Contractor's Pricing Guide:

Residential Repair & Remodeling Costs

Bob Mewis, CCP, Senior Editor

RS Means
FROM THE GORDIAN GROUP®

2016

Engineering Director
Bob Mewis, CCP (*1, 2, 4, 13, 14, 31, 32, 33, 34, 35, 41, 44, 46*)

Contributing Editors
Christopher Babbitt
Adrian C. Charest, PE
Cheryl Elsmore
Wafaa Hamitou
Joseph Kelble
Charles Kibbee
Robert J. Kuchta (*8*)
Michael Landry
Thomas Lane (*6, 7*)
Genevieve Medeiros
Elisa Mello
Chris Morris (*26, 27, 28, 48*)

Melville J. Mossman, PE (*21, 22, 23*)
Marilyn Phelan, AIA (*9, 10, 11, 12*)
Stephen C. Plotner (*3, 5*)
Stephen Rosenberg
Kevin Souza
Keegan Spraker
Tim Tonello
David Yazbek

Vice President Data & Engineering
Chris Anderson

Product Manager
Andrea Sillah

Production Manager
Debbie Panarelli

Production
Sharon Larsen
Jonathan Forgit
Sheryl Rose
Mary Lou Geary

Technical Support
Gary L. Hoitt
Kathryn S. Rodriguez

Cover Design
Blaire Gaddis

Numbers in italics are the divisional responsibilities for each editor. Please contact the designated editor directly with any questions.

RSMeans
FROM THE GORDIAN GROUP®

The most trusted source for construction cost estimating data

2016

Contractor's Pricing Guide:
Residential Repair & Remodeling Costs

RSMeans
Construction Publishers & Consultants
1099 Hingham Street, Suite 201
Rockland, MA 02370
United States of America
1-888-606-7279
www.RSMeans.com

Printed in the United States of America
ISSN 1542-9156
ISBN 978-1-943215-19-5

0346 $49.95 per copy (in United States)
Price is subject to change without prior notice.

Table of Contents

Foreword

Who We Are

Since 1942, RSMeans has delivered construction cost estimating information and consulting throughout North America. In 2014, RSMeans was acquired by The Gordian Group, combining two industry-leading construction cost databases. Through the RSMeans line of products and services, The Gordian Group provides innovative construction cost estimating data to organizations pursuing efficient and effective construction planning, estimating, procurement, and information solutions.

Our Offerings

With RSMeans' construction cost estimating data, contractors, architects, engineers, facility owners, and managers can utilize the most up-to-date data available in many formats to meet their specific planning and estimating needs.

When you purchase information from RSMeans, you are, in effect, hiring the services of a full-time staff of construction and engineering professionals.

Our thoroughly-experienced and highly-qualified staff works daily to collect, analyze, and disseminate comprehensive cost information to meet your needs.

These staff members have years of practical construction experience and engineering training prior to joining the firm. As a result, you can count on them not only for accurate cost figures, but also for additional background reference information that will help you create a realistic estimate.

The RSMeans organization is equipped to help you solve construction problems through its variety of data solutions.

Access our comprehensive database electronically with RSMeans Online. Quick, intuitive, easy to use, and updated continuously throughout the year, RSMeans Online is the most accurate and up-to-date cost estimating data.

This up-to-date, accurate, and localized data can be leveraged for a broad range of applications, such as Custom Cost Engineering solutions, Market and Custom Analytics, and Third-Party Legal Resources.

To ensure you are getting the most from your cost estimating data, we also offer a myriad of reference guides, training, and professional seminars (learn more at **www.RSMeans.com/ Learn**).

In short, RSMeans can provide you with the tools and expertise for developing accurate and dependable construction estimates and budgets in a variety of ways.

Our Commitment

Today at RSMeans, we do more than talk about the quality of our data and the usefulness of the information. We stand behind all of our data—from historical cost indexes to construction materials and techniques—to craft current costs and predict future trends.

If you have any questions about our products or services, please call us toll-free at 1-888-606-7279. You can also visit our website at: **www.RSMeans.com**.

How the Cost Data Is Built: An Overview

A Powerful Construction Tool

You have in your hands one of the most powerful construction tools available today. A successful project is built on the foundation of an accurate and dependable estimate. This data will enable you to construct just such an estimate.

For the casual user the data is designed to be:

- quickly and easily understood so you can get right to your estimate.
- filled with valuable information so you can understand the necessary factors that go into the cost estimate.

For the regular user, the book is designed to be:

- a handy reference that can be quickly referred to for key costs.
- a comprehensive, fully reliable source of current construction costs, so you'll be prepared to estimate any project.
- a source book for project cost, product selections, and alternate materials and methods.

To meet all of these requirements we have organized the data into the following clearly defined sections.

Unit Price Section

The cost data has been divided into 23 sections representing the order of construction. Within each section is a listing of components and variations to those components applicable to that section. Costs are shown for the various actions that must be done to that component.

Reference Section

This section includes information on Location Factors and a listing of Abbreviations.

Location Factors

You can adjust total project costs to over 900 locations throughout the U.S. and Canada by using the data in this section.

Abbreviations:

A listing of the abbreviations used throughout, along with the terms they represent, is included.

Index

A comprehensive listing of all terms and subjects will help you quickly find what you need when you are not sure where it falls in the order of construction.

The Scope of This Data

This data is designed to be as comprehensive and as easy to use as possible. To that end we have made certain assumptions and limited its scope in two key ways:

1. We have established material prices based on a national average.
2. We have computed labor costs based on residential wage rates.

Project Size

This data is intended for use by those involved primarily in residential repair and remodeling construction costing between $10,000–$100,000.

With reasonable exercise of judgment, the figures can be used for any building work. For other types of projects, such as new home construction or commercial buildings, consult the appropriate RSMeans data set for more information.

How to Use the Cost Data: The Details

What's Behind the Numbers? The Development of Cost Data

The staff at RSMeans continually monitors developments in the construction industry in order to ensure reliable, thorough, and up-to-date cost information. While overall construction costs may vary relative to general economic conditions, price fluctuations within the industry are dependent upon many factors. Individual price variations may, in fact, be opposite to overall economic trends. Therefore, costs are constantly tracked, and complete updates are performed yearly. Also, new items are frequently added in response to changes in materials and methods.

Costs—$ (U.S.)

All costs represent U.S. national averages and are given in U.S. dollars. The RSMeans City Cost Indexes (CCI) or Location Factors can be used to adjust costs to a particular location. The CCI or Location Factors for Canada can be used to adjust U.S. national averages to local costs in Canadian dollars. No exchange rate conversion is necessary because it has already been factored in.

Material Costs

The RSMeans staff contacts manufacturers, dealers, distributors, and contractors all across the U.S. and Canada to determine national average material costs. If you have access to current material costs for your specific location, you may wish to make adjustments to reflect differences from the national average. Included within material costs are fasteners for a normal installation. RSMeans engineers use manufacturers' recommendations, written specifications, and/or standard construction practices for size and spacing of fasteners. Adjustments to material costs may be required for your specific application or location. The manufacturer's warranty is assumed. Extended warranties are not included in the material costs. **Material costs do not include sales tax.**

Labor Costs

Labor costs are based upon a mathematical average of trade-specific wages in 30 major U.S. cities. The type of wage (union, open shop, or residential) is identified on the inside back cover of printed publications or is selected by the estimator when using the electronic products. Markups for the wages can also be found on the inside back cover of printed publications and/or under the labor references found in the electronic products.

- If wage rates in your area vary from those used, or if rate increases are expected within a given year, labor costs should be adjusted accordingly.

Labor costs reflect productivity based on actual working conditions. In addition to actual installation, these figures include time spent during a normal weekday on tasks, such as material receiving and handling, mobilization at site, site movement, breaks, and cleanup.

Productivity data is developed over an extended period so as not to be influenced by abnormal variations and reflects a typical average.

Equipment Costs

Equipment costs include not only rental, but also operating costs for equipment under normal use. The operating costs include parts and labor for routine servicing, such as repair and replacement of pumps, filters, and worn lines. Normal operating expendables, such as fuel, lubricants, tires, and electricity (where applicable), are also included. Extraordinary operating expendables with highly variable wear patterns, such as diamond bits and blades, are excluded. These costs are included under materials. Equipment rental rates are obtained from industry sources throughout North America—contractors, suppliers, dealers, manufacturers, and distributors.

Rental rates can also be treated as reimbursement costs for contractor-owned equipment. Owned equipment costs include depreciation, loan payments, interest, taxes, insurance, storage, and major repairs.

Equipment costs do not include operators' wages.

Factors Affecting Costs

Costs can vary depending upon a number of variables. Here's a listing of some factors that affect costs and points to consider.

Quality—The prices for materials and the workmanship upon which productivity is based represent sound construction work. They are also in line with industry standard and manufacturer specifications and are frequently used by federal, state, and local governments.

Overtime—We have made no allowance for overtime. If you anticipate premium time or work beyond normal working hours, be sure to make an appropriate adjustment to your labor costs.

Productivity—The productivity, daily output, and labor-hour figures for each line item are based on working an eight-hour day in daylight hours in moderate temperatures. For work that extends beyond normal work hours or is performed under adverse conditions, productivity may decrease.

Size of Project—The size, scope of work, and type of construction project will have a significant impact on cost. Economies of scale can reduce costs for large projects. Unit costs can often run higher for small projects.

Location—Material prices are for metropolitan areas. However, in dense urban areas, traffic and site storage limitations may increase costs. Beyond a 20-mile radius of metropolitan areas, extra trucking or transportation charges may also increase the material costs slightly.

On the other hand, lower wage rates may be in effect. Be sure to consider both of these factors when preparing an estimate, particularly if the job site is located in a central city or remote rural location. In addition, highly specialized subcontract items may require travel and per-diem expenses for mechanics.

Other Factors—

- season of year
- contractor management
- weather conditions
- local union restrictions
- building code requirements
- availability of:
  - adequate energy
  - skilled labor
  - building materials
- owner's special requirements/restrictions
- safety requirements
- environmental considerations
- access

Unpredictable Factors—General business conditions influence "in-place" costs of all items. Substitute materials and construction methods may have to be employed. These may affect the installed cost and/or life cycle costs. Such factors may be difficult to evaluate and cannot necessarily be predicted on the basis of the job's location in a particular section of the country. Thus, where these factors apply, you may find significant but unavoidable cost variations for which you will have to apply a measure of judgment to your estimate.

Final Checklist

Estimating can be a straightforward process provided you remember the basics. Here's a checklist of some of the steps you should remember to complete before finalizing your estimate.

Did you remember to:

- factor in the City Cost Index or Location Factors for your locale?
- take into consideration which items have been marked up and by how much?
- mark up the entire estimate sufficiently for your purposes?

- read the background information on techniques and technical matters that could impact your project time span and cost?
- include all components of your project in the final estimate?
- double check your figures for accuracy?
- call RSMeans if you have any questions about your estimate or the data you've used? Remember, RSMeans stands behind its information. If you have any questions about your estimate, about the costs you've used from our data, or even about the technical aspects of the job that may affect your estimate, feel free to call the RSMeans editors at 1-888-606-7279.

Free Quarterly Updates

Stay up-to-date throughout the year with RSMeans' free cost data updates four times a year. Sign up online to make sure you have access to the newest data. Every quarter we provide the city cost adjustment factors for hundreds of cities and key materials.

Register at: **http://info.TheGordianGroup.com/ 2016Updates.html.**

Unit Price Section

Table of Contents

Table of Contents (cont.)

How RSMeans Unit Price Data Works

The prices in this publication include all overhead and profit mark-ups for the installing contractor. If a general contractor is involved, an additional 10% should be added to the cost shown in the "Total" column.

The costs for material, labor and any equipment are shown here. Material costs include delivery and a 10% mark-up. No sales taxes are included. Labor costs include all mark-ups as shown on the inside back cover of this data set. Equipment costs also include a 10% mark-up.

The cost in the total column should be multiplied by the quantity of each item to determine the total cost for that item. The sum of all items' total costs becomes the project's total cost.

The data set is arranged into 23 categories that follow the order of construction for residential projects. The category is shown as white letters on a black background at the top of each page.

A brief specification of the cost item is shown here.

Graphics representing some of the cost items are found at the beginning of each category.

The cost data is arranged in a tree structure format. Major cost sections are left justified and are shown as bold headings within a grey background. Cost items and sub-cost items are found under the bold headings. The complete description should read as "bold heading, cost item, sub-cost item."

For example, the identified item's description is "Beam/Girder, Solid Wood Beam, 2" x 8"."

Each action performed to the cost item is conveniently listed to the right of the description. See the explanation below for the scope of work included as part of the action.

The unit of measure upon which all the costs are based is shown here. An abbreviations list is included within the reference section found at the back of this data set.

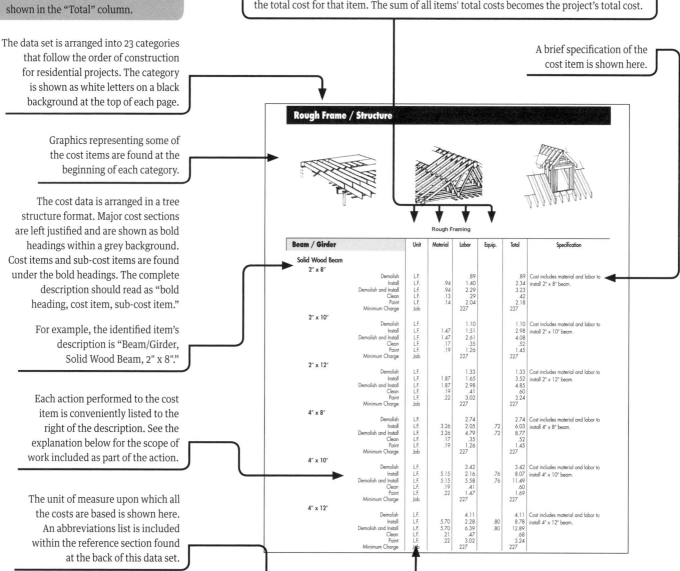

Rough Frame / Structure

Rough Framing

Beam / Girder	Action	Unit	Material	Labor	Equip.	Total	Specification
Solid Wood Beam							
2" x 8"							
	Demolish	L.F.		.89		.89	Cost includes material and labor to
	Install	L.F.	.94	1.40		2.34	install 2" x 8" beam.
	Demolish and Install	L.F.	.94	2.29		3.23	
	Clean	L.F.	.13	.29		.42	
	Paint	L.F.	.14	2.04		2.18	
	Minimum Charge	Job		227		227	
2" x 10"							
	Demolish	L.F.		1.10		1.10	Cost includes material and labor to
	Install	L.F.	1.47	1.51		2.98	install 2" x 10" beam.
	Demolish and Install	L.F.	1.47	2.61		4.08	
	Clean	L.F.	.17	.35		.52	
	Paint	L.F.	.19	1.26		1.45	
	Minimum Charge	Job		227		227	
2" x 12"							
	Demolish	L.F.		1.33		1.33	Cost includes material and labor to
	Install	L.F.	1.87	1.65		3.52	install 2" x 12" beam.
	Demolish and Install	L.F.	1.87	2.98		4.85	
	Clean	L.F.	.19	.41		.60	
	Paint	L.F.	.22	3.02		3.24	
	Minimum Charge	Job		227		227	
4" x 8"							
	Demolish	L.F.		2.74		2.74	Cost includes material and labor to
	Install	L.F.	3.26	2.05	.72	6.03	install 4" x 8" beam.
	Demolish and Install	L.F.	3.26	4.79	.72	8.77	
	Clean	L.F.	.17	.35		.52	
	Paint	L.F.	.19	1.26		1.45	
	Minimum Charge	Job		227		227	
4" x 10"							
	Demolish	L.F.		3.42		3.42	Cost includes material and labor to
	Install	L.F.	5.15	2.16	.76	8.07	install 4" x 10" beam.
	Demolish and Install	L.F.	5.15	5.58	.76	11.49	
	Clean	L.F.	.19	.41		.60	
	Paint	L.F.	.22	1.47		1.69	
	Minimum Charge	Job		227		227	
4" x 12"							
	Demolish	L.F.		4.11		4.11	Cost includes material and labor to
	Install	L.F.	5.70	2.28	.80	8.78	install 4" x 12" beam.
	Demolish and Install	L.F.	5.70	6.39	.80	12.89	
	Clean	L.F.	.21	.47		.68	
	Paint	L.F.	.22	3.02		3.24	
	Minimum Charge	Job		227		227	

Action	Scope
Demolish	Includes removal of the item and hauling the debris to a truck, dumpster, or storage area.
Install	Includes the installation of the item with normal placement and spacing of fasteners.
Demolish and Install	Includes the demolish action and the install action.
Reinstall	Includes the reinstallation of a removed item.
Clean	Includes cleaning of the item.
Paint	Includes up to two coats of finish on the item (unless otherwise indicated).
Minimum Charge	Includes the minimum labor or the minimum labor and equipment cost. Use this charge for small quantities of material.

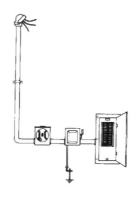

| Scaffolding | Temporary Electrical Service | Air Compressor |

Construction Fees		Unit	Material	Labor	Equip.	Total	Specification
Asbestos Test							
	Minimum Charge	Job		266		266	Minimum charge for testing.
Architectural							
	Minimum Charge	Day	2075			2075	Minimum labor charge for architect.
Engineering							
	Minimum Charge	Day	1900			1900	Minimum labor charge for field engineer.

Construction Permits		Unit	Material	Labor	Equip.	Total	Specification
Job (less than 5K)							
	Minimum Charge	Job	65.50			65.50	Permit fees, job less than 5K.
Job (5K to 10K)							
	Minimum Charge	Job	110			110	Permit fees, job more than 5K, but less than 10K.
Job (10K to 25K)							
	Minimum Charge	Job	244			244	Permit fees, job more than 10K, but less than 25K.
Job (25K to 50K)							
	Minimum Charge	Job	470			470	Permit fees, job more than 25K, but less than 50K.
Job (50K to 100K)							
	Minimum Charge	Job	860			860	Permit fees, job more than 50K, but less than 100K.
Job (100K to 250K)							
	Minimum Charge	Job	1800			1800	Permit fees, job more than 100K, but less than 250K.
Job (250K to 500K)							
	Minimum Charge	Job	3350			3350	Permit fees, job more than 250K, but less than 500K.
Job (500K to 1M)							
	Minimum Charge	Job	6000			6000	Permit fees, job more than 500K, but less than 1 million.

Trade Labor		Unit	Material	Labor	Equip.	Total	Specification
Carpenter							
Daily (8 hours)							
	Install	Day		455		455	Daily labor rate for one carpenter.
	Minimum Charge	Job		227		227	
Weekly (40 hours)							
	Install	Week		2275		2275	Weekly labor rate for one carpenter.
	Minimum Charge	Job		227		227	
Drywaller							
Daily (8 hours)							
	Install	Day		455		455	Daily labor rate for one carpenter.
	Minimum Charge	Job		227		227	
Weekly (40 hours)							
	Install	Week		2275		2275	Weekly labor rate for one carpenter.
	Minimum Charge	Job		227		227	
Roofer							
Daily (8 hours)							
	Install	Day		425		425	Daily labor rate for one skilled roofer.
	Minimum Charge	Job		212		212	
Weekly (40 hours)							
	Install	Week		2125		2125	Weekly labor rate for one skilled
	Minimum Charge	Job		212		212	roofer.
Painter							
Daily (8 hours)							
	Install	Day		380		380	Daily labor rate for one painter.
	Minimum Charge	Job		189		189	
Weekly (40 hours)							
	Install	Week		1900		1900	Weekly labor rate for one painter.
	Minimum Charge	Job		189		189	
Mason							
Daily (8 hours)							
	Install	Day		430		430	Daily labor rate for one skilled mason.
	Minimum Charge	Job		216		216	
Weekly (40 hours)							
	Install	Week		2150		2150	Weekly labor rate for one skilled
	Minimum Charge	Job		216		216	mason.
Electrician							
Daily (8 hours)							
	Install	Day		485		485	Daily labor rate for an electrician.
	Minimum Charge	Job		241		241	
Weekly (40 hours)							
	Install	Week		2425		2425	Weekly labor rate for an electrician.
	Minimum Charge	Job		241		241	
Plumber							
Daily (8 hours)							
	Install	Day		515		515	Daily labor rate for a plumber.
	Minimum Charge	Job		258		258	
Weekly (40 hours)							
	Install	Week		2575		2575	Weekly labor rate for a plumber.
	Minimum Charge	Job		258		258	
Common Laborer							
Daily (8 hours)							
	Install	Day		330		330	Daily labor rate for a common
	Minimum Charge	Job		165		165	laborer.

For customer support on your Contractor's Pricing Guide: Residential Repair & Remodeling, call 888.606.7279.

Job Costs

Trade Labor

	Unit	Material	Labor	Equip.	Total	Specification
Weekly (40 hours)						
Install	Week		1650		1650	Weekly labor rate for a common
Minimum Charge	Job		165		165	laborer.

Temporary Utilities

	Unit	Material	Labor	Equip.	Total	Specification
Power Pole						
Job Site Electricity						
Install	Month	57			57	Includes average cost of electricity per
Minimum Charge	Job		241		241	month.
Water Service						
Job Site Water						
Install	Month	75			75	Includes average costs for water for
Minimum Charge	Month	75			75	one month.
Telephone						
Install	Month	93.50			93.50	Includes average cost for local and
Minimum Charge	Month	93.50			93.50	long distance telephone service for
						one month.
Chemical Toilet						
Install	Month			305	305	Monthly rental of portable toilet.
Minimum Charge	Day			15.20	15.20	
Office Trailer						
Place and Remove						
Install	Job	385			385	Includes cost to a contractor to deliver,
Minimum Charge	Job		202		202	place and level an office trailer and
						remove at end of job.
Storage Container						
Place and Remove						
Install	Job	218			218	Includes cost to a contractor to deliver
Minimum Charge	Job		202		202	and place a storage container and
						remove at end of job.
Fencing						
Install	L.F.	3.18			3.18	Includes material costs for temporary
Minimum Charge	Job		165		165	fencing.
Railing						
Install	L.F.	17.05	12.05	.92	30.02	Includes material and labor to install
Minimum Charge	Job		165		165	primed steel pipe.

Equipment Rental

	Unit	Material	Labor	Equip.	Total	Specification
Flatbed Truck						
Daily Rental						
Install	Day			273	273	Daily rental of flatbed truck.
Minimum Charge	Job			273	273	
Weekly Rental						
Install	Week			1375	1375	Weekly rental of flatbed truck.
Minimum Charge	Job			273	273	
Monthly Rental						
Install	Month			5450	5450	Monthly rental of flatbed truck.
Minimum Charge	Job			273	273	
Dump Truck						
Daily Rental						
Install	Day			460	460	Daily rental of dump truck.
Minimum Charge	Day			460	460	

3

Job Costs

Equipment Rental		Unit	Material	Labor	Equip.	Total	Specification
Weekly Rental							
	Install	Week			2300	2300	Weekly rental of dump truck.
	Minimum Charge	Day			460	460	
Monthly Rental							
	Install	Month			9175	9175	Monthly rental of dump truck.
	Minimum Charge	Day			460	460	
Forklift							
Daily Rental							
	Install	Day			365	365	Daily rental of forklift.
	Minimum Charge	Day			365	365	
Weekly Rental							
	Install	Week			1825	1825	Weekly rental of forklift.
	Minimum Charge	Day			365	365	
Monthly Rental							
	Install	Month			7350	7350	Monthly rental of forklift.
	Minimum Charge	Day			365	365	
Bobcat							
Daily Rental							
	Install	Day			191	191	Daily rental of wheeled, skid steer
	Minimum Charge	Day			191	191	loader.
Weekly Rental							
	Install	Week			955	955	Weekly rental of wheeled, skid steer
	Minimum Charge	Day			191	191	loader.
Monthly Rental							
	Install	Month			3800	3800	Monthly rental of wheeled, skid steer
	Minimum Charge	Day			191	191	loader.
Air Compressor							
Daily Rental							
	Install	Day			157	157	Daily rental of air compressor.
	Minimum Charge	Day			157	157	
Weekly Rental							
	Install	Week			785	785	Weekly rental of air compressor.
	Minimum Charge	Day			157	157	
Monthly Rental							
	Install	Month			3150	3150	Monthly rental of air compressor.
	Minimum Charge	Day			157	157	
Jackhammer							
Daily Rental							
	Install	Day			11.20	11.20	Daily rental of air tool, jackhammer.
	Minimum Charge	Day			11.20	11.20	
Weekly Rental							
	Install	Week			56	56	Weekly rental of air tool, jackhammer.
	Minimum Charge	Day			11.20	11.20	
Monthly Rental							
	Install	Month			224	224	Monthly rental of air tool,
	Minimum Charge	Day			11.20	11.20	jackhammer.
Concrete Mixer							
Daily Rental							
	Install	Day			154	154	Daily rental of concrete mixer.
	Minimum Charge	Day			154	154	
Weekly Rental							
	Install	Week			770	770	Weekly rental of concrete mixer.
	Minimum Charge	Day			154	154	

Job Costs

Equipment Rental		Unit	Material	Labor	Equip.	Total	Specification
Monthly Rental							
	Install	Month			3075	3075	Monthly rental of concrete mixer.
	Minimum Charge	Day			154	154	
Concrete Bucket							
Daily Rental							
	Install	Day			22.50	22.50	Daily rental of concrete bucket.
	Minimum Charge	Day			22.50	22.50	
Weekly Rental							
	Install	Week			112	112	Weekly rental of concrete bucket.
	Minimum Charge	Day			22.50	22.50	
Monthly Rental							
	Install	Month			450	450	Monthly rental of concrete bucket.
	Minimum Charge	Day			22.50	22.50	
Concrete Pump							
Daily Rental							
	Install	Day			800	800	Daily rental of concrete pump.
	Minimum Charge	Day			800	800	
Weekly Rental							
	Install	Week			4000	4000	Weekly rental of concrete pump.
	Minimum Charge	Day			800	800	
Monthly Rental							
	Install	Month			16000	16000	Monthly rental of concrete pump.
	Minimum Charge	Day			800	800	
Generator							
Daily Rental							
	Install	Day			40	40	Daily rental of generator.
	Minimum Charge	Day			40	40	
Weekly Rental							
	Install	Week			201	201	Weekly rental of generator.
	Minimum Charge	Day			40	40	
Monthly Rental							
	Install	Month			805	805	Monthly rental of generator.
	Minimum Charge	Day			40	40	
Welder							
Daily Rental							
	Install	Day			45	45	Daily rental of welder.
	Minimum Charge	Day			45	45	
Weekly Rental							
	Install	Week			225	225	Weekly rental of welder.
	Minimum Charge	Day			45	45	
Monthly Rental							
	Install	Month			900	900	Monthly rental of welder.
	Minimum Charge	Day			45	45	
Sandblaster							
Daily Rental							
	Install	Day			23	23	Daily rental of sandblaster.
	Minimum Charge	Day			23	23	
Weekly Rental							
	Install	Week			114	114	Weekly rental of sandblaster.
	Minimum Charge	Day			23	23	
Monthly Rental							
	Install	Month			460	460	Monthly rental of sandblaster.
	Minimum Charge	Day			23	23	

Job Costs

Equipment Rental

Equipment Rental		Unit	Material	Labor	Equip.	Total	Specification
Space Heater							
Daily Rental							
	Install	Day			19.05	19.05	Daily rental of space heater.
	Minimum Charge	Day			23	23	
Weekly Rental							
	Install	Day			95	95	Weekly rental of space heater.
	Minimum Charge	Day			23	23	
Monthly Rental							
	Install	Day			380	380	Monthly rental of space heater.
	Minimum Charge	Day			23	23	

Move / Reset Contents

Move / Reset Contents		Unit	Material	Labor	Equip.	Total	Specification
Small Room							
	Install	Room		40		40	Includes cost to remove and reset the
	Minimum Charge	Job		165		165	contents of a small room.
Average Room							
	Install	Room		48.50		48.50	Includes cost to remove and reset the
	Minimum Charge	Job		165		165	contents of an average room.
Large Room							
	Install	Room		67		67	Includes cost to remove and reset the
	Minimum Charge	Job		165		165	contents of a large room.
Extra Large Room							
	Install	Room		106		106	Includes cost to remove and reset the
	Minimum Charge	Job		165		165	contents of an extra large room.
Store Contents							
Weekly							
	Install	Week	287			287	Storage fee, per week.
Monthly							
	Install	Month	860			860	Storage fee, per month.

Cover and Protect

Cover and Protect		Unit	Material	Labor	Equip.	Total	Specification
Cover / Protect Walls							
	Install	SF Wall		.13		.13	Includes labor costs to cover and
	Minimum Charge	Job		165		165	protect walls.
Cover / Protect Floors							
	Install	SF Flr.		.13		.13	Includes labor costs to cover and
	Minimum Charge	Job		165		165	protect floors.

Temporary Bracing

Temporary Bracing		Unit	Material	Labor	Equip.	Total	Specification
Shim Support Piers							
	Install	Ea.	4.51	131		135.51	Includes labor and material to install
	Minimum Charge	Job		395		395	shims support for piers.

Job Costs

Jack and Relevel

		Unit	Material	Labor	Equip.	Total	Specification
Single Story							
	Install	S.F.	1.87	1.58		3.45	Includes labor and equipment to jack
	Minimum Charge	Job		395		395	and relevel single-story building.
Two-story							
	Install	S.F.	1.87	2.63		4.50	Includes labor and equipment to jack
	Minimum Charge	Job		395		395	and relevel two-story building.

Dumpster Rental

		Unit	Material	Labor	Equip.	Total	Specification
Small Load							
	Install	Week			615	615	Dumpster, weekly rental, 2 dumps, 5
	Minimum Charge	Week			615	615	yard container.
Medium Load							
	Install	Week			715	715	Dumpster, weekly rental, 2 dumps, 10
	Minimum Charge	Week			715	715	yard container.
Large Load							
	Install	Week			1075	1075	Dumpster, weekly rental, 2 dumps, 30
	Minimum Charge	Week			1075	1075	yard container.

Debris Hauling

		Unit	Material	Labor	Equip.	Total	Specification
Per Ton							
	Install	Ton		105	56.50	161.50	Includes hauling of debris by trailer or
	Minimum Charge	Ton		790	420	1210	dump truck.
Per Cubic Yard							
	Install	C.Y.		35	18.75	53.75	Includes hauling of debris by trailer or
	Minimum Charge	Job		790	420	1210	dump truck.
Per Pick-up Truck Load							
	Install	Ea.		183		183	Includes hauling of debris by pick-up
	Minimum Charge	Ea.		183		183	truck.

Dump Fees

		Unit	Material	Labor	Equip.	Total	Specification
Per Cubic Yard							
	Install	C.Y.	36.50			36.50	Dump charges, building materials.
	Minimum Charge	C.Y.	36.50			36.50	
Per Ton							
	Install	Ton	138			138	Dump charges, building materials.
	Minimum Charge	Ton	138			138	

Scaffolding

		Unit	Material	Labor	Equip.	Total	Specification
4' to 6' High							
	Install	Month	73.50			73.50	Includes monthly rental of scaffolding,
	Minimum Charge	Month	73.50			73.50	steel tubular, 30" wide, 7' long, 5' high.
7' to 11' High							
	Install	Month	87			87	Includes monthly rental of scaffolding,
	Minimum Charge	Month	87			87	steel tubular, 30" wide, 7' long, 10' high.

For customer support on your Contractor's Pricing Guide: Residential Repair & Remodeling, call 888.606.7279.

Job Costs

Scaffolding

Scaffolding		Unit	Material	Labor	Equip.	Total	Specification
12' to 16' High							
	Install	Month	100			100	Includes monthly rental of scaffolding,
	Minimum Charge	Month	100			100	steel tubular, 30" wide, 7' long, 15' high.
17' to 21' High							
	Install	Month	129			129	Includes monthly rental of scaffolding,
	Minimum Charge	Month	129			129	steel tubular, 30" wide, 7' long, 20' high.
22' to 26' High							
	Install	Month	142			142	Includes monthly rental of scaffolding,
	Minimum Charge	Month	142			142	steel tubular, 30" wide, 7' long, 25' high.
27' to 30' High							
	Install	Month	155			155	Includes monthly rental of scaffolding,
	Minimum Charge	Month	155			155	steel tubular, 30" wide, 7' long, 30' high.

Construction Clean-up

Construction Clean-up		Unit	Material	Labor	Equip.	Total	Specification
Initial							
	Clean	S.F.		.41		.41	Includes labor for general clean-up.
	Minimum Charge	Job		165		165	
Progressive							
	Clean	S.F.		.21		.21	Includes labor for progressive site/job
	Minimum Charge	Job		165		165	clean-up on a weekly basis.

For customer support on your Contractor's Pricing Guide: Residential Repair & Remodeling, call 888.606.7279.

Building Demolition		Unit	Material	Labor	Equip.	Total	Specification
Wood-framed Building							
1 Story Demolition							
	Demolish	S.F.		1.61	1.99	3.60	Includes demolition of 1 floor only
	Minimum Charge	Job		2550	3150	5700	light wood-framed building, including hauling of debris. Dump fees not included.
2nd Story Demolition							
	Demolish	S.F.		1.64	2.03	3.67	Includes demolition of 2nd floor only
	Minimum Charge	Job		2550	3150	5700	light wood-framed building, including hauling of debris. Dump fees not included.
3rd Story Demolition							
	Demolish	S.F.		2.13	2.64	4.77	Includes demolition of 3rd floor only
	Minimum Charge	Job		2550	3150	5700	light wood-framed building, including hauling of debris. Dump fees not included.
Masonry Building							
	Demolish	S.F.		2.29	2.84	5.13	Includes demolition of one story
	Minimum Charge	Job		2550	3150	5700	masonry building, including hauling of debris. Dump fees not included.
Concrete Building							
Unreinforced							
	Demolish	S.F.		4.64	1.29	5.93	Includes demolition of a one story
	Minimum Charge	Job		2550	3150	5700	non-reinforced concrete building, including hauling of debris. Dump fees not included.
Reinforced							
	Demolish	S.F.		5.35	1.48	6.83	Includes demolition of a one story
	Minimum Charge	Job		2550	3150	5700	reinforced concrete building, including hauling of debris. Dump fees not included.
Steel Building							
	Demolish	S.F.		3.06	3.78	6.84	Includes demolition of a metal
	Minimum Charge	Job		2550	3150	5700	building, including hauling of debris. Dump fees not included. No salvage value assumed for scrap metal.

9

Job Preparation

General Clean-up

General Clean-up		Unit	Material	Labor	Equip.	Total	Specification
Remove Debris	Clean	S.F.		1.05	.27	1.32	Includes removal of debris caused by the loss including contents items and loose debris.
	Minimum Charge	Job		525	137	662	
Muck-out	Clean	S.F.		.64		.64	Includes labor to remove 2″ of muck and mud after a flood loss.
	Minimum Charge	Job		165		165	
Deodorize / Disinfect	Clean	SF Flr.		.22		.22	Includes deodorizing building with application of mildicide and disinfectant to floors and walls - will vary depending on the extent of deodorization required.
	Minimum Charge	Job		165		165	
Clean Walls							
Heavy	Clean	S.F.		.15		.15	Includes labor and material for heavy cleaning (multiple applications) with detergent and solvent.
	Minimum Charge	Job		165		165	
Clean Ceiling							
Light	Clean	S.F.		.25		.25	Includes labor and material to clean light to moderate smoke and soot.
	Minimum Charge	Job		165		165	
Heavy	Clean	S.F.		.38		.38	Includes labor and material to clean heavy smoke and soot.
	Minimum Charge	Job		165		165	

Flood Clean-up

Flood Clean-up		Unit	Material	Labor	Equip.	Total	Specification
Emergency Service Call							
After Hours / Weekend	Minimum Charge	Job		220		220	Includes minimum charges for deflooding clean-up after hours or on weekends.
Grey Water / Sewage	Minimum Charge	Job		165		165	Includes minimum charges for deflooding clean-up.
Water Extraction							
Clean (non-grey)	Clean	S.F.	.03	.66		.69	Includes labor to remove clean uncontaminated water from the loss.
	Minimum Charge	Job		165		165	
Grey (non-solids)	Clean	S.F.	.07	1.32		1.39	Includes labor to remove non-solid grey water from the loss.
	Minimum Charge	Job		165		165	
Sewage / Solids	Clean	SF Flr.	.11	1.65		1.76	Includes labor to remove water from the loss that may contain sewage or solids.
	Minimum Charge	Job		165		165	

For customer support on your Contractor's Pricing Guide: Residential Repair & Remodeling, call 888.606.7279.

Job Preparation

Flood Clean-up

Flood Clean-up		Unit	Material	Labor	Equip.	Total	Specification
Mildicide Walls							
Topographical	Clean	SF Wall		.30		.30	Clean-up of deflooding including
	Minimum Charge	Job		165		165	topographical mildicide application.
Injection Treatments	Clean	SF Flr.	44.50	5.10		49.60	Clean-up of deflooding including
	Minimum Charge	Job		165		165	mildicide injection.
Carpet Cleaning							
Uncontaminated	Clean	SF Flr.		.21		.21	Includes labor to clean carpet.
	Minimum Charge	Job		165		165	
Contaminated	Clean	SF Flr.		.33		.33	Includes labor to clean contaminated
	Minimum Charge	Job		165		165	carpet.
Stairway	Clean	Ea.	.12	3.88		4	Includes labor to clean contaminated
	Minimum Charge	Job		165		165	carpet on a stairway.
Carpet Treatment							
Disinfect / Deodorize	Clean	SF Flr.		.21		.21	Includes disinfecting and deodorizing
	Minimum Charge	Job		165		165	of carpet.
Disinfect For Sewage	Clean	SF Flr.		.53		.53	Includes labor to disinfect and
	Minimum Charge	Job		165		165	deodorize carpet exposed to sewage.
Mildicide	Clean	SF Flr.	.02	.21		.23	Includes labor and materials to apply
	Minimum Charge	Job		165		165	mildicide to carpet.
Thermal Fog Area	Clean	SF Flr.	.01	.51		.52	Includes labor and materials to
	Minimum Charge	Job		165		165	perform thermal fogging during
							deflooding clean-up.
Wet Fog / ULV Area	Clean	SF Flr.	.03	.29		.32	Includes labor and materials to
	Minimum Charge	Job		165		165	perform wet fogging / ULV during
							deflooding clean-up.
Steam Clean Fixtures	Clean	Ea.	.01	33		33.01	Includes labor and materials to steam
	Minimum Charge	Job		165		165	clean fixtures during deflooding
							clean-up.
Pressure Wash Sewage	Clean	SF Flr.		3		3	Includes labor to pressure wash
	Minimum Charge	Job		165		165	sewage contamination.
Air Mover	Minimum Charge	Day			46	46	Daily rental of air mover.
Dehumidifier							
Medium Size	Minimum Charge	Day			54	54	Daily rental of dehumidifier, medium.
Large Size	Minimum Charge	Day			103	103	Daily rental of dehumidifier, large.

11

Job Preparation

Flood Clean-up		Unit	Material	Labor	Equip.	Total	Specification
Air Purifier							
HEPA Filter-Bacteria							
	Minimum Charge	Ea.	360			360	HEPA filter for air purifier.

Asbestos Removal		Unit	Material	Labor	Equip.	Total	Specification
Asbestos Analysis							
	Minimum Charge	Job		266		266	Minimum charge for testing.
Asbestos Pipe Insulation							
1/2" to 3/4" Diameter							
	Demolish	L.F.	.34	6.85		7.19	Minimum charge to remove asbestos. Includes full Tyvek suits for workers changed four times per eight-hour shift with respirators, air monitoring and supervision by qualified professionals.
	Minimum Charge	Job		4250		4250	
1" to 3" Diameter							
	Demolish	L.F.	1.05	21.50		22.55	Minimum charge to remove asbestos. Includes full Tyvek suits for workers changed four times per eight-hour shift with respirators, air monitoring and supervision by qualified professionals.
	Minimum Charge	Job		4250		4250	
Asbestos-based Siding							
	Demolish	S.F.	2.33	47.50		49.83	Minimum charge to remove asbestos. Includes full Tyvek suits for workers changed four times per eight-hour shift with respirators, air monitoring and supervision by qualified professionals.
	Minimum Charge	Job		4250		4250	
Asbestos-based Plaster							
	Demolish	S.F.	1.79	36.50		38.29	Minimum charge to remove asbestos. Includes full Tyvek suits for workers changed four times per eight-hour shift with respirators, air monitoring and supervision by qualified professionals.
	Minimum Charge	Job		4250		4250	
Encapsulate Asbestos Ceiling							
	Install	S.F.	.44	.20		.64	Includes encapsulation with penetrating sealant sprayed onto ceiling and walls with airless sprayer.
	Minimum Charge	Job		4250		4250	
Scrape Acoustical Ceiling							
	Demolish	S.F.	.07	1.20		1.27	Minimum charge to remove asbestos. Includes full Tyvek suits for workers changed four times per eight-hour shift with respirators, air monitoring and supervision by qualified professionals.
	Minimum Charge	Job		4250		4250	

Job Preparation

Asbestos Removal

Asbestos Removal		Unit	Material	Labor	Equip.	Total	Specification
Seal Ceiling Asbestos	Install	S.F.	2.42	4.01		6.43	Includes attaching plastic cover to ceiling area and sealing where asbestos exists, taping all seams, caulk or foam insulation in cracks and negative air system.
	Minimum Charge	Job		4250		4250	
Seal Wall Asbestos	Install	S.F.	2.42	3.56		5.98	Includes attaching plastic cover to wall area and sealing where asbestos exists, taping all seams, caulk or foam insulation in cracks and negative air system.
	Minimum Charge	Job		4250		4250	
Seal Floor Asbestos	Install	S.F.	3.87	7.10		10.97	Includes attaching plastic cover to floor area and sealing where asbestos exists, taping all seams, caulk or foam insulation in cracks and negative air system.
	Minimum Charge	Job		4250		4250	
Negative Air Vent System	Install	Day			47.50	47.50	Daily rental of negative air vent system.
	Minimum Charge	Week			238	238	
HEPA Vacuum Cleaner	Install	Day			19.35	19.35	Daily rental of HEPA vacuum, 16 gallon.
	Minimum Charge	Week			97	97	
Airless Sprayer	Install	Day			87	87	Daily rental of airless sprayer.
	Minimum Charge	Week			435	435	
Decontamination Unit	Install	Day			152	152	Daily rental of decontamination unit.
	Minimum Charge	Week			760	760	
Light Stand	Install	Day			17.50	17.50	Daily rental of floodlight w/tripod.
	Minimum Charge	Week			87.50	87.50	

Room Demolition

Room Demolition		Unit	Material	Labor	Equip.	Total	Specification
Strip Typical Room	Demolish	S.F.		3.30		3.30	Includes 4 labor hours for general demolition.
	Minimum Charge	Job		183		183	
Strip Typical Bathroom	Demolish	S.F.		4.40		4.40	Includes 4 labor hours for general demolition.
	Minimum Charge	Job		183		183	
Strip Typical Kitchen	Demolish	S.F.		4.23		4.23	Includes 4 labor hours for general demolition.
	Minimum Charge	Job		183		183	
Strip Typical Laundry	Demolish	S.F.		3.77		3.77	Includes 4 labor hours for general demolition.
	Minimum Charge	Job		183		183	

13

Job Preparation

Selective Demolition		Unit	Material	Labor	Equip.	Total	Specification
Concrete Slab							
4" Unreinforced							
	Demolish	S.F.		.24	.56	.80	Includes labor and equipment to
	Minimum Charge	Job		530	1250	1780	remove unreinforced concrete slab and haul debris to truck or dumpster.
6" Unreinforced							
	Demolish	S.F.		.26	.63	.89	Includes labor and equipment to
	Minimum Charge	Job		530	1250	1780	remove unreinforced concrete slab and haul debris to truck or dumpster.
8" Unreinforced							
	Demolish	S.F.		.30	.72	1.02	Includes labor and equipment to
	Minimum Charge	Job		530	1250	1780	remove unreinforced concrete slab and haul debris to truck or dumpster.
4" Reinforced							
	Demolish	S.F.		.24	.56	.80	Includes labor and equipment to
	Minimum Charge	Job		530	1250	1780	remove reinforced concrete slab and haul debris to truck or dumpster.
6" Reinforced							
	Demolish	S.F.		.26	.63	.89	Includes labor and equipment to
	Minimum Charge	Job		530	1250	1780	remove reinforced concrete slab and haul debris to truck or dumpster.
8" Reinforced							
	Demolish	S.F.		.30	.72	1.02	Includes labor and equipment to
	Minimum Charge	Job		530	1250	1780	remove reinforced concrete slab and haul debris to truck or dumpster.
Brick Wall							
4" Thick							
	Demolish	S.F.		3.05	.46	3.51	Includes labor and equipment to
	Minimum Charge	Job		930	140	1070	remove brick wall and haul debris to a central location for disposal.
8" Thick							
	Demolish	S.F.		4.19	.63	4.82	Includes labor and equipment to
	Minimum Charge	Job		930	140	1070	remove brick wall and haul debris to a central location for disposal.
12" Thick							
	Demolish	S.F.		5.60	.84	6.44	Includes labor and equipment to
	Minimum Charge	Job		930	140	1070	remove brick wall and haul debris to a central location for disposal.
Concrete Masonry Wall							
4" Thick							
	Demolish	S.F.		3.39	.51	3.90	Includes minimum labor and
	Minimum Charge	Job		930	140	1070	equipment to remove 4" block wall and haul debris to a central location for disposal.
6" Thick							
	Demolish	S.F.		4.19	.63	4.82	Includes minimum labor and
	Minimum Charge	Job		930	140	1070	equipment to remove 6" block wall and haul debris to a central location for disposal.

Job Preparation

Selective Demolition		Unit	Material	Labor	Equip.	Total	Specification
8" Thick							
	Demolish	S.F.		5.15	.77	5.92	Includes minimum labor and
	Minimum Charge	Job		930	140	1070	equipment to remove 8" block wall
							and haul debris to a central location
							for disposal.
12" Thick							
	Demolish	S.F.		7.45	1.12	8.57	Includes minimum labor and
	Minimum Charge	Job		930	140	1070	equipment to remove 12" block wall
							and haul debris to a central location
							for disposal.
Tree Removal							
Small, 4" - 8"							
	Demolish	Ea.		119	102	221	Minimum charge, common labor
	Minimum Charge	Job		165		165	trade.
Medium, 9" - 12"							
	Demolish	Ea.		179	154	333	Minimum charge, common labor
	Minimum Charge	Job		165		165	trade.
Large, 13" - 18"							
	Demolish	Ea.		214	184	398	Minimum charge, tree removal.
	Minimum Charge	Job		555	229	784	
Very Large, 19" - 24"							
	Demolish	Ea.		268	230	498	Minimum charge, tree removal.
	Minimum Charge	Job		555	229	784	
Stump Removal							
	Demolish	Ea.		79.50	166	245.50	Includes minimum charges for a site
	Minimum Charge	Job		310	112	422	demolition crew and backhoe /
							loader.

For customer support on your Contractor's Pricing Guide: Residential Repair & Remodeling, call 888.606.7279.

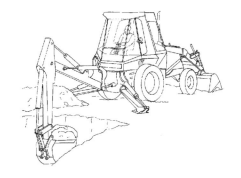

Backhoe/Loader - Wheel Type   Tractor Loader - Wheel Type   Backhoe - Crawler Type

Excavation		Unit	Material	Labor	Equip.	Total	Specification
Footings							
Hand Trenching							
	Install	L.F.		2.06		2.06	Includes hand excavation for footing
	Minimum Charge	Job		82.50		82.50	or trench up to 10″ deep and 12″
							wide.
Machine Trenching							
	Install	L.F.		.80	.40	1.20	Includes machine excavation for
	Minimum Charge	Job		400	201	601	footing or trench up to 10″ deep and
							12″ wide.
Trenching w / Backfill							
Hand							
	Install	L.F.		3.30		3.30	Includes hand excavation, backfill and
	Minimum Charge	Job		82.50		82.50	compaction.
Machine							
	Install	L.F.		3.98	15.75	19.73	Includes machine excavation, backfill
	Minimum Charge	Job		165	37.50	202.50	and compaction.

Backfill		Unit	Material	Labor	Equip.	Total	Specification
Hand (no compaction)							
	Install	L.C.Y.		27.50		27.50	Includes backfill by hand in average
	Minimum Charge	Job		165		165	soil without compaction.
Hand (w / compaction)							
	Install	E.C.Y.		39		39	Includes backfill by hand in average
	Minimum Charge	Job		165		165	soil with compaction.
Machine (no compaction)							
	Install	L.C.Y.		3.98	2.01	5.99	Includes backfill, by 55 HP wheel
	Minimum Charge	Job		400	201	601	loader, of trenches from loose material
							piled adjacent without compaction.
Machine (w / compaction)							
	Install	L.C.Y.		10.60	5.35	15.95	Includes backfill, by 55 HP wheel
	Minimum Charge	Job		400	201	601	loader, of trenches from loose material
							piled adjacent with compaction by
							vibrating tampers.

17

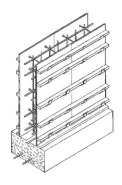

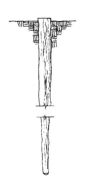

Concrete Footing and Formwork   **Wood Pile**

Piles	Unit	Material	Labor	Equip.	Total	Specification
Treated Wood, 12" Butt						
To 30'						
Demolish	L.F.		5.35	1.94	7.29	Includes material, labor and
Install	V.L.F.	14.40	5.10	3	22.50	equipment to install up to 30' long
Demolish and Install	V.L.F.	14.40	10.45	4.94	29.79	treated wood piling.
Paint	L.F.	.43	2.16		2.59	
Minimum Charge	Job		3175	1875	5050	
30' to 40'						
Demolish	L.F.		5.35	1.94	7.29	Includes material, labor and
Install	V.L.F.	15.95	4.54	2.68	23.17	equipment to install 30' to 40' long
Demolish and Install	V.L.F.	15.95	9.89	4.62	30.46	treated wood piling.
Paint	L.F.	.43	2.16		2.59	
Minimum Charge	Job		3175	1875	5050	
40' to 50'						
Demolish	L.F.		5.35	1.94	7.29	Includes material, labor and
Install	V.L.F.	17.05	4.41	2.61	24.07	equipment to install 40' to 50' long
Demolish and Install	V.L.F.	17.05	9.76	4.55	31.36	treated wood piling.
Paint	L.F.	.43	2.16		2.59	
Minimum Charge	Job		3175	1875	5050	
50' to 60'						
Demolish	L.F.		5.35	1.94	7.29	Includes material, labor and
Install	V.L.F.	19.10	3.97	2.35	25.42	equipment to install 50' to 60' long
Demolish and Install	V.L.F.	19.10	9.32	4.29	32.71	treated wood piling.
Paint	L.F.	.43	2.16		2.59	
Minimum Charge	Job		3175	1875	5050	
60' to 80'						
Demolish	L.F.		5.35	1.94	7.29	Includes material, labor and
Install	V.L.F.	25.50	3.78	2.82	32.10	equipment to install 60' to 80' long
Demolish and Install	V.L.F.	25.50	9.13	4.76	39.39	treated wood piling.
Paint	L.F.	.43	2.16		2.59	
Minimum Charge	Job		3175	1875	5050	
Untreated Wood, 12" Butt						
To 30'						
Demolish	L.F.		5.35	1.94	7.29	Includes material, labor and
Install	V.L.F.	14.85	5.10	3	22.95	equipment to install up to 30' long
Demolish and Install	V.L.F.	14.85	10.45	4.94	30.24	untreated wood piling.
Paint	L.F.	.43	2.16		2.59	
Minimum Charge	Job		3175	1875	5050	

Foundation

Piles	Unit	Material	Labor	Equip.	Total	Specification
30' to 40'						
Demolish	L.F.		5.35	1.94	7.29	Includes material, labor and
Install	V.L.F.	14.85	4.54	2.68	22.07	equipment to install 30' to 40' long
Demolish and Install	V.L.F.	14.85	9.89	4.62	29.36	untreated wood piling.
Paint	L.F.	.43	2.16		2.59	
Minimum Charge	Job		3175	1875	5050	
40' to 50'						
Demolish	L.F.		5.35	1.94	7.29	Includes material, labor and
Install	V.L.F.	14.85	4.41	2.61	21.87	equipment to install 40' to 50' long
Demolish and Install	V.L.F.	14.85	9.76	4.55	29.16	untreated wood piling.
Paint	L.F.	.43	2.16		2.59	
Minimum Charge	Job		3175	1875	5050	
50' to 60'						
Demolish	L.F.		5.35	1.94	7.29	Includes material, labor and
Install	V.L.F.	14.85	3.97	2.35	21.17	equipment to install 50' to 60' long
Demolish and Install	V.L.F.	14.85	9.32	4.29	28.46	untreated wood piling.
Paint	L.F.	.43	2.16		2.59	
Minimum Charge	Job		3175	1875	5050	
60' to 80'						
Demolish	L.F.		5.35	1.94	7.29	Includes material, labor and
Install	V.L.F.	18.60	3.78	2.23	24.61	equipment to install 60' to 80' long
Demolish and Install	V.L.F.	18.60	9.13	4.17	31.90	untreated wood piling.
Paint	L.F.	.43	2.16		2.59	
Minimum Charge	Job		3175	1875	5050	
Precast Concrete						
10" Square						
Demolish	L.F.		5.45	1.99	7.44	Includes material, labor and
Install	V.L.F.	14.25	4.54	2.68	21.47	equipment to install a precast,
Demolish and Install	V.L.F.	14.25	9.99	4.67	28.91	prestressed, 40' long, 10" square
Paint	L.F.	.43	2.16		2.59	concrete pile.
Minimum Charge	Job		3175	1875	5050	
12" Square						
Demolish	L.F.		5.85	2.13	7.98	Includes material, labor and
Install	V.L.F.	20.50	4.67	2.76	27.93	equipment to install a precast,
Demolish and Install	V.L.F.	20.50	10.52	4.89	35.91	prestressed, 40' long, 12" square
Paint	L.F.	.43	2.16		2.59	concrete pile.
Minimum Charge	Job		3175	1875	5050	
14" Square						
Demolish	L.F.		6	2.18	8.18	Includes material, labor and
Install	V.L.F.	24	5.30	3.13	32.43	equipment to install a precast,
Demolish and Install	V.L.F.	24	11.30	5.31	40.61	prestressed, 40' long, 14" square
Paint	L.F.	.43	2.16		2.59	concrete pile.
Minimum Charge	Job		3175	1875	5050	
16" Square						
Demolish	L.F.		6.30	2.29	8.59	Includes material, labor and
Install	V.L.F.	33.50	5.65	3.35	42.50	equipment to install a precast,
Demolish and Install	V.L.F.	33.50	11.95	5.64	51.09	prestressed, 40' long, 16" square
Paint	L.F.	.43	2.16		2.59	concrete pile.
Minimum Charge	Job		3175	1875	5050	
18" Square						
Demolish	L.F.		6.85	2.48	9.33	Includes material, labor and
Install	V.L.F.	40	6.10	4.56	50.66	equipment to install a precast,
Demolish and Install	V.L.F.	40	12.95	7.04	59.99	prestressed, 40' long, 18" square
Paint	L.F.	.43	2.16		2.59	concrete pile.
Minimum Charge	Job		3175	1875	5050	

Foundation

Piles	Unit	Material	Labor	Equip.	Total	Specification
Cross Bracing						
Treated 2″ x 6″						
Demolish	L.F.		1.08		1.08	Cost includes material and labor to
Install	L.F.	.85	6.35		7.20	install 2″ x 6″ lumber for cross-bracing
Demolish and Install	L.F.	.85	7.43		8.28	of foundation piling or posts.
Reinstall	L.F.		5.09		5.09	
Clean	S.F.		.41		.41	
Paint	L.F.	.31	1.45		1.76	
Minimum Charge	Job		790		790	
Treated 2″ x 8″						
Demolish	L.F.		1.08		1.08	Cost includes material and labor to
Install	L.F.	1.17	6.55		7.72	install 2″ x 8″ lumber for cross-bracing
Demolish and Install	L.F.	1.17	7.63		8.80	of foundation piling or posts.
Reinstall	L.F.		5.26		5.26	
Clean	S.F.		.41		.41	
Paint	L.F.	.41	1.51		1.92	
Minimum Charge	Job		790		790	
Treated 2″ x 10″						
Demolish	L.F.		1.11		1.11	Cost includes material and labor to
Install	L.F.	1.49	6.80		8.29	install 2″ x 10″ lumber for
Demolish and Install	L.F.	1.49	7.91		9.40	cross-bracing of foundation piling or
Reinstall	L.F.		5.44		5.44	posts.
Clean	S.F.		.41		.41	
Paint	L.F.	.48	1.54		2.02	
Minimum Charge	Job		790		790	
Treated 2″ x 12″						
Demolish	L.F.		1.33		1.33	Cost includes material and labor to
Install	L.F.	2.09	7.05		9.14	install 2″ x 12″ lumber for
Demolish and Install	L.F.	2.09	8.38		10.47	cross-bracing of foundation piling or
Reinstall	L.F.		5.63		5.63	posts.
Clean	S.F.		.41		.41	
Paint	L.F.	.57	1.57		2.14	
Minimum Charge	Job		790		790	
Metal Pipe						
6″ Concrete Filled						
Demolish	L.F.		7.60	2.76	10.36	Includes material, labor and
Install	V.L.F.	21	6.60	3.91	31.51	equipment to install concrete filled
Demolish and Install	V.L.F.	21	14.20	6.67	41.87	steel pipe pile.
Paint	L.F.	.32	2.16		2.48	
Minimum Charge	Job		3175	1875	5050	
6″ Unfilled						
Demolish	L.F.		7.60	2.76	10.36	Includes material, labor and
Install	V.L.F.	18.30	6.05	3.57	27.92	equipment to install unfilled steel pipe
Demolish and Install	V.L.F.	18.30	13.65	6.33	38.28	pile.
Paint	L.F.	.32	2.16		2.48	
Minimum Charge	Job		3175	1875	5050	
8″ Concrete Filled						
Demolish	L.F.		7.60	2.76	10.36	Includes material, labor and
Install	V.L.F.	23	6.90	4.08	33.98	equipment to install concrete filled
Demolish and Install	V.L.F.	23	14.50	6.84	44.34	steel pipe pile.
Paint	L.F.	.32	2.16		2.48	
Minimum Charge	Job		3175	1875	5050	
8″ Unfilled						
Demolish	L.F.		7.60	2.76	10.36	Includes material, labor and
Install	V.L.F.	21	6.35	3.75	31.10	equipment to install unfilled steel pipe
Demolish and Install	V.L.F.	21	13.95	6.51	41.46	pile.
Paint	L.F.	.32	2.16		2.48	
Minimum Charge	Job		3175	1875	5050	

Piles	Unit	Material	Labor	Equip.	Total	Specification
10″ Concrete Filled						
Demolish	L.F.		7.85	2.84	10.69	Includes material, labor and
Install	V.L.F.	27.50	7.05	4.17	38.72	equipment to install concrete filled
Demolish and Install	V.L.F.	27.50	14.90	7.01	49.41	steel pipe pile.
Paint	L.F.	.32	2.16		2.48	
Minimum Charge	Job		3175	1875	5050	
10″ Unfilled						
Demolish	L.F.		7.85	2.84	10.69	Includes material, labor and
Install	V.L.F.	24	6.35	3.75	34.10	equipment to install unfilled steel pipe
Demolish and Install	V.L.F.	24	14.20	6.59	44.79	pile.
Paint	L.F.	.32	2.16		2.48	
Minimum Charge	Job		3175	1875	5050	
12″ Concrete Filled						
Demolish	L.F.		7.85	2.84	10.69	Includes material, labor and
Install	V.L.F.	35	7.65	4.52	47.17	equipment to install concrete filled
Demolish and Install	V.L.F.	35	15.50	7.36	57.86	steel pipe pile.
Paint	L.F.	.32	2.16		2.48	
Minimum Charge	Job		3175	1875	5050	
12″ Unfilled						
Demolish	L.F.		7.85	2.84	10.69	Includes material, labor and
Install	V.L.F.	29.50	6.70	3.95	40.15	equipment to install unfilled steel pipe
Demolish and Install	V.L.F.	29.50	14.55	6.79	50.84	pile.
Paint	L.F.	.32	2.16		2.48	
Minimum Charge	Job		3175	1875	5050	
Steel H Section						
H Section 8 x 8 x 36#						
Demolish	L.F.		5.30	1.93	7.23	Includes material, labor and
Install	V.L.F.	18.20	4.96	2.93	26.09	equipment to install steel H section
Demolish and Install	V.L.F.	18.20	10.26	4.86	33.32	pile.
Paint	S.F.	.12	.33		.45	
Minimum Charge	Job		3175	1875	5050	
H Section 10 x 10 x 57#						
Demolish	L.F.		5.80	2.11	7.91	Includes material, labor and
Install	V.L.F.	28.50	5.20	3.08	36.78	equipment to install steel H section
Demolish and Install	V.L.F.	28.50	11	5.19	44.69	pile.
Paint	S.F.	.12	.33		.45	
Minimum Charge	Job		3175	1875	5050	
H Section 12 x 12 x 74#						
Demolish	L.F.		6	2.18	8.18	Includes material, labor and
Install	V.L.F.	37.50	5.40	4.02	46.92	equipment to install steel H section
Demolish and Install	V.L.F.	37.50	11.40	6.20	55.10	pile.
Paint	S.F.	.12	.33		.45	
Minimum Charge	Job		3175	1875	5050	
H Section 14 x 14 x 89#						
Demolish	L.F.		6.30	2.29	8.59	Includes material, labor and
Install	V.L.F.	45	5.90	4.39	55.29	equipment to install steel H section
Demolish and Install	V.L.F.	45	12.20	6.68	63.88	pile.
Paint	S.F.	.12	.33		.45	
Minimum Charge	Job		3175	1875	5050	

For customer support on your Contractor's Pricing Guide: Residential Repair & Remodeling, call 888.606.7279.

Foundation

Foundation Post

	Unit	Material	Labor	Equip.	Total	Specification
Wood Foundation Post						
4″ x 4″						
Demolish	L.F.		.86	.31	1.17	Cost includes material and labor to
Install	L.F.	1.36	5.50		6.86	install 4″ x 4″ treated wood post by
Demolish and Install	L.F.	1.36	6.36	.31	8.03	hand in normal soil conditions.
Reinstall	L.F.		4.40		4.40	
Minimum Charge	Job		790		790	
4″ x 6″						
Demolish	L.F.		.86	.31	1.17	Cost includes material and labor to
Install	L.F.	1.98	5.50		7.48	install 4″ x 6″ treated wood post by
Demolish and Install	L.F.	1.98	6.36	.31	8.65	hand in normal soil conditions.
Reinstall	L.F.		4.40		4.40	
Minimum Charge	Job		790		790	
6″ x 6″						
Demolish	L.F.		.86	.31	1.17	Cost includes material and labor to
Install	L.F.	3.19	5.50		8.69	install 6″ x 6″ treated wood post by
Demolish and Install	L.F.	3.19	6.36	.31	9.86	hand in normal soil conditions.
Reinstall	L.F.		4.40		4.40	
Minimum Charge	Job		790		790	
Water Jet						
Demolish	Ea.		98		98	Includes material, labor and
Install	Ea.	73.50	755		828.50	equipment to water jet a 6″ x 6″ wood
Demolish and Install	Ea.	73.50	853		926.50	post under an existing building. Cost
Reinstall	Ea.		754.29		754.29	may vary depending upon location
Minimum Charge	Job		660		660	and access.
Shim						
Install	Ea.	4.40	79		83.40	Includes labor and material to trim or
Minimum Charge	Job		790		790	readjust existing foundation posts under an existing building. Access of 2 - 4 foot assumed.
Cross-bracing						
Treated 2″ x 6″						
Demolish	L.F.		1.08		1.08	Cost includes material and labor to
Install	L.F.	.85	6.35		7.20	install 2″ x 6″ lumber for cross-bracing
Demolish and Install	L.F.	.85	7.43		8.28	of foundation piling or posts.
Reinstall	L.F.		5.09		5.09	
Clean	S.F.		.41		.41	
Paint	L.F.	.31	1.45		1.76	
Minimum Charge	Job		790		790	
Treated 2″ x 8″						
Demolish	L.F.		1.08		1.08	Cost includes material and labor to
Install	L.F.	1.17	6.55		7.72	install 2″ x 8″ lumber for cross-bracing
Demolish and Install	L.F.	1.17	7.63		8.80	of foundation piling or posts.
Reinstall	L.F.		5.26		5.26	
Clean	S.F.		.41		.41	
Paint	L.F.	.41	1.51		1.92	
Minimum Charge	Job		790		790	
Treated 2″ x 10″						
Demolish	L.F.		1.11		1.11	Cost includes material and labor to
Install	L.F.	1.49	6.80		8.29	install 2″ x 10″ lumber for
Demolish and Install	L.F.	1.49	7.91		9.40	cross-bracing of foundation piling or
Reinstall	L.F.		5.44		5.44	posts.
Clean	S.F.		.41		.41	
Paint	L.F.	.48	1.54		2.02	
Minimum Charge	Job		790		790	

For customer support on your Contractor's Pricing Guide: Residential Repair & Remodeling, call 888.606.7279.

Foundation

Foundation Post

	Unit	Material	Labor	Equip.	Total	Specification
Treated 2″ x 12″						
Demolish	L.F.		1.33		1.33	Cost includes material and labor to
Install	L.F.	2.09	7.05		9.14	install 2″ x 12″ lumber for
Demolish and Install	L.F.	2.09	8.38		10.47	cross-bracing of foundation piling or
Reinstall	L.F.		5.63		5.63	posts.
Clean	S.F.		.41		.41	
Paint	L.F.	.57	1.57		2.14	
Minimum Charge	Job		790		790	

Concrete Footing

	Unit	Material	Labor	Equip.	Total	Specification
Continuous						
12″ w x 6″ d						
Demolish	L.F.		.59	1.39	1.98	Includes material and labor to install
Install	L.F.	4.43	3.69	.02	8.14	continuous reinforced concrete footing
Demolish and Install	L.F.	4.43	4.28	1.41	10.12	poured by chute cast against earth
Minimum Charge	Job		910		910	including reinforcing, forming and
						finishing.
12″ w x 12″ d						
Demolish	L.F.		.71	1.67	2.38	Includes material and labor to install
Install	L.F.	9.90	7.35	.04	17.29	continuous reinforced concrete footing
Demolish and Install	L.F.	9.90	8.06	1.71	19.67	poured by chute cast against earth
Minimum Charge	Job		910		910	including reinforcing, forming and
						finishing.
18″ w x 10″ d						
Demolish	L.F.		.71	1.67	2.38	Includes material and labor to install
Install	L.F.	9.25	9.10	.05	18.40	continuous reinforced concrete footing
Demolish and Install	L.F.	9.25	9.81	1.72	20.78	poured by chute cast against earth
Minimum Charge	Job		910		910	including reinforcing, forming and
						finishing.
24″ w x 24″ d						
Demolish	L.F.		1.06	2.51	3.57	Includes material and labor to install
Install	L.F.	27.50	29.50	.17	57.17	continuous reinforced concrete footing
Demolish and Install	L.F.	27.50	30.56	2.68	60.74	poured by chute cast against earth
Minimum Charge	Job		910		910	including reinforcing, forming and
						finishing.
Stem Wall						
Single Story						
Demolish	L.F.		14.90	12.90	27.80	Includes material and labor to install
Install	L.F.	23	79	68	170	stem wall 24″ above and 18″ below
Demolish and Install	L.F.	23	93.90	80.90	197.80	grade for a single story structure.
Minimum Charge	Job		910		910	Includes 12″ wide and 8″ deep
						footing.
Two Story						
Demolish	L.F.		14.90	12.90	27.80	Includes material and labor to install
Install	L.F.	28	113	98	239	stem wall 24″ above and 18″ below
Demolish and Install	L.F.	28	127.90	110.90	266.80	grade typical for a two story structure,
Minimum Charge	Job		910		910	includes 18″ wide and 10″ deep
						footing.

Foundation

Concrete Slab	Unit	Material	Labor	Equip.	Total	Specification
Reinforced						
4″						
Demolish	S.F.		.24	.56	.80	Cost includes material and labor to
Install	S.F.	2.31	.42		2.73	install 4″ reinforced concrete
Demolish and Install	S.F.	2.31	.66	.56	3.53	slab-on-grade poured by chute
Clean	S.F.	.09	.37		.46	including forms, vapor barrier, wire
Paint	S.F.	.20	.82		1.02	mesh, 3000 PSI concrete, float finish,
Minimum Charge	Job		207		207	and curing.
6″						
Demolish	S.F.		.26	.63	.89	Cost includes material and labor to
Install	S.F.	3.15	.64	.01	3.80	install 6″ reinforced concrete
Demolish and Install	S.F.	3.15	.90	.64	4.69	slab-on-grade poured by chute
Clean	S.F.	.09	.37		.46	including forms, vapor barrier, wire
Paint	S.F.	.20	.82		1.02	mesh, 3000 PSI concrete, float finish,
Minimum Charge	Job		207		207	and curing.
8″						
Demolish	S.F.		.30	.72	1.02	Cost includes material and labor to
Install	S.F.	4.10	.83	.01	4.94	install 8″ reinforced concrete
Demolish and Install	S.F.	4.10	1.13	.73	5.96	slab-on-grade poured by chute
Clean	S.F.	.09	.37		.46	including forms, vapor barrier, wire
Paint	S.F.	.20	.82		1.02	mesh, 3000 PSI concrete, float finish,
Minimum Charge	Job		207		207	and curing.
Unreinforced						
4″						
Demolish	S.F.		.24	.56	.80	Cost includes material and labor to
Install	S.F.	1.57	1.02	.01	2.60	install 4″ concrete slab-on-grade
Demolish and Install	S.F.	1.57	1.26	.57	3.40	poured by chute, including forms,
Clean	S.F.	.09	.37		.46	vapor barrier, 3000 PSI concrete, float
Paint	S.F.	.20	.82		1.02	finish, and curing.
Minimum Charge	Job		207		207	
6″						
Demolish	S.F.		.26	.63	.89	Cost includes material and labor to
Install	S.F.	2.29	1.05	.01	3.35	install 6″ concrete slab-on-grade
Demolish and Install	S.F.	2.29	1.31	.64	4.24	poured by chute, including forms,
Clean	S.F.	.09	.37		.46	vapor barrier, 3000 PSI concrete, float
Paint	S.F.	.20	.82		1.02	finish, and curing.
Minimum Charge	Job		207		207	
8″						
Demolish	S.F.		.30	.72	1.02	Cost includes material and labor to
Install	S.F.	3.14	1.10	.01	4.25	install 8″ concrete slab-on-grade
Demolish and Install	S.F.	3.14	1.40	.73	5.27	poured by chute, including forms,
Clean	S.F.	.09	.37		.46	vapor barrier, 3000 PSI concrete, float
Paint	S.F.	.20	.82		1.02	finish, and curing.
Minimum Charge	Job		207		207	
Lightweight						
Demolish	S.F.		.98	.27	1.25	Includes material and labor to install
Install	S.F.	2.23	3.09		5.32	lightweight concrete, 3″ - 4″ thick.
Demolish and Install	S.F.	2.23	4.07	.27	6.57	
Clean	S.F.	.09	.37		.46	
Paint	S.F.	.20	.82		1.02	
Minimum Charge	Job		207		207	

For customer support on your Contractor's Pricing Guide: Residential Repair & Remodeling, call 888.606.7279.

Foundation

Concrete Slab		Unit	Material	Labor	Equip.	Total	Specification
Gunite							
	Demolish	S.F.		1.23	.45	1.68	Includes material and labor to install
	Install	S.F.	.41	1.32	.22	1.95	gunite on flat plane per inch of
	Demolish and Install	S.F.	.41	2.55	.67	3.63	thickness using 1" as base price. No
	Clean	S.F.	.09	.37		.46	forms or reinforcing are included.
	Paint	S.F.	.20	.82		1.02	
	Minimum Charge	Job		1150		1150	
Epoxy Inject							
	Install	L.F.	5.45	6.90		12.35	Includes labor and material to inject
	Minimum Charge	Job		207		207	epoxy into cracks.
Slabjacking							
	Install	S.F.	.51	1.52	.12	2.15	Includes labor, material and
	Minimum Charge	Job		207		207	equipment to install slabjacking under
							an existing building.
Scrape and Paint							
	Paint	S.F.	.21	.94		1.15	Includes labor and materials to scrape
	Minimum Charge	Job		189		189	and paint a concrete floor.
Vapor Barrier							
	Install	S.F.	.03	.12		.15	Cost includes material and labor to
	Minimum Charge	Job		227		227	install 4 mil polyethylene vapor barrier
							10' wide sheets with 6" overlaps.

For customer support on your Contractor's Pricing Guide: Residential Repair & Remodeling, call 888.606.7279.

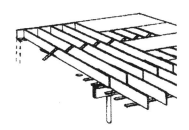  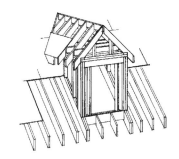

Rough Framing

Beam / Girder		Unit	Material	Labor	Equip.	Total	Specification
Solid Wood Beam							
2″ x 8″							
	Demolish	L.F.		.89		.89	Cost includes material and labor to
	Install	L.F.	.94	1.40		2.34	install 2″ x 8″ beam.
	Demolish and Install	L.F.	.94	2.29		3.23	
	Clean	L.F.	.13	.29		.42	
	Paint	L.F.	.14	2.04		2.18	
	Minimum Charge	Job		227		227	
2″ x 10″							
	Demolish	L.F.		1.10		1.10	Cost includes material and labor to
	Install	L.F.	1.47	1.51		2.98	install 2″ x 10″ beam.
	Demolish and Install	L.F.	1.47	2.61		4.08	
	Clean	L.F.	.17	.35		.52	
	Paint	L.F.	.19	1.26		1.45	
	Minimum Charge	Job		227		227	
2″ x 12″							
	Demolish	L.F.		1.33		1.33	Cost includes material and labor to
	Install	L.F.	1.87	1.65		3.52	install 2″ x 12″ beam.
	Demolish and Install	L.F.	1.87	2.98		4.85	
	Clean	L.F.	.19	.41		.60	
	Paint	L.F.	.22	3.02		3.24	
	Minimum Charge	Job		227		227	
4″ x 8″							
	Demolish	L.F.		2.74		2.74	Cost includes material and labor to
	Install	L.F.	3.26	2.05	.72	6.03	install 4″ x 8″ beam.
	Demolish and Install	L.F.	3.26	4.79	.72	8.77	
	Clean	L.F.	.17	.35		.52	
	Paint	L.F.	.19	1.26		1.45	
	Minimum Charge	Job		227		227	
4″ x 10″							
	Demolish	L.F.		3.42		3.42	Cost includes material and labor to
	Install	L.F.	5.15	2.16	.76	8.07	install 4″ x 10″ beam.
	Demolish and Install	L.F.	5.15	5.58	.76	11.49	
	Clean	L.F.	.19	.41		.60	
	Paint	L.F.	.22	1.47		1.69	
	Minimum Charge	Job		227		227	
4″ x 12″							
	Demolish	L.F.		4.11		4.11	Cost includes material and labor to
	Install	L.F.	5.70	2.28	.80	8.78	install 4″ x 12″ beam.
	Demolish and Install	L.F.	5.70	6.39	.80	12.89	
	Clean	L.F.	.21	.47		.68	
	Paint	L.F.	.22	3.02		3.24	
	Minimum Charge	Job		227		227	

Beam / Girder		Unit	Material	Labor	Equip.	Total	Specification
6″ x 8″							
	Demolish	L.F.		4.11		4.11	Cost includes material and labor to
	Install	L.F.	8.25	3.91	1.38	13.54	install 6″ x 8″ beam.
	Demolish and Install	L.F.	8.25	8.02	1.38	17.65	
	Clean	L.F.	.19	.41		.60	
	Paint	L.F.	.22	1.47		1.69	
	Minimum Charge	Job		227		227	
6″ x 10″							
	Demolish	L.F.		5.15		5.15	Cost includes material and labor to
	Install	L.F.	4.41	1.82		6.23	install 6″ x 10″ triple beam.
	Demolish and Install	L.F.	4.41	6.97		11.38	
	Clean	L.F.	.21	.47		.68	
	Paint	L.F.	.25	1.68		1.93	
	Minimum Charge	Job		227		227	
6″ x 12″							
	Demolish	L.F.		6.10		6.10	Cost includes material and labor to
	Install	L.F.	5.60	1.91		7.51	install 6″ x 12″ triple beam.
	Demolish and Install	L.F.	5.60	8.01		13.61	
	Clean	L.F.	.24	.53		.77	
	Paint	L.F.	.29	1.89		2.18	
	Minimum Charge	Job		227		227	
Steel I-type							
W8 x 31							
	Demolish	L.F.		9.55	3.66	13.21	Includes material and labor to install
	Install	L.F.	49.50	2.89	1.55	53.94	W shaped steel beam / girder.
	Demolish and Install	L.F.	49.50	12.44	5.21	67.15	
	Minimum Charge	Job		525		525	
W8 x 48							
	Demolish	L.F.		9.55	3.66	13.21	Includes material and labor to install
	Install	L.F.	77	3.03	1.62	81.65	W shaped steel beam / girder.
	Demolish and Install	L.F.	77	12.58	5.28	94.86	
	Minimum Charge	Job		525		525	
W8 x 67							
	Demolish	L.F.		9.55	3.66	13.21	Includes material and labor to install
	Install	L.F.	107	3.17	1.70	111.87	W shaped steel beam / girder.
	Demolish and Install	L.F.	107	12.72	5.36	125.08	
	Minimum Charge	Job		525		525	
W10 x 45							
	Demolish	L.F.		9.55	3.66	13.21	Includes material and labor to install
	Install	L.F.	72	3.03	1.62	76.65	W shaped steel beam / girder.
	Demolish and Install	L.F.	72	12.58	5.28	89.86	
	Minimum Charge	Job		525		525	
W10 x 68							
	Demolish	L.F.		9.55	3.66	13.21	Includes material and labor to install
	Install	L.F.	109	3.17	1.70	113.87	W shaped steel beam / girder.
	Demolish and Install	L.F.	109	12.72	5.36	127.08	
	Minimum Charge	Job		525		525	
W12 x 50							
	Demolish	L.F.		9.55	3.66	13.21	Includes material and labor to install
	Install	L.F.	80.50	3.03	1.62	85.15	W shaped steel beam / girder.
	Demolish and Install	L.F.	80.50	12.58	5.28	98.36	
	Minimum Charge	Job		525		525	
W12 x 87							
	Demolish	L.F.		14.10	5.40	19.50	Includes material and labor to install
	Install	L.F.	140	3.17	1.70	144.87	W shaped steel beam / girder.
	Demolish and Install	L.F.	140	17.27	7.10	164.37	
	Minimum Charge	Job		525		525	

Rough Frame / Structure

Beam / Girder

	Unit	Material	Labor	Equip.	Total	Specification
W12 x 120						
Demolish	L.F.		14.10	5.40	19.50	Includes material and labor to install
Install	L.F.	193	3.25	1.74	197.99	W shaped steel beam / girder.
Demolish and Install	L.F.	193	17.35	7.14	217.49	
Minimum Charge	Job		525		525	
W14 x 74						
Demolish	L.F.		14.10	5.40	19.50	Includes material and labor to install
Install	L.F.	119	3.17	1.70	123.87	W shaped steel beam / girder.
Demolish and Install	L.F.	119	17.27	7.10	143.37	
Minimum Charge	Job		525		525	
W14 x 120						
Demolish	L.F.		14.10	5.40	19.50	Includes material and labor to install
Install	L.F.	193	3.25	1.74	197.99	W shaped steel beam / girder.
Demolish and Install	L.F.	193	17.35	7.14	217.49	
Minimum Charge	Job		525		525	

Column

	Unit	Material	Labor	Equip.	Total	Specification
Concrete						
Small Diameter						
Demolish	L.F.		15.55	2.33	17.88	Includes material and labor to install
Install	L.F.	163	14.40	2.73	180.13	small diameter concrete column.
Demolish and Install	L.F.	163	29.95	5.06	198.01	
Minimum Charge	Job		720	136	856	
Large Diameter						
Demolish	L.F.		46.50	7	53.50	Includes material and labor to install
Install	L.F.	289	16.90	3.21	309.11	large diameter concrete column.
Demolish and Install	L.F.	289	63.40	10.21	362.61	
Minimum Charge	Job		720	136	856	

Floor Framing System

	Unit	Material	Labor	Equip.	Total	Specification
12" O.C.						
2" x 6" Joists						
Demolish	S.F.		.76		.76	Cost includes material and labor to
Install	S.F.	.78	.73		1.51	install 2" x 6" joists, 12" O.C.
Demolish and Install	S.F.	.78	1.49		2.27	including box or band joist. Blocking
Reinstall	S.F.		.58		.58	or bridging not included.
Minimum Charge	Job		227		227	
2" x 8" Joists						
Demolish	S.F.		.78		.78	Cost includes material and labor to
Install	S.F.	1.09	.83		1.92	install 2" x 8" joists, 12" O.C.
Demolish and Install	S.F.	1.09	1.61		2.70	including box or band joist. Blocking
Reinstall	S.F.		.66		.66	or bridging not included.
Minimum Charge	Job		227		227	
2" x 10" Joists						
Demolish	S.F.		.80		.80	Cost includes material and labor to
Install	S.F.	1.72	1.01		2.73	install 2" x 10" joists, 12" O.C.
Demolish and Install	S.F.	1.72	1.81		3.53	including box or band joist. Blocking
Reinstall	S.F.		.81		.81	or bridging not included.
Minimum Charge	Job		227		227	

Rough Frame / Structure

Floor Framing System

	Unit	Material	Labor	Equip.	Total	Specification
2" x 12" Joists						
Demolish	S.F.		.83		.83	Cost includes material and labor to
Install	S.F.	2.19	1.04		3.23	install 2" x 12" joists, 12" O.C.
Demolish and Install	S.F.	2.19	1.87		4.06	including box or band joist. Blocking
Reinstall	S.F.		.83		.83	or bridging not included.
Minimum Charge	Job		227		227	
Block / Bridge						
Demolish	Ea.		1.15		1.15	Includes material and labor to install
Install	Ea.	1.96	2.55		4.51	set of cross bridging or per block of
Demolish and Install	Ea.	1.96	3.70		5.66	solid bridging for 2" x 10" joists cut to
Minimum Charge	Job		227		227	size on site.

16" O.C.

	Unit	Material	Labor	Equip.	Total	Specification
2" x 6" Joists						
Demolish	S.F.		.57		.57	Cost includes material and labor to
Install	S.F.	.61	.54		1.15	install 2" x 6" joists including box or
Demolish and Install	S.F.	.61	1.11		1.72	band joist installed 16" O.C. Does not
Reinstall	S.F.		.43		.43	include beams, blocking or bridging.
Minimum Charge	Job		227		227	
2" x 8" Joists						
Demolish	S.F.		.59		.59	Cost includes material and labor to
Install	S.F.	.84	.62		1.46	install 2" x 8" joists including box or
Demolish and Install	S.F.	.84	1.21		2.05	band joist installed 16" O.C. Does not
Reinstall	S.F.		.50		.50	include beams, blocking or bridging.
Minimum Charge	Job		227		227	
2" x 10" Joists						
Demolish	S.F.		.60		.60	Cost includes material and labor to
Install	S.F.	1.32	.76		2.08	install 2" x 10" joists including box or
Demolish and Install	S.F.	1.32	1.36		2.68	band joist 16" O.C. Does not include
Reinstall	S.F.		.61		.61	beams, blocking or bridging.
Minimum Charge	Job		227		227	
2" x 12" Joists						
Demolish	S.F.		.62		.62	Cost includes material and labor to
Install	S.F.	1.68	.78		2.46	install 2" x 12" joists including box or
Demolish and Install	S.F.	1.68	1.40		3.08	band joist installed 16" O.C. Does not
Reinstall	S.F.		.62		.62	include beams, blocking or bridging.
Minimum Charge	Job		227		227	
Block / Bridge						
Demolish	Ea.		1.15		1.15	Includes material and labor to install
Install	Ea.	1.96	2.55		4.51	set of cross bridging or per block of
Demolish and Install	Ea.	1.96	3.70		5.66	solid bridging for 2" x 10" joists cut to
Minimum Charge	Job		227		227	size on site.

24" O.C.

	Unit	Material	Labor	Equip.	Total	Specification
2" x 6" Joists						
Demolish	S.F.		.38		.38	Cost includes material and labor to
Install	S.F.	.43	.36		.79	install 2" x 6" joists including box or
Demolish and Install	S.F.	.43	.74		1.17	band joist installed 24" O.C. Does not
Reinstall	S.F.		.29		.29	include beams, blocking or bridging.
Minimum Charge	Job		227		227	
2" x 8" Joists						
Demolish	S.F.		.39		.39	Cost includes material and labor to
Install	S.F.	.59	.41		1	install 2" x 8" joists including box or
Demolish and Install	S.F.	.59	.80		1.39	band joist installed 24" O.C. Does not
Reinstall	S.F.		.33		.33	include beams, blocking or bridging.
Minimum Charge	Job		227		227	

Floor Framing System		Unit	Material	Labor	Equip.	Total	Specification
2″ x 10″ Joists							
	Demolish	S.F.		.40		.40	Cost includes material and labor to
	Install	S.F.	.94	.50		1.44	install 2″ x 10″ joists including box or
	Demolish and Install	S.F.	.94	.90		1.84	band joist installed 24″ O.C. Does not
	Reinstall	S.F.		.40		.40	include beams, blocking or bridging.
	Minimum Charge	Job		227		227	
2″ x 12″ Joists							
	Demolish	S.F.		.42		.42	Cost includes material and labor to
	Install	S.F.	1.20	.52		1.72	install 2″ x 12″ joists including box or
	Demolish and Install	S.F.	1.20	.94		2.14	band joist installed 24″ O.C. Does not
	Reinstall	S.F.		.42		.42	include beams, blocking or bridging.
	Minimum Charge	Job		227		227	
Block / Bridge							
	Demolish	Ea.		1.15		1.15	Includes material and labor to install
	Install	Ea.	1.96	2.55		4.51	set of cross bridging or per block of
	Demolish and Install	Ea.	1.96	3.70		5.66	solid bridging for 2″ x 10″ joists cut to
	Minimum Charge	Job		227		227	size on site.
Engineered Lumber, Joist							
9-1/2″							
	Demolish	S.F.		.60		.60	Includes material and labor to install
	Install	S.F.	1.84	.45		2.29	engineered lumber truss / joists per
	Demolish and Install	S.F.	1.84	1.05		2.89	S.F. of floor area based on joists 16″
	Reinstall	S.F.		.36		.36	O.C. Beams, supports and bridging
	Minimum Charge	Job		227		227	are not included.
11-7/8″							
	Demolish	S.F.		.62		.62	Includes material and labor to install
	Install	S.F.	2.02	.46		2.48	engineered lumber truss / joists per
	Demolish and Install	S.F.	2.02	1.08		3.10	S.F. of floor area based on joists 16″
	Reinstall	S.F.		.37		.37	O.C. Beams, supports and bridging
	Minimum Charge	Job		227		227	are not included.
14″							
	Demolish	S.F.		.65		.65	Includes material and labor to install
	Install	S.F.	2.10	.50		2.60	engineered lumber truss / joists per
	Demolish and Install	S.F.	2.10	1.15		3.25	S.F. of floor area based on joists 16″
	Reinstall	S.F.		.40		.40	O.C. Beams, supports and bridging
	Minimum Charge	Job		227		227	are not included.
16″							
	Demolish	S.F.		.67		.67	Includes material and labor to install
	Install	S.F.	3.54	.52		4.06	engineered lumber truss / joists per
	Demolish and Install	S.F.	3.54	1.19		4.73	S.F. of floor area based on joists 16″
	Reinstall	S.F.		.42		.42	O.C. Beams, supports and bridging
	Minimum Charge	Job		227		227	are not included.
Block / Bridge							
	Demolish	Pr.		1.15		1.15	Includes material and labor to install
	Install	Pr.	1.03	3.36		4.39	set of steel, one-nail type cross
	Demolish and Install	Pr.	1.03	4.51		5.54	bridging for trusses placed 16″ O.C.
	Minimum Charge	Job		227		227	
Block / Bridge							
2″ x 6″							
	Demolish	Ea.		1.15		1.15	Includes material and labor to install
	Install	Ea.	.89	2.05		2.94	set of cross bridging or per block of
	Demolish and Install	Ea.	.89	3.20		4.09	solid bridging for 2″ x 6″ joists cut to
	Minimum Charge	Job		227		227	size on site.

31

Floor Framing System		Unit	Material	Labor	Equip.	Total	Specification
2" x 8"							
	Demolish	Ea.		1.15		1.15	Includes material and labor to install
	Install	Ea.	1.24	2.27		3.51	set of cross bridging or per block of
	Demolish and Install	Ea.	1.24	3.42		4.66	solid bridging for 2" x 8" joists cut to
	Minimum Charge	Job		227		227	size on site.
2" x 10"							
	Demolish	Ea.		1.15		1.15	Includes material and labor to install
	Install	Ea.	1.96	2.55		4.51	set of cross bridging or per block of
	Demolish and Install	Ea.	1.96	3.70		5.66	solid bridging for 2" x 10" joists cut to
	Minimum Charge	Job		227		227	size on site.
2" x 12"							
	Demolish	Ea.		1.15		1.15	Includes material and labor to install
	Install	Ea.	2.50	3.01		5.51	set of cross bridging or per block of
	Demolish and Install	Ea.	2.50	4.16		6.66	solid bridging for 2" x 12" joists cut to
	Minimum Charge	Job		227		227	size on site.
Ledger Strips							
1" x 2"							
	Demolish	L.F.		.31		.31	Cost includes material and labor to
	Install	L.F.	.26	1.32		1.58	install 1" x 2" ledger strip nailed to the
	Demolish and Install	L.F.	.26	1.63		1.89	face of studs, beams or joist.
	Minimum Charge	Job		227		227	
1" x 3"							
	Demolish	L.F.		.31		.31	Includes material and labor to install
	Install	L.F.	.51	1.51		2.02	up to 1" x 4" ledger strip nailed to the
	Demolish and Install	L.F.	.51	1.82		2.33	face of studs, beams or joist.
	Minimum Charge	Job		227		227	
1" x 4"							
	Demolish	L.F.		.31		.31	Includes material and labor to install
	Install	L.F.	.51	1.51		2.02	up to 1" x 4" ledger strip nailed to the
	Demolish and Install	L.F.	.51	1.82		2.33	face of studs, beams or joist.
	Minimum Charge	Job		227		227	
2" x 2"							
	Demolish	L.F.		.33		.33	Cost includes material and labor to
	Install	L.F.	.42	1.38		1.80	install 2" x 2" ledger strip nailed to the
	Demolish and Install	L.F.	.42	1.71		2.13	face of studs, beams or joist.
	Minimum Charge	Job		227		227	
2" x 4"							
	Demolish	L.F.		.37		.37	Cost includes material and labor to
	Install	L.F.	.44	1.82		2.26	install 2" x 4" ledger strip nailed to the
	Demolish and Install	L.F.	.44	2.19		2.63	face of studs, beams or joist.
	Reinstall	L.F.		1.45		1.45	
	Minimum Charge	Job		227		227	
Ledger Boards							
2" x 6"							
	Demolish	L.F.		.37		.37	Cost includes material and labor to
	Install	L.F.	5.15	2.84		7.99	install 2" x 6" ledger board fastened to
	Demolish and Install	L.F.	5.15	3.21		8.36	a wall, joists or studs.
	Reinstall	L.F.		2.27		2.27	
	Minimum Charge	Job		227		227	
4" x 6"							
	Demolish	L.F.		.61		.61	Cost includes material and labor to
	Install	L.F.	9.10	3.24		12.34	install 4" x 6" ledger board fastened to
	Demolish and Install	L.F.	9.10	3.85		12.95	a wall, joists or studs.
	Reinstall	L.F.		2.59		2.59	
	Minimum Charge	Job		227		227	

Rough Frame / Structure

Floor Framing System

	Unit	Material	Labor	Equip.	Total	Specification
4" x 8"						
Demolish	L.F.		.81		.81	Cost includes material and labor to
Install	L.F.	9.15	3.78		12.93	install 4" x 8" ledger board fastened to
Demolish and Install	L.F.	9.15	4.59		13.74	a wall, joists or studs.
Reinstall	L.F.		3.03		3.03	
Minimum Charge	Job		227		227	

Sill Plate Per L.F.

	Unit	Material	Labor	Equip.	Total	Specification
2" x 4"						
Demolish	L.F.		.31		.31	Cost includes material and labor to
Install	L.F.	.57	1.65		2.22	install 2" x 4" pressure treated lumber,
Demolish and Install	L.F.	.57	1.96		2.53	drilled and installed on foundation
Reinstall	L.F.		1.32		1.32	bolts 48" O.C. Bolts, nuts and washers
Minimum Charge	Job		227		227	are not included.
2" x 6"						
Demolish	L.F.		.47		.47	Cost includes material and labor to
Install	L.F.	.84	1.82		2.66	install 2" x 6" pressure treated lumber,
Demolish and Install	L.F.	.84	2.29		3.13	drilled and installed on foundation
Reinstall	L.F.		1.45		1.45	bolts 48" O.C. Bolts, nuts and washers
Minimum Charge	Job		227		227	not included.
2" x 8"						
Demolish	L.F.		.63		.63	Cost includes material and labor to
Install	L.F.	1.14	2.02		3.16	install 2" x 8" pressure treated lumber,
Demolish and Install	L.F.	1.14	2.65		3.79	drilled and installed on foundation
Reinstall	L.F.		1.61		1.61	bolts 48" O.C. Bolts, nuts and washers
Minimum Charge	Job		227		227	not included.

Earthquake Strapping

	Unit	Material	Labor	Equip.	Total	Specification
Install	Ea.	3.23	2.84		6.07	Includes labor and material to replace
Minimum Charge	Job		227		227	an earthquake strap.

Subflooring

	Unit	Material	Labor	Equip.	Total	Specification
Plywood						
1/2"						
Demolish	S.F.		.95		.95	Cost includes material and labor to
Install	SF Flr.	.57	.49		1.06	install 1/2" plywood subfloor, CD
Demolish and Install	SF Flr.	.57	1.44		2.01	standard interior grade, plugged and
Minimum Charge	Job		227		227	touch sanded.
5/8"						
Demolish	S.F.		.96		.96	Cost includes material and labor to
Install	SF Flr.	.64	.54		1.18	install 5/8" plywood subfloor, CD
Demolish and Install	SF Flr.	.64	1.50		2.14	standard interior grade, plugged and
Reinstall	SF Flr.		.43		.43	touch sanded.
Minimum Charge	Job		227		227	
3/4"						
Demolish	S.F.		.98		.98	Cost includes material and labor to
Install	SF Flr.	.76	.59		1.35	install 3/4" plywood subfloor, CD
Demolish and Install	SF Flr.	.76	1.57		2.33	standard interior grade, plugged and
Reinstall	SF Flr.		.47		.47	touch sanded.
Minimum Charge	Job		227		227	

Rough Frame / Structure

Subflooring

Subflooring	Unit	Material	Labor	Equip.	Total	Specification
Particle Board						
1/2″						
Demolish	S.F.		.48		.48	Cost includes material and labor to
Install	SF Flr.	.50	.61		1.11	install 1/2″ particle board subfloor.
Demolish and Install	SF Flr.	.50	1.09		1.59	
Minimum Charge	Job		227		227	
5/8″						
Demolish	S.F.		.48		.48	Cost includes material and labor to
Install	SF Flr.	.59	.66		1.25	install 5/8″ particle board subfloor.
Demolish and Install	SF Flr.	.59	1.14		1.73	
Minimum Charge	Job		227		227	
3/4″						
Demolish	S.F.		.49		.49	Cost includes material and labor to
Install	SF Flr.	.75	.73		1.48	install 3/4″ particle board subfloor.
Demolish and Install	SF Flr.	.75	1.22		1.97	
Minimum Charge	Job		227		227	
Plank Board						
Demolish	S.F.		1.55		1.55	Cost includes material and labor to
Install	SF Flr.	1.69	1.01		2.70	install 1″ x 6″ standard grade plank
Demolish and Install	SF Flr.	1.69	2.56		4.25	flooring.
Reinstall	SF Flr.		1.01		1.01	
Minimum Charge	Job		227		227	
Felt Underlay						
Demolish	S.F.		.09		.09	Includes material and labor to install
Install	S.F.	.06	.12		.18	#15 felt building paper.
Demolish and Install	S.F.	.06	.21		.27	
Minimum Charge	Job		227		227	
Prep (for flooring)						
Install	S.F.		.38		.38	Preparation of subflooring for
Minimum Charge	Job		202		202	installation of new finished flooring. Cost reflects average time to prep area.

Wall Framing System

Wall Framing System	Unit	Material	Labor	Equip.	Total	Specification
2″ x 4″						
8′ High						
Demolish	L.F.		4.89		4.89	Cost includes material and labor to
Install	L.F.	4.51	7.55		12.06	install 2″ x 4″ x 8′ high wall system,
Demolish and Install	L.F.	4.51	12.44		16.95	including studs, treated bottom plate,
Reinstall	L.F.		6.05		6.05	double top plate and one row of fire
Minimum Charge	Job		227		227	blocking.
9′ High						
Demolish	L.F.		5.45		5.45	Cost includes material and labor to
Install	L.F.	4.84	7.55		12.39	install 2″ x 4″ x 9′ wall system
Demolish and Install	L.F.	4.84	13		17.84	including studs, treated bottom plate,
Reinstall	L.F.		6.05		6.05	double top plate and one row of fire
Minimum Charge	Job		227		227	blocking.
10′ High						
Demolish	L.F.		6.10		6.10	Cost includes material and labor to
Install	L.F.	5.15	7.55		12.70	install 2″ x 4″ x 10′ wall system
Demolish and Install	L.F.	5.15	13.65		18.80	including studs, treated bottom plate,
Reinstall	L.F.		6.05		6.05	double top plate and one row of fire
Minimum Charge	Job		227		227	blocking.

Wall Framing System	Unit	Material	Labor	Equip.	Total	Specification
12' High						
Demolish	L.F.		7.35		7.35	Cost includes material and labor to
Install	L.F.	5.85	9.45		15.30	install 2" x 4" x 12' wall system
Demolish and Install	L.F.	5.85	16.80		22.65	including studs, treated bottom plate,
Reinstall	L.F.		7.57		7.57	double top plate and one row of fire
Minimum Charge	Job		227		227	blocking.
2" x 6"						
8' High						
Demolish	L.F.		4.58		4.58	Cost includes material and labor to
Install	L.F.	6.90	8.40		15.30	install 2" x 6" x 8' wall system
Demolish and Install	L.F.	6.90	12.98		19.88	including studs, treated bottom plate,
Reinstall	L.F.		6.73		6.73	double top plate and one row of fire
Minimum Charge	Job		227		227	blocking.
9' High						
Demolish	L.F.		5.15		5.15	Cost includes material and labor to
Install	L.F.	7.45	8.40		15.85	install 2" x 6" x 9' wall system
Demolish and Install	L.F.	7.45	13.55		21	including studs, treated bottom plate,
Reinstall	L.F.		6.73		6.73	double top plate and one row of fire
Minimum Charge	Job		227		227	blocking.
10' High						
Demolish	L.F.		5.65		5.65	Cost includes material and labor to
Install	L.F.	7.90	8.40		16.30	install 2" x 6" x 10' wall system
Demolish and Install	L.F.	7.90	14.05		21.95	including studs, treated bottom plate,
Reinstall	L.F.		6.73		6.73	double top plate and one row of fire
Minimum Charge	Job		227		227	blocking.
12' High						
Demolish	L.F.		6.65		6.65	Cost includes material and labor to
Install	L.F.	8.90	10.30		19.20	install 2" x 6" x 12' wall system
Demolish and Install	L.F.	8.90	16.95		25.85	including studs, treated bottom plate,
Reinstall	L.F.		8.25		8.25	double top plate and one row of fire
Minimum Charge	Job		227		227	blocking.
Fireblock						
2" x 4", 16" O.C. System						
Demolish	L.F.		.61		.61	Cost includes material and labor to
Install	L.F.	.44	1.51		1.95	install 2" x 4" fireblocks in wood frame
Demolish and Install	L.F.	.44	2.12		2.56	walls per L.F. of wall to be blocked.
Reinstall	L.F.		1.21		1.21	
Minimum Charge	Job		227		227	
2" x 4", 24" O.C. System						
Demolish	L.F.		.61		.61	Cost includes material and labor to
Install	L.F.	.44	1.51		1.95	install 2" x 4" fireblocks in wood frame
Demolish and Install	L.F.	.44	2.12		2.56	walls per L.F. of wall to be blocked.
Reinstall	L.F.		1.21		1.21	
Minimum Charge	Job		227		227	
2" x 6", 16" O.C. System						
Demolish	L.F.		.92		.92	Cost includes material and labor to
Install	L.F.	.67	1.51		2.18	install 2" x 6" fireblocks in wood frame
Demolish and Install	L.F.	.67	2.43		3.10	walls per L.F. of wall to be blocked.
Reinstall	L.F.		1.21		1.21	
Minimum Charge	Job		227		227	
2" x 6", 24" O.C. System						
Demolish	L.F.		.92		.92	Cost includes material and labor to
Install	L.F.	.67	1.51		2.18	install 2" x 6" fireblocks in wood frame
Demolish and Install	L.F.	.67	2.43		3.10	walls per L.F. of wall to be blocked.
Reinstall	L.F.		1.21		1.21	
Minimum Charge	Job		227		227	

Wall Framing System	Unit	Material	Labor	Equip.	Total	Specification
Bracing						
1" x 3"						
Demolish	L.F.		.35		.35	Cost includes material and labor to
Install	L.F.	.43	2.84		3.27	install 1" x 3" let-in bracing.
Demolish and Install	L.F.	.43	3.19		3.62	
Minimum Charge	Job		227		227	
1" x 4"						
Demolish	L.F.		.35		.35	Cost includes material and labor to
Install	L.F.	.51	2.84		3.35	install 1" x 4" let-in bracing.
Demolish and Install	L.F.	.51	3.19		3.70	
Minimum Charge	Job		227		227	
1" x 6"						
Demolish	L.F.		.35		.35	Cost includes material and labor to
Install	L.F.	.78	3.03		3.81	install 1" x 6" let-in bracing.
Demolish and Install	L.F.	.78	3.38		4.16	
Minimum Charge	Job		227		227	
2" x 3"						
Demolish	L.F.		.46		.46	Cost includes material and labor to
Install	L.F.	.42	3.03		3.45	install 2" x 3" let-in bracing.
Demolish and Install	L.F.	.42	3.49		3.91	
Reinstall	L.F.		2.42		2.42	
Minimum Charge	Job		227		227	
2" x 4"						
Demolish	L.F.		.46		.46	Cost includes material and labor to
Install	L.F.	.44	3.03		3.47	install 2" x 4" let-in bracing.
Demolish and Install	L.F.	.44	3.49		3.93	
Reinstall	L.F.		2.42		2.42	
Minimum Charge	Job		227		227	
2" x 6"						
Demolish	L.F.		.46		.46	Cost includes material and labor to
Install	L.F.	.67	3.24		3.91	install 2" x 6" let-in bracing.
Demolish and Install	L.F.	.67	3.70		4.37	
Reinstall	L.F.		2.59		2.59	
Minimum Charge	Job		227		227	
2" x 8"						
Demolish	L.F.		.46		.46	Cost includes material and labor to
Install	L.F.	.94	3.24		4.18	install 2" x 8" let-in bracing.
Demolish and Install	L.F.	.94	3.70		4.64	
Reinstall	L.F.		2.59		2.59	
Minimum Charge	Job		227		227	
Earthquake Strapping						
Install	Ea.	3.23	2.84		6.07	Includes labor and material to replace
Minimum Charge	Job		227		227	an earthquake strap.
Hurricane Clips						
Install	Ea.	1.23	3.13		4.36	Includes labor and material to install a
Minimum Charge	Job		227		227	hurricane clip.
Stud						
2" x 4"						
Demolish	L.F.		.37		.37	Cost includes material and labor to
Install	L.F.	.43	.83		1.26	install 2" x 4" wall stud per L.F. of stud.
Demolish and Install	L.F.	.43	1.20		1.63	
Reinstall	L.F.		.66		.66	
Minimum Charge	Job		227		227	

Wall Framing System

Wall Framing System	Unit	Material	Labor	Equip.	Total	Specification
2" x 6"						
Demolish	L.F.		.46		.46	Cost includes material and labor to
Install	L.F.	.67	.91		1.58	install 2" x 6" stud per L.F. of stud.
Demolish and Install	L.F.	.67	1.37		2.04	
Reinstall	L.F.		.73		.73	
Minimum Charge	Job		227		227	
Plates						
2" x 4"						
Demolish	L.F.		.33		.33	Cost includes material and labor to
Install	L.F.	.44	1.14		1.58	install 2" x 4" plate per L.F. of plate.
Demolish and Install	L.F.	.44	1.47		1.91	
Reinstall	L.F.		.91		.91	
Minimum Charge	Job		227		227	
2" x 6"						
Demolish	L.F.		.35		.35	Cost includes material and labor to
Install	L.F.	.67	1.21		1.88	install 2" x 6" plate per L.F. of plate.
Demolish and Install	L.F.	.67	1.56		2.23	
Reinstall	L.F.		.97		.97	
Minimum Charge	Job		227		227	
Headers						
2" x 6"						
Demolish	L.F.		.33		.33	Includes material and labor to install
Install	L.F.	.67	2.52		3.19	header over wall openings and
Demolish and Install	L.F.	.67	2.85		3.52	around floor, ceiling and roof
Clean	L.F.	.02	.42		.44	openings or flush beam.
Paint	L.F.	.11	1.51		1.62	
Minimum Charge	Job		227		227	
2" x 8"						
Demolish	L.F.		.35		.35	Includes material and labor to install
Install	L.F.	.94	2.67		3.61	header over wall openings and
Demolish and Install	L.F.	.94	3.02		3.96	around floor, ceiling and roof
Clean	L.F.	.02	.42		.44	openings or flush beam.
Paint	L.F.	.14	2.04		2.18	
Minimum Charge	Job		227		227	
2" x 10"						
Demolish	L.F.		.35		.35	Includes material and labor to install
Install	L.F.	1.47	2.84		4.31	header over wall openings and
Demolish and Install	L.F.	1.47	3.19		4.66	around floor, ceiling and roof
Clean	L.F.	.03	.42		.45	openings or flush beam.
Paint	L.F.	.18	2.52		2.70	
Minimum Charge	Job		227		227	
2" x 12"						
Demolish	L.F.		.37		.37	Includes material and labor to install
Install	L.F.	1.87	3.03		4.90	header over wall openings and
Demolish and Install	L.F.	1.87	3.40		5.27	around floor, ceiling and roof
Clean	L.F.	.04	.42		.46	openings or flush beam.
Paint	L.F.	.22	3.02		3.24	
Minimum Charge	Job		227		227	
4" x 8"						
Demolish	L.F.		.65		.65	Includes material and labor to install
Install	L.F.	3.26	3.49		6.75	header over wall openings and
Demolish and Install	L.F.	3.26	4.14		7.40	around floor, ceiling and roof
Clean	L.F.	.02	.42		.44	openings or flush beam.
Paint	L.F.	.14	2.04		2.18	
Minimum Charge	Job		227		227	

Rough Frame / Structure

Wall Framing System

Wall Framing System	Unit	Material	Labor	Equip.	Total	Specification
4″ x 10″						
Demolish	L.F.		.70		.70	Includes material and labor to install
Install	L.F.	5.15	3.78		8.93	header over wall openings and
Demolish and Install	L.F.	5.15	4.48		9.63	around floor, ceiling and roof
Clean	L.F.	.03	.42		.45	openings or flush beam.
Paint	L.F.	.18	2.52		2.70	
Minimum Charge	Job		227		227	
4″ x 12″						
Demolish	L.F.		.73		.73	Includes material and labor to install
Install	L.F.	5.70	4.78		10.48	header over wall openings and
Demolish and Install	L.F.	5.70	5.51		11.21	around floor, ceiling and roof
Clean	L.F.	.04	.42		.46	openings or flush beam.
Paint	L.F.	.22	3.02		3.24	
Minimum Charge	Job		227		227	
6″ x 8″						
Demolish	L.F.		.65		.65	Includes material and labor to install
Install	L.F.	8.25	2.52		10.77	header over wall openings and
Demolish and Install	L.F.	8.25	3.17		11.42	around floor, ceiling and roof
Clean	L.F.	.02	.42		.44	openings or flush beam.
Paint	L.F.	.14	2.04		2.18	
Minimum Charge	Job		227		227	
6″ x 10″						
Demolish	L.F.		.70		.70	Includes material and labor to install
Install	L.F.	7.20	5.50		12.70	header over wall openings and
Demolish and Install	L.F.	7.20	6.20		13.40	around floor, ceiling and roof
Clean	L.F.	.03	.42		.45	openings or flush beam.
Paint	L.F.	.18	2.52		2.70	
Minimum Charge	Job		227		227	
6″ x 12″						
Demolish	L.F.		.73		.73	Includes material and labor to install
Install	L.F.	9.10	6.50		15.60	header over wall openings and
Demolish and Install	L.F.	9.10	7.23		16.33	around floor, ceiling and roof
Clean	L.F.	.04	.42		.46	openings or flush beam.
Paint	L.F.	.22	1.05		1.27	
Minimum Charge	Job		227		227	

Rough-in Opening

Rough-in Opening	Unit	Material	Labor	Equip.	Total	Specification
Door w / 2″ x 4″ Lumber						
3′ Wide						
Demolish	Ea.		11.45		11.45	Includes material and labor to install
Install	Ea.	18.30	14.20		32.50	header, double studs each side,
Demolish and Install	Ea.	18.30	25.65		43.95	cripples, blocking and nails, up to 3′
Reinstall	Ea.		11.35		11.35	opening in 2″ x 4″ stud wall 8′ high.
Minimum Charge	Job		227		227	
4′ Wide						
Demolish	Ea.		11.45		11.45	Includes material and labor to install
Install	Ea.	19.70	14.20		33.90	header, double studs each side,
Demolish and Install	Ea.	19.70	25.65		45.35	cripples, blocking and nails, up to 4′
Reinstall	Ea.		11.35		11.35	opening in 2″ x 4″ stud wall 8′ high.
Minimum Charge	Job		227		227	
5′ Wide						
Demolish	Ea.		11.45		11.45	Includes material and labor to install
Install	Ea.	23.50	14.20		37.70	header, double studs each side,
Demolish and Install	Ea.	23.50	25.65		49.15	cripples, blocking and nails, up to 5′
Reinstall	Ea.		11.35		11.35	opening in 2″ x 4″ stud wall 8′ high.
Minimum Charge	Job		227		227	

Rough Frame / Structure

Rough-in Opening		Unit	Material	Labor	Equip.	Total	Specification
6' Wide							
	Demolish	Ea.		11.45		11.45	Includes material and labor to install
	Install	Ea.	25.50	14.20		39.70	header, double studs each side,
	Demolish and Install	Ea.	25.50	25.65		51.15	cripples, blocking and nails, up to 6'
	Reinstall	Ea.		11.35		11.35	opening in 2" x 4" stud wall 8' high.
	Minimum Charge	Job		227		227	
8' Wide							
	Demolish	Ea.		12.20		12.20	Includes material and labor to install
	Install	Ea.	38	15.15		53.15	header, double studs each side,
	Demolish and Install	Ea.	38	27.35		65.35	cripples, blocking and nails, up to 8'
	Reinstall	Ea.		12.11		12.11	opening in 2" x 4" stud wall 8' high.
	Minimum Charge	Job		227		227	
10' Wide							
	Demolish	Ea.		12.20		12.20	Includes material and labor to install
	Install	Ea.	52.50	15.15		67.65	header, double studs each side,
	Demolish and Install	Ea.	52.50	27.35		79.85	cripples, blocking and nails, up to 10'
	Reinstall	Ea.		12.11		12.11	opening in 2" x 4" stud wall 8' high.
	Minimum Charge	Job		227		227	
12' Wide							
	Demolish	Ea.		12.20		12.20	Includes material and labor to install
	Install	Ea.	71.50	15.15		86.65	header, double studs each side,
	Demolish and Install	Ea.	71.50	27.35		98.85	cripples, blocking and nails, up to 12'
	Reinstall	Ea.		12.11		12.11	opening in 2" x 4" stud wall 8' high.
	Minimum Charge	Job		227		227	

Door w / 2" x 6" Lumber

		Unit	Material	Labor	Equip.	Total	Specification
3' Wide							
	Demolish	Ea.		11.45		11.45	Includes material and labor to install
	Install	Ea.	26	14.20		40.20	header, double studs each side,
	Demolish and Install	Ea.	26	25.65		51.65	cripples, blocking and nails, up to 3'
	Reinstall	Ea.		11.35		11.35	opening in 2" x 6" stud wall 8' high.
	Minimum Charge	Job		227		227	
4' Wide							
	Demolish	Ea.		11.45		11.45	Includes material and labor to install
	Install	Ea.	27	14.20		41.20	header, double studs each side,
	Demolish and Install	Ea.	27	25.65		52.65	cripples, blocking and nails, up to 4'
	Reinstall	Ea.		11.35		11.35	opening in 2" x 6" stud wall 8' high.
	Minimum Charge	Job		227		227	
5' Wide							
	Demolish	Ea.		11.45		11.45	Includes material and labor to install
	Install	Ea.	31.50	14.20		45.70	header, double studs each side,
	Demolish and Install	Ea.	31.50	25.65		57.15	cripples, blocking and nails up to 5'
	Reinstall	Ea.		11.35		11.35	opening in 2" x 6" stud wall 8' high.
	Minimum Charge	Job		227		227	
6' Wide							
	Demolish	Ea.		11.45		11.45	Includes material and labor to install
	Install	Ea.	33	14.20		47.20	header, double studs each side,
	Demolish and Install	Ea.	33	25.65		58.65	cripples, blocking and nails up to 6'
	Reinstall	Ea.		11.35		11.35	opening in 2" x 6" stud wall 8' high.
	Minimum Charge	Job		227		227	
8' Wide							
	Demolish	Ea.		12.20		12.20	Includes material and labor to install
	Install	Ea.	45.50	15.15		60.65	header, double studs each side,
	Demolish and Install	Ea.	45.50	27.35		72.85	cripples, blocking and nails up to 8'
	Reinstall	Ea.		12.11		12.11	opening in 2" x 6" stud wall 8' high.
	Minimum Charge	Job		227		227	
10' Wide							
	Demolish	Ea.		12.20		12.20	Includes material and labor to install
	Install	Ea.	59.50	15.15		74.65	header, double studs each side,
	Demolish and Install	Ea.	59.50	27.35		86.85	cripples, blocking and nails up to 10'
	Reinstall	Ea.		12.11		12.11	opening in 2" x 6" stud wall 8' high.
	Minimum Charge	Job		227		227	

Rough Frame / Structure

Rough-in Opening

	Unit	Material	Labor	Equip.	Total	Specification
12' Wide						Includes material and labor to install
Demolish	Ea.		12.20		12.20	header, double studs each side,
Install	Ea.	79	15.15		94.15	cripples, blocking and nails, up to 12'
Demolish and Install	Ea.	79	27.35		106.35	opening in 2" x 6" stud wall 8' high.
Reinstall	Ea.		12.11		12.11	
Minimum Charge	Job		227		227	

Rough-in Opening

	Unit	Material	Labor	Equip.	Total	Specification
Window w / 2" x 4" Lumber						
2' Wide						Includes material and labor to install
Demolish	Ea.		15.30		15.30	header, double studs each side of
Install	Ea.	19.45	18.90		38.35	opening, cripples, blocking, nails and
Demolish and Install	Ea.	19.45	34.20		53.65	sub-sills for opening up to 2' in a 2" x
Reinstall	Ea.		15.13		15.13	4" stud wall 8' high.
Minimum Charge	Job		227		227	
3' Wide						Includes material and labor to install
Demolish	Ea.		15.30		15.30	header, double studs each side of
Install	Ea.	22.50	18.90		41.40	opening, cripples, blocking, nails and
Demolish and Install	Ea.	22.50	34.20		56.70	sub-sills for opening up to 3' in a 2" x
Reinstall	Ea.		15.13		15.13	4" stud wall 8' high.
Minimum Charge	Job		227		227	
4' Wide						Includes material and labor to install
Demolish	Ea.		15.30		15.30	header, double studs each side of
Install	Ea.	25	18.90		43.90	opening, cripples, blocking, nails and
Demolish and Install	Ea.	25	34.20		59.20	sub-sills for opening up to 4' in a 2" x
Reinstall	Ea.		15.13		15.13	4" stud wall 8' high.
Minimum Charge	Job		227		227	
5' Wide						Includes material and labor to install
Demolish	Ea.		15.30		15.30	header, double studs each side of
Install	Ea.	29	18.90		47.90	opening, cripples, blocking, nails and
Demolish and Install	Ea.	29	34.20		63.20	sub-sills for opening up to 5' in a 2" x
Reinstall	Ea.		15.13		15.13	4" stud wall 8' high.
Minimum Charge	Job		227		227	
6' Wide						Includes material and labor to install
Demolish	Ea.		15.30		15.30	header, double studs each side of
Install	Ea.	32	18.90		50.90	opening, cripples, blocking, nails and
Demolish and Install	Ea.	32	34.20		66.20	sub-sills for opening up to 6' in a 2" x
Reinstall	Ea.		15.13		15.13	4" stud wall 8' high.
Minimum Charge	Job		227		227	
7' Wide						Includes material and labor to install
Demolish	Ea.		15.30		15.30	header, double studs each side of
Install	Ea.	42.50	18.90		61.40	opening, cripples, blocking, nails and
Demolish and Install	Ea.	42.50	34.20		76.70	sub-sills for opening up to 7' in a 2" x
Reinstall	Ea.		15.13		15.13	4" stud wall 8' high.
Minimum Charge	Job		227		227	
8' Wide						Includes material and labor to install
Demolish	Ea.		16.65		16.65	header, double studs each side of
Install	Ea.	47.50	20.50		68	opening, cripples, blocking, nails and
Demolish and Install	Ea.	47.50	37.15		84.65	sub-sills for opening up to 8' in a 2" x
Reinstall	Ea.		16.51		16.51	4" stud wall 8' high.
Minimum Charge	Job		227		227	

Rough Frame / Structure

Rough-in Opening	Unit	Material	Labor	Equip.	Total	Specification
10' Wide						
Demolish	Ea.		16.65		16.65	Includes material and labor to install
Install	Ea.	62.50	20.50		83	header, double studs each side of
Demolish and Install	Ea.	62.50	37.15		99.65	opening, cripples, blocking, nails and
Reinstall	Ea.		16.51		16.51	sub-sills for opening up to 10' in a 2"
Minimum Charge	Job		227		227	x 4" stud wall 8' high.
12' Wide						
Demolish	Ea.		16.65		16.65	Includes material and labor to install
Install	Ea.	84.50	20.50		105	header, double studs each side of
Demolish and Install	Ea.	84.50	37.15		121.65	opening, cripples, blocking, nails and
Reinstall	Ea.		16.51		16.51	sub-sills for opening up to 12' in a 2"
Minimum Charge	Job		227		227	x 4" stud wall 8' high.

Window w / 2" x 6" Lumber

Rough-in Opening	Unit	Material	Labor	Equip.	Total	Specification
2' Wide						
Demolish	Ea.		15.30		15.30	Includes material and labor to install
Install	Ea.	28.50	18.90		47.40	header, double studs each side of
Demolish and Install	Ea.	28.50	34.20		62.70	opening, cripples, blocking, nails and
Reinstall	Ea.		15.13		15.13	sub-sills for opening up to 2' in a 2" x
Minimum Charge	Job		227		227	6" stud wall 8' high.
3' Wide						
Demolish	Ea.		15.30		15.30	Includes material and labor to install
Install	Ea.	32	18.90		50.90	header, double studs each side of
Demolish and Install	Ea.	32	34.20		66.20	opening, cripples, blocking, nails and
Reinstall	Ea.		15.13		15.13	sub-sills for opening up to 3' in a 2" x
Minimum Charge	Job		227		227	6" stud wall 8' high.
4' Wide						
Demolish	Ea.		15.30		15.30	Includes material and labor to install
Install	Ea.	35	18.90		53.90	header, double studs each side of
Demolish and Install	Ea.	35	34.20		69.20	opening, cripples, blocking, nails and
Reinstall	Ea.		15.13		15.13	sub-sills for opening up to 4' in a 2" x
Minimum Charge	Job		227		227	6" stud wall 8' high.
5' Wide						
Demolish	Ea.		15.30		15.30	Includes material and labor to install
Install	Ea.	39.50	18.90		58.40	header, double studs each side of
Demolish and Install	Ea.	39.50	34.20		73.70	opening, cripples, blocking, nails and
Reinstall	Ea.		15.13		15.13	sub-sills for opening up to 5' in a 2" x
Minimum Charge	Job		227		227	6" stud wall 8' high.
6' Wide						
Demolish	Ea.		15.30		15.30	Includes material and labor to install
Install	Ea.	43.50	18.90		62.40	header, double studs each side of
Demolish and Install	Ea.	43.50	34.20		77.70	opening, cripples, blocking, nails and
Reinstall	Ea.		15.13		15.13	sub-sills for opening up to 6' in a 2" x
Minimum Charge	Job		227		227	6" stud wall 8' high.
7' Wide						
Demolish	Ea.		15.30		15.30	Includes material and labor to install
Install	Ea.	55	18.90		73.90	header, double studs each side of
Demolish and Install	Ea.	55	34.20		89.20	opening, cripples, blocking, nails and
Reinstall	Ea.		15.13		15.13	sub-sills for opening up to 7' in a 2" x
Minimum Charge	Job		227		227	6" stud wall 8' high.

Rough Frame / Structure

Rough-in Opening

	Unit	Material	Labor	Equip.	Total	Specification
8' Wide						
Demolish	Ea.		16.65		16.65	Includes material and labor to install
Install	Ea.	60	20.50		80.50	header, double studs each side of
Demolish and Install	Ea.	60	37.15		97.15	opening, cripples, blocking, nails and
Reinstall	Ea.		16.51		16.51	sub-sills for opening up to 8' in a 2" x
Minimum Charge	Job		227		227	6" stud wall 8' high.
10' Wide						
Demolish	Ea.		16.65		16.65	Includes material and labor to install
Install	Ea.	76	20.50		96.50	header, double studs each side of
Demolish and Install	Ea.	76	37.15		113.15	opening, cripples, blocking, nails and
Reinstall	Ea.		16.51		16.51	sub-sills for opening up to 10' in a 2"
Minimum Charge	Job		227		227	x 6" stud wall 8' high.
12' Wide						
Demolish	Ea.		16.65		16.65	Includes material and labor to install
Install	Ea.	99	20.50		119.50	header, double studs each side of
Demolish and Install	Ea.	99	37.15		136.15	opening, cripples, blocking, nails and
Reinstall	Ea.		16.51		16.51	sub-sills for opening up to 12' in a 2"
Minimum Charge	Job		227		227	x 6" stud wall 8' high.

Glue-Laminated Beams

	Unit	Material	Labor	Equip.	Total	Specification
3-1/2" x 6"						
Demolish	L.F.		2.74		2.74	Includes material and labor to install
Install	L.F.	5.60	3.42	1.21	10.23	glue-laminated wood beam.
Demolish and Install	L.F.	5.60	6.16	1.21	12.97	
Reinstall	L.F.		2.74	.97	3.71	
Clean	L.F.	.13	.41		.54	
Paint	L.F.	.37	1.11		1.48	
Minimum Charge	Job		227		227	
3-1/2" x 9"						
Demolish	L.F.		2.74		2.74	Includes material and labor to install
Install	L.F.	7.55	3.42	1.21	12.18	glue-laminated wood beam.
Demolish and Install	L.F.	7.55	6.16	1.21	14.92	
Reinstall	L.F.		2.74	.97	3.71	
Clean	L.F.	.18	.43		.61	
Paint	L.F.	.48	1.44		1.92	
Minimum Charge	Job		227		227	
3-1/2" x 12"						
Demolish	L.F.		4.11		4.11	Includes material and labor to install
Install	L.F.	10.05	3.42	1.21	14.68	glue-laminated wood beam.
Demolish and Install	L.F.	10.05	7.53	1.21	18.79	
Reinstall	L.F.		2.74	.97	3.71	
Clean	L.F.	.22	.44		.66	
Paint	L.F.	.59	1.80		2.39	
Minimum Charge	Job		227		227	
3-1/2" x 15"						
Demolish	L.F.		4.11		4.11	Includes material and labor to install
Install	L.F.	12.55	3.54	1.25	17.34	glue-laminated wood beam.
Demolish and Install	L.F.	12.55	7.65	1.25	21.45	
Reinstall	L.F.		2.83	1	3.83	
Clean	L.F.	.25	.46		.71	
Paint	L.F.	.69	2.10		2.79	
Minimum Charge	Job		227		227	

Rough Frame / Structure

Glue-Laminated Beams	Unit	Material	Labor	Equip.	Total	Specification
3-1/2" x 18"						
Demolish	L.F.		5.15		5.15	Includes material and labor to install
Install	L.F.	16.90	3.54	1.25	21.69	glue-laminated wood beam.
Demolish and Install	L.F.	16.90	8.69	1.25	26.84	
Reinstall	L.F.		2.83	1	3.83	
Clean	L.F.	.30	.47		.77	
Paint	L.F.	.83	2.44		3.27	
Minimum Charge	Job		227		227	
5-1/8" x 6"						
Demolish	L.F.		4.11		4.11	Includes material and labor to install
Install	L.F.	8.25	3.42	1.21	12.88	glue-laminated wood beam.
Demolish and Install	L.F.	8.25	7.53	1.21	16.99	
Reinstall	L.F.		2.74	.97	3.71	
Clean	L.F.	.14	.41		.55	
Paint	L.F.	.40	1.22		1.62	
Minimum Charge	Job		227		227	
5-1/8" x 9"						
Demolish	L.F.		5.15		5.15	Includes material and labor to install
Install	L.F.	12.30	3.42	1.21	16.93	glue-laminated wood beam.
Demolish and Install	L.F.	12.30	8.57	1.21	22.08	
Reinstall	L.F.		2.74	.97	3.71	
Clean	L.F.	.19	.43		.62	
Paint	L.F.	.52	1.56		2.08	
Minimum Charge	Job		227		227	
5-1/8" x 12"						
Demolish	L.F.		6.10		6.10	Includes material and labor to install
Install	L.F.	16.45	3.42	1.21	21.08	glue-laminated wood beam.
Demolish and Install	L.F.	16.45	9.52	1.21	27.18	
Reinstall	L.F.		2.74	.97	3.71	
Clean	L.F.	.23	.44		.67	
Paint	L.F.	.64	1.89		2.53	
Minimum Charge	Job		227		227	
5-1/8" x 18"						
Demolish	L.F.		10.25		10.25	Includes material and labor to install
Install	L.F.	25	3.54	1.25	29.79	glue-laminated wood beam.
Demolish and Install	L.F.	25	13.79	1.25	40.04	
Reinstall	L.F.		2.83	1	3.83	
Clean	L.F.	.31	.47		.78	
Paint	L.F.	.74	2.55		3.29	
Minimum Charge	Job		227		227	
6-3/4" x 12"						
Demolish	L.F.		6.10		6.10	Includes material and labor to install
Install	L.F.	21.50	3.54	1.25	26.29	glue-laminated wood beam.
Demolish and Install	L.F.	21.50	9.64	1.25	32.39	
Reinstall	L.F.		2.83	1	3.83	
Clean	L.F.	.25	.46		.71	
Paint	L.F.	.69	2.10		2.79	
Minimum Charge	Job		227		227	
6-3/4" x 15"						
Demolish	L.F.		10.25		10.25	Includes material and labor to install
Install	L.F.	27	3.54	1.25	31.79	glue-laminated wood beam.
Demolish and Install	L.F.	27	13.79	1.25	42.04	
Reinstall	L.F.		2.83	1	3.83	
Clean	L.F.	.30	.47		.77	
Paint	L.F.	.83	2.44		3.27	
Minimum Charge	Job		227		227	

For customer support on your Contractor's Pricing Guide: Residential Repair & Remodeling, call 888.606.7279.

Glue-Laminated Beams

Glue-Laminated Beams	Unit	Material	Labor	Equip.	Total	Specification
6-3/4″ x 18″						
Demolish	L.F.		10.25		10.25	Includes material and labor to install
Install	L.F.	32.50	3.67	1.29	37.46	glue-laminated wood beam.
Demolish and Install	L.F.	32.50	13.92	1.29	47.71	
Reinstall	L.F.		2.93	1.03	3.96	
Clean	L.F.	.34	.49		.83	
Paint	L.F.	.94	2.78		3.72	
Minimum Charge	Job		227		227	
Hardware						
5-1/4″ Glue-Lam Seat						
Install	Ea.	230	2.52		232.52	Includes labor and material to install
Minimum Charge	Job		227		227	beam hangers.
6-3/4″ Glue-Lam Seat						
Install	Ea.	235	2.52		237.52	Includes labor and material to install
Minimum Charge	Job		227		227	beam hangers.
Earthquake Strapping						
Install	Ea.	3.23	2.84		6.07	Includes labor and material to replace
Minimum Charge	Job		227		227	an earthquake strap.
Hurricane Clips						
Install	Ea.	1.23	3.13		4.36	Includes labor and material to install a
Minimum Charge	Job		227		227	hurricane clip.

Metal Stud Framing

Metal Stud Framing	Unit	Material	Labor	Equip.	Total	Specification
16″ O.C. System						
4″, 16 Ga.						
Demolish	S.F.		1.35		1.35	Includes material and labor to install
Install	S.F.	1.41	1.40		2.81	load bearing cold rolled metal stud
Demolish and Install	S.F.	1.41	2.75		4.16	walls, to 10′ high, including studs, top
Reinstall	S.F.		1.12		1.12	and bottom track and screws.
Minimum Charge	Job		227		227	
6″, 16 Ga.						
Demolish	S.F.		1.35		1.35	Includes material and labor to install
Install	S.F.	1.78	1.42		3.20	load bearing cold rolled metal stud
Demolish and Install	S.F.	1.78	2.77		4.55	walls, to 10′ high, including studs, top
Reinstall	S.F.		1.14		1.14	and bottom track and screws.
Minimum Charge	Job		227		227	
4″, 25 Ga.						
Demolish	S.F.		.93		.93	Includes material and labor to install
Install	S.F.	.45	.96		1.41	cold rolled metal stud walls, to 10′
Demolish and Install	S.F.	.45	1.89		2.34	high, including studs, top and bottom
Reinstall	S.F.		.76		.76	track and screws.
Minimum Charge	Job		227		227	
6″, 25 Ga.						
Demolish	S.F.		.93		.93	Includes material and labor to install
Install	S.F.	.54	.97		1.51	cold rolled metal stud walls, to 10′
Demolish and Install	S.F.	.54	1.90		2.44	high, including studs, top and bottom
Reinstall	S.F.		.77		.77	track and screws.
Minimum Charge	Job		227		227	
24″ O.C. System						
4″, 25 Ga.						
Demolish	S.F.		.79		.79	Includes material and labor to install
Install	S.F.	.33	.61		.94	cold rolled metal stud walls, to 10′
Demolish and Install	S.F.	.33	1.40		1.73	high, including studs, top and bottom
Reinstall	S.F.		.49		.49	track and screws.
Minimum Charge	Job		227		227	

Rough Frame / Structure

Metal Stud Framing

		Unit	Material	Labor	Equip.	Total	Specification
6", 25 Ga.							
	Demolish	S.F.		.79		.79	Includes material and labor to install
	Install	S.F.	.40	.63		1.03	cold rolled metal stud walls, to 10'
	Demolish and Install	S.F.	.40	1.42		1.82	high, including studs, top and bottom
	Reinstall	S.F.		.50		.50	track and screws.
	Minimum Charge	Job		227		227	
4", 16 Ga.							
	Demolish	L.F.		9.05		9.05	Includes material and labor to install
	Install	S.F.	1.02	1.01		2.03	load bearing cold rolled metal stud
	Demolish and Install	S.F.	1.02	10.06		11.08	walls, to 10' high, including studs, top
	Reinstall	S.F.		.81		.81	and bottom track and screws.
	Minimum Charge	Job		227		227	
6", 16 Ga.							
	Demolish	L.F.		9.05		9.05	Includes material and labor to install
	Install	S.F.	1.30	1.03		2.33	load bearing cold rolled metal stud
	Demolish and Install	S.F.	1.30	10.08		11.38	walls, to 10' high, including studs, top
	Reinstall	S.F.		.83		.83	and bottom track and screws.
	Minimum Charge	Job		227		227	

Metal Joist

		Unit	Material	Labor	Equip.	Total	Specification
Bar Joist							
18K9							
	Demolish	L.F.		1.09	.42	1.51	Includes material, labor and
	Install	L.F.	9.30	2.35	1	12.65	equipment to install open web joist.
	Demolish and Install	L.F.	9.30	3.44	1.42	14.16	
	Reinstall	L.F.		1.88	.80	2.68	
	Minimum Charge	Job		2050	725	2775	
16K6							
	Demolish	L.F.		1.09	.42	1.51	Includes material, labor and
	Install	L.F.	5.45	1.57	.12	7.14	equipment to install open web joist.
	Demolish and Install	L.F.	5.45	2.66	.54	8.65	
	Reinstall	L.F.		1.26	.10	1.36	
	Minimum Charge	Job		2050	725	2775	

Exterior Sheathing

		Unit	Material	Labor	Equip.	Total	Specification
CDX Plywood							
5/16"							
	Demolish	S.F.		.62		.62	Cost includes material and labor to
	Install	S.F.	.68	.57		1.25	install 5/16" CDX plywood sheathing.
	Demolish and Install	S.F.	.68	1.19		1.87	
	Minimum Charge	Job		227		227	
3/8"							
	Demolish	S.F.		.64		.64	Cost includes material and labor to
	Install	S.F.	.68	.76		1.44	install 3/8" CDX plywood sheathing.
	Demolish and Install	S.F.	.68	1.40		2.08	
	Minimum Charge	Job		227		227	
1/2"							
	Demolish	S.F.		.65		.65	Cost includes material and labor to
	Install	S.F.	.75	.81		1.56	install 1/2" CDX plywood sheathing.
	Demolish and Install	S.F.	.75	1.46		2.21	
	Minimum Charge	Job		227		227	

Rough Frame / Structure

Exterior Sheathing

Exterior Sheathing	Unit	Material	Labor	Equip.	Total	Specification
5/8"						
Demolish	S.F.		.67		.67	Cost includes material and labor to
Install	S.F.	.84	.86		1.70	install 5/8" CDX plywood sheathing.
Demolish and Install	S.F.	.84	1.53		2.37	
Minimum Charge	Job		227		227	
3/4"						
Demolish	S.F.		.68		.68	Cost includes material and labor to
Install	S.F.	1	.93		1.93	install 3/4" CDX plywood sheathing.
Demolish and Install	S.F.	1	1.61		2.61	
Minimum Charge	Job		227		227	
OSB						
1/2"						
Demolish	S.F.		.65		.65	Cost includes material and labor to
Install	S.F.	.55	.65		1.20	install 4' x 8' x 1/2" OSB sheathing.
Demolish and Install	S.F.	.55	1.30		1.85	
Minimum Charge	Job		227		227	
5/8"						
Demolish	S.F.		.67		.67	Cost includes material and labor to
Install	S.F.	.55	.52		1.07	install 4' x 8' x 5/8" OSB sheathing.
Demolish and Install	S.F.	.55	1.19		1.74	
Minimum Charge	Job		227		227	
Vapor Barrier						
Black Paper						
Install	S.F.	.06	.12		.18	Includes material and labor to install
Minimum Charge	Job		227		227	#15 felt building paper.

Plywood Sheathing

Plywood Sheathing	Unit	Material	Labor	Equip.	Total	Specification
Finish Plywood						
5/16"						
Demolish	S.F.		.62		.62	Includes material and labor to install
Install	S.F.	1.09	.28		1.37	exterior 5/16" AC plywood on walls.
Demolish and Install	S.F.	1.09	.90		1.99	
Minimum Charge	Job		227		227	
3/8"						
Demolish	S.F.		.64		.64	Includes material and labor to install
Install	S.F.	1.09	.38		1.47	exterior 3/8" AC plywood on walls.
Demolish and Install	S.F.	1.09	1.02		2.11	
Minimum Charge	Job		227		227	
1/2"						
Demolish	S.F.		.65		.65	Includes material and labor to install
Install	S.F.	1.29	.40		1.69	exterior 1/2" AC plywood on walls.
Demolish and Install	S.F.	1.29	1.05		2.34	
Minimum Charge	Job		227		227	
5/8"						
Demolish	S.F.		.67		.67	Includes material and labor to install
Install	S.F.	1.42	.40		1.82	exterior 5/8" AC plywood on walls.
Demolish and Install	S.F.	1.42	1.07		2.49	
Reinstall	S.F.		.40		.40	
Minimum Charge	Job		227		227	
3/4"						
Demolish	S.F.		.68		.68	Includes material and labor to install
Install	S.F.	1.63	.40		2.03	exterior 3/4" AC plywood on walls.
Demolish and Install	S.F.	1.63	1.08		2.71	
Reinstall	S.F.		.40		.40	
Minimum Charge	Job		227		227	

For customer support on your Contractor's Pricing Guide: Residential Repair & Remodeling, call 888.606.7279.

Rough Frame / Structure

Stairs

	Unit	Material	Labor	Equip.	Total	Specification
Job-built						
Treads and Risers						
Demolish	Ea.		46		46	Includes material and labor to install
Install	Ea.	620	455		1075	three 2" x 12" stringers, treads and
Demolish and Install	Ea.	620	501		1121	risers of 3/4" CDX plywood, installed
Clean	Flight	4.07	28.50		32.57	in a straight or "L" shaped run
Paint	Ea.	.12	75.50		75.62	including carpet.
Minimum Charge	Job		227		227	
Landing						
Demolish	S.F.		.37		.37	Includes material and labor to install
Install	S.F.	76	3.15		79.15	landing framing, 3/4" CDX plywood
Demolish and Install	S.F.	76	3.52		79.52	surface and carpet.
Clean	S.F.		.41		.41	
Paint	S.F.	.44	.93		1.37	
Minimum Charge	Job		227		227	

Ceiling Framing

	Unit	Material	Labor	Equip.	Total	Specification
16" O.C. System						
2" x 6" Joists						
Demolish	S.F.		.57		.57	Cost includes material and labor to
Install	S.F.	.51	.75		1.26	install 2" x 6" joists including end and
Demolish and Install	S.F.	.51	1.32		1.83	header joist installed 16" O.C. Does
Reinstall	S.F.		.60		.60	not include beams, ledger strips,
Minimum Charge	Job		227		227	blocking or bridging.
2" x 8" Joists						
Demolish	S.F.		.59		.59	Cost includes material and labor to
Install	S.F.	.70	.92		1.62	install 2" x 8" joists including end and
Demolish and Install	S.F.	.70	1.51		2.21	header joist installed 16" O.C. Does
Reinstall	S.F.		.74		.74	not include beams, ledger strips,
Minimum Charge	Job		227		227	blocking or bridging.
2" x 10" Joists						
Demolish	S.F.		.60		.60	Cost includes material and labor to
Install	S.F.	1.12	1.09		2.21	install 2" x 10" joists including end
Demolish and Install	S.F.	1.12	1.69		2.81	and header joist installed 16" O.C.
Reinstall	S.F.		.87		.87	Does not include beams, ledger strips,
Minimum Charge	Job		227		227	blocking or bridging.
2" x 12" Joists						
Demolish	S.F.		.62		.62	Cost includes material and labor to
Install	S.F.	1.42	1.33		2.75	install 2" x 12" joists including end
Demolish and Install	S.F.	1.42	1.95		3.37	and header joist installed 16" O.C.
Reinstall	S.F.		1.06		1.06	Does not include beams, ledger strips,
Minimum Charge	Job		227		227	blocking or bridging.
Block / Bridge						
Demolish	Ea.		1.15		1.15	Includes material and labor to install
Install	Ea.	1.96	2.55		4.51	set of cross bridging or per block of
Demolish and Install	Ea.	1.96	3.70		5.66	solid bridging for 2" x 10" joists cut to
Minimum Charge	Job		227		227	size on site.

Ceiling Framing	Unit	Material	Labor	Equip.	Total	Specification
24" O.C. System						
2" x 6" Joists						
Demolish	S.F.		.38		.38	Cost includes material and labor to
Install	S.F.	.34	.50		.84	install 2" x 6" joists including end and
Demolish and Install	S.F.	.34	.88		1.22	header joist installed 24" O.C. Does
Reinstall	S.F.		.40		.40	not include beams, ledger strips,
Minimum Charge	Job		227		227	blocking or bridging.
2" x 8" Joists						
Demolish	S.F.		.39		.39	Cost includes material and labor to
Install	S.F.	.47	.61		1.08	install 2" x 8" joists including end and
Demolish and Install	S.F.	.47	1		1.47	header joist installed 24" O.C. Does
Reinstall	S.F.		.49		.49	not include beams, ledger strips,
Minimum Charge	Job		227		227	blocking or bridging.
2" x 10" Joists						
Demolish	S.F.		.40		.40	Cost includes material and labor to
Install	S.F.	.75	.72		1.47	install 2" x 10" joists including end
Demolish and Install	S.F.	.75	1.12		1.87	and header joist installed 24" O.C.
Reinstall	S.F.		.58		.58	Does not include beams, ledger strips,
Minimum Charge	Job		227		227	blocking or bridging.
2" x 12" Joists						
Demolish	S.F.		.42		.42	Cost includes material and labor to
Install	S.F.	.96	.88		1.84	install 2" x 12" joists including end
Demolish and Install	S.F.	.96	1.30		2.26	and header joist installed 24" O.C.
Reinstall	S.F.		.71		.71	Does not include beams, ledger strips,
Minimum Charge	Job		227		227	blocking or bridging.
Block / Bridge						
Demolish	Ea.		1.15		1.15	Includes material and labor to install
Install	Ea.	1.96	2.55		4.51	set of cross bridging or per block of
Demolish and Install	Ea.	1.96	3.70		5.66	solid bridging for 2" x 10" joists cut to
Minimum Charge	Job		227		227	size on site.
Ledger Strips						
1" x 2"						
Demolish	L.F.		.31		.31	Cost includes material and labor to
Install	L.F.	.26	1.32		1.58	install 1" x 2" ledger strip nailed to the
Demolish and Install	L.F.	.26	1.63		1.89	face of studs, beams or joist.
Minimum Charge	Job		227		227	
1" x 3"						
Demolish	L.F.		.31		.31	Includes material and labor to install
Install	L.F.	.51	1.51		2.02	up to 1" x 4" ledger strip nailed to the
Demolish and Install	L.F.	.51	1.82		2.33	face of studs, beams or joist.
Minimum Charge	Job		227		227	
1" x 4"						
Demolish	L.F.		.31		.31	Includes material and labor to install
Install	L.F.	.51	1.51		2.02	up to 1" x 4" ledger strip nailed to the
Demolish and Install	L.F.	.51	1.82		2.33	face of studs, beams or joist.
Minimum Charge	Job		227		227	
2" x 2"						
Demolish	L.F.		.33		.33	Cost includes material and labor to
Install	L.F.	.42	1.38		1.80	install 2" x 2" ledger strip nailed to the
Demolish and Install	L.F.	.42	1.71		2.13	face of studs, beams or joist.
Minimum Charge	Job		227		227	
2" x 4"						
Demolish	L.F.		.37		.37	Cost includes material and labor to
Install	L.F.	.44	1.82		2.26	install 2" x 4" ledger strip nailed to the
Demolish and Install	L.F.	.44	2.19		2.63	face of studs, beams or joist.
Reinstall	L.F.		1.45		1.45	
Minimum Charge	Job		227		227	

Ceiling Framing

	Unit	Material	Labor	Equip.	Total	Specification
Ledger Boards						
2" x 4"						
Demolish	L.F.		.37		.37	Cost includes material and labor to
Install	L.F.	5.15	2.52		7.67	install 2" x 4" ledger board fastened to
Demolish and Install	L.F.	5.15	2.89		8.04	a wall, joists or studs.
Reinstall	L.F.		2.02		2.02	
Minimum Charge	Job		227		227	
4" x 6"						
Demolish	L.F.		.61		.61	Cost includes material and labor to
Install	L.F.	9.10	3.24		12.34	install 4" x 6" ledger board fastened to
Demolish and Install	L.F.	9.10	3.85		12.95	a wall, joists or studs.
Reinstall	L.F.		2.59		2.59	
Minimum Charge	Job		227		227	
4" x 8"						
Demolish	L.F.		.81		.81	Cost includes material and labor to
Install	L.F.	9.15	3.78		12.93	install 4" x 8" ledger board fastened to
Demolish and Install	L.F.	9.15	4.59		13.74	a wall, joists or studs.
Reinstall	L.F.		3.03		3.03	
Minimum Charge	Job		227		227	
Earthquake Strapping						
Install	Ea.	3.23	2.84		6.07	Includes labor and material to replace
Minimum Charge	Job		227		227	an earthquake strap.
Hurricane Clips						
Install	Ea.	1.23	3.13		4.36	Includes labor and material to install a
Minimum Charge	Job		227		227	hurricane clip.

Roof Framing

	Unit	Material	Labor	Equip.	Total	Specification
16" O.C. System						
2" x 4" Rafters						
Demolish	S.F.		.83		.83	Cost includes material and labor to
Install	S.F.	.33	.81		1.14	install 2" x 4", 16" O.C. rafter framing
Demolish and Install	S.F.	.33	1.64		1.97	for flat, shed or gable roofs, 25' span,
Reinstall	S.F.		.65		.65	up to 5/12 slope, per S.F. of roof.
Minimum Charge	Job		227		227	
2" x 6" Rafters						
Demolish	S.F.		.87		.87	Cost includes material and labor to
Install	S.F.	.51	.85		1.36	install 2" x 6", 16" O.C. rafter framing
Demolish and Install	S.F.	.51	1.72		2.23	for flat, shed or gable roofs, 25' span,
Reinstall	S.F.		.68		.68	up to 5/12 slope, per S.F. of roof.
Minimum Charge	Job		227		227	
2" x 8" Rafters						
Demolish	S.F.		.89		.89	Cost includes material and labor to
Install	S.F.	.70	.91		1.61	install 2" x 8", 16" O.C. rafter framing
Demolish and Install	S.F.	.70	1.80		2.50	for flat, shed or gable roofs, 25' span,
Reinstall	S.F.		.73		.73	up to 5/12 slope, per S.F. of roof.
Minimum Charge	Job		227		227	
2" x 10" Rafters						
Demolish	S.F.		.89		.89	Cost includes material and labor to
Install	S.F.	1.10	1.38		2.48	install 2" x 10", 16" O.C. rafter
Demolish and Install	S.F.	1.10	2.27		3.37	framing for flat, shed or gable roofs,
Reinstall	S.F.		1.10		1.10	25' span, up to 5/12 slope, per S.F.
Minimum Charge	Job		227		227	of roof.

Rough Frame / Structure

Roof Framing		Unit	Material	Labor	Equip.	Total	Specification
2" x 12" Rafters							
	Demolish	S.F.		.91		.91	Cost includes material and labor to
	Install	S.F.	1.41	1.50		2.91	install 2" x 12", 16" O.C. rafter
	Demolish and Install	S.F.	1.41	2.41		3.82	framing for flat, shed or gable roofs,
	Reinstall	S.F.		1.20		1.20	25' span, up to 5/12 slope, per S.F.
	Minimum Charge	Job		227		227	of roof.
24" O.C. System							
2" x 4" Rafters							
	Demolish	S.F.		.63		.63	Cost includes material and labor to
	Install	S.F.	.22	.61		.83	install 2" x 4", 24" O.C. rafter framing
	Demolish and Install	S.F.	.22	1.24		1.46	for flat, shed or gable roofs, 25' span,
	Reinstall	S.F.		.49		.49	up to 5/12 slope, per S.F. of roof.
	Minimum Charge	Job		227		227	
2" x 6" Rafters							
	Demolish	S.F.		.66		.66	Cost includes material and labor to
	Install	S.F.	.34	.64		.98	install 2" x 6", 24" O.C. rafter framing
	Demolish and Install	S.F.	.34	1.30		1.64	for flat, shed or gable roofs, 25' span,
	Reinstall	S.F.		.51		.51	up to 5/12 slope, per S.F. of roof.
	Minimum Charge	Job		227		227	
2" x 8" Rafters							
	Demolish	S.F.		.67		.67	Cost includes material and labor to
	Install	S.F.	.46	.68		1.14	install 2" x 8", 24" O.C. rafter framing
	Demolish and Install	S.F.	.46	1.35		1.81	for flat, shed or gable roofs, 25' span,
	Reinstall	S.F.		.55		.55	up to 5/12 slope, per S.F. of roof.
	Minimum Charge	Job		227		227	
2" x 10" Rafters							
	Demolish	S.F.		.67		.67	Cost includes material and labor to
	Install	S.F.	.74	1.03		1.77	install 2" x 10", 24" O.C. rafter
	Demolish and Install	S.F.	.74	1.70		2.44	framing for flat, shed or gable roofs,
	Reinstall	S.F.		.83		.83	25' span, up to 5/12 slope, per S.F.
	Minimum Charge	Job		227		227	of roof.
2" x 12" Rafters							
	Demolish	S.F.		.68		.68	Cost includes material and labor to
	Install	S.F.	.94	1.13		2.07	install 2" x 12", 24" O.C. rafter
	Demolish and Install	S.F.	.94	1.81		2.75	framing for flat, shed or gable roofs,
	Reinstall	S.F.		.90		.90	25' span, up to 5/12 slope, per S.F.
	Minimum Charge	Job		227		227	of roof.
Rafter							
2" x 4"							
	Demolish	L.F.		.85		.85	Cost includes material and labor to
	Install	L.F.	.44	1.14		1.58	install 2" x 4" rafters for flat, shed or
	Demolish and Install	L.F.	.44	1.99		2.43	gable roofs, up to 5/12 slope, 25'
	Reinstall	L.F.		.91		.91	span, per L.F.
	Minimum Charge	Job		227		227	
2" x 6"							
	Demolish	L.F.		.86		.86	Cost includes material and labor to
	Install	L.F.	.67	1.14		1.81	install 2" x 6" rafters for flat, shed or
	Demolish and Install	L.F.	.67	2		2.67	gable roofs, up to 5/12 slope, 25'
	Reinstall	L.F.		.91		.91	span, per L.F.
	Minimum Charge	Job		227		227	
2" x 8"							
	Demolish	L.F.		.88		.88	Cost includes material and labor to
	Install	L.F.	.94	1.21		2.15	install 2" x 8" rafters for flat, shed or
	Demolish and Install	L.F.	.94	2.09		3.03	gable roofs, up to 5/12 slope, 25'
	Reinstall	L.F.		.97		.97	span, per L.F.
	Minimum Charge	Job		227		227	

Rough Frame / Structure

Roof Framing	Unit	Material	Labor	Equip.	Total	Specification
2" x 10"						
Demolish	L.F.		.89		.89	Cost includes material and labor to
Install	L.F.	1.47	1.83		3.30	install 2" x 10" rafters for flat, shed or
Demolish and Install	L.F.	1.47	2.72		4.19	gable roofs, up to 5/12 slope, 25'
Reinstall	L.F.		1.47		1.47	span, per L.F.
Minimum Charge	Job		227		227	
2" x 12"						
Demolish	L.F.		.90		.90	Cost includes material and labor to
Install	L.F.	1.87	2		3.87	install 2" x 12" rafters for flat, shed or
Demolish and Install	L.F.	1.87	2.90		4.77	gable roofs, up to 5/12 slope, 25'
Reinstall	L.F.		1.60		1.60	span, per L.F.
Minimum Charge	Job		227		227	
2"x 4" Valley / Jack						
Demolish	L.F.		.85		.85	Includes material and labor to install
Install	L.F.	.67	1.91		2.58	up to 2" x 6" valley/jack rafters for
Demolish and Install	L.F.	.67	2.76		3.43	flat, shed or gable roofs, up to 5/12
Reinstall	L.F.		1.53		1.53	slope, 25' span, per L.F.
Minimum Charge	Job		227		227	
2"x 6" Valley / Jack						
Demolish	L.F.		.86		.86	Includes material and labor to install
Install	L.F.	.67	1.91		2.58	up to 2" x 6" valley/jack rafters for
Demolish and Install	L.F.	.67	2.77		3.44	flat, shed or gable roofs, up to 5/12
Reinstall	L.F.		1.53		1.53	slope, 25' span, per L.F.
Minimum Charge	Job		227		227	
Ridgeboard						
2" x 4"						
Demolish	L.F.		.81		.81	Cost includes material and labor to
Install	L.F.	.44	1.65		2.09	install 2" x 4" ridgeboard for flat, shed
Demolish and Install	L.F.	.44	2.46		2.90	or gable roofs, up to 5/12 slope, 25'
Reinstall	L.F.		1.32		1.32	span, per L.F.
Minimum Charge	Job		227		227	
2" x 6"						
Demolish	L.F.		.84		.84	Cost includes material and labor to
Install	L.F.	.67	1.82		2.49	install 2" x 6" ridgeboard for flat, shed
Demolish and Install	L.F.	.67	2.66		3.33	or gable roofs, up to 5/12 slope, 25'
Reinstall	L.F.		1.45		1.45	span, per L.F.
Minimum Charge	Job		227		227	
2" x 8"						
Demolish	L.F.		.86		.86	Cost includes material and labor to
Install	L.F.	.94	2.02		2.96	install 2" x 8" ridgeboard for flat, shed
Demolish and Install	L.F.	.94	2.88		3.82	or gable roofs, up to 5/12 slope, 25'
Reinstall	L.F.		1.61		1.61	span, per L.F.
Minimum Charge	Job		227		227	
2" x 10"						
Demolish	L.F.		.89		.89	Cost includes material and labor to
Install	L.F.	1.47	2.27		3.74	install 2" x 10" ridgeboard for flat,
Demolish and Install	L.F.	1.47	3.16		4.63	shed or gable roofs, up to 5/12
Reinstall	L.F.		1.82		1.82	slope, 25' span, per L.F.
Minimum Charge	Job		227		227	
2" x 12"						
Demolish	L.F.		.92		.92	Cost includes material and labor to
Install	L.F.	1.87	1.30		3.17	install 2" x 12" ridgeboard for flat,
Demolish and Install	L.F.	1.87	2.22		4.09	shed or gable roofs, up to 5/12
Reinstall	L.F.		1.04		1.04	slope, 25' span, per L.F.
Minimum Charge	Job		227		227	

For customer support on your Contractor's Pricing Guide: Residential Repair & Remodeling, call 888.606.7279.

Rough Frame / Structure

Roof Framing

		Unit	Material	Labor	Equip.	Total	Specification
Collar Beam							
1″ x 6″							
	Demolish	L.F.		.65		.65	Cost includes material and labor to
	Install	L.F.	.78	.57		1.35	install 1″ x 6″ collar beam or tie for
	Demolish and Install	L.F.	.78	1.22		2	roof framing.
	Reinstall	L.F.		.45		.45	
	Minimum Charge	Job		227		227	
2″ x 6″							
	Demolish	L.F.		.76		.76	Cost includes material and labor to
	Install	L.F.	.67	.57		1.24	install 2″ x 6″ collar beam or tie for
	Demolish and Install	L.F.	.67	1.33		2	roof framing.
	Reinstall	L.F.		.45		.45	
	Minimum Charge	Job		227		227	
Purlins							
2″ x 6″							
	Demolish	L.F.		.73		.73	Cost includes material and labor to
	Install	L.F.	.67	.50		1.17	install 2″ x 6″ purlins below roof
	Demolish and Install	L.F.	.67	1.23		1.90	rafters.
	Reinstall	L.F.		.40		.40	
	Minimum Charge	Job		227		227	
2″ x 8″							
	Demolish	L.F.		.74		.74	Cost includes material and labor to
	Install	L.F.	.94	.51		1.45	install 2″ x 8″ purlins below roof
	Demolish and Install	L.F.	.94	1.25		2.19	rafters.
	Reinstall	L.F.		.41		.41	
	Minimum Charge	Job		227		227	
2″ x 10″							
	Demolish	L.F.		.75		.75	Cost includes material and labor to
	Install	L.F.	1.47	.52		1.99	install 2″ x 10″ purlins below roof
	Demolish and Install	L.F.	1.47	1.27		2.74	rafters.
	Reinstall	L.F.		.41		.41	
	Minimum Charge	Job		227		227	
2″ x 12″							
	Demolish	L.F.		.76		.76	Cost includes material and labor to
	Install	L.F.	1.87	.52		2.39	install 2″ x 12″ purlins below roof
	Demolish and Install	L.F.	1.87	1.28		3.15	rafters.
	Reinstall	L.F.		.42		.42	
	Minimum Charge	Job		227		227	
4″ x 6″							
	Demolish	L.F.		.76		.76	Cost includes material and labor to
	Install	L.F.	3.19	.52		3.71	install 4″ x 6″ purlins below roof
	Demolish and Install	L.F.	3.19	1.28		4.47	rafters.
	Reinstall	L.F.		.42		.42	
	Minimum Charge	Job		227		227	
4″ x 8″							
	Demolish	L.F.		.76		.76	Cost includes material and labor to
	Install	L.F.	3.26	.53		3.79	install 4″ x 8″ purlins below roof
	Demolish and Install	L.F.	3.26	1.29		4.55	rafters.
	Reinstall	L.F.		.43		.43	
	Minimum Charge	Job		227		227	
Ledger Board							
2″ x 6″							
	Demolish	L.F.		.37		.37	Cost includes material and labor to
	Install	L.F.	.67	2.05		2.72	install 2″ x 6″ ledger board nailed to
	Demolish and Install	L.F.	.67	2.42		3.09	the face of a rafter.
	Reinstall	L.F.		1.64		1.64	
	Minimum Charge	Job		227		227	

Rough Frame / Structure

Roof Framing	Unit	Material	Labor	Equip.	Total	Specification
4" x 6"						
Demolish	L.F.		.61		.61	Cost includes material and labor to
Install	L.F.	3.19	3.01		6.20	install 4" x 6" ledger board nailed to
Demolish and Install	L.F.	3.19	3.62		6.81	the face of a rafter.
Reinstall	L.F.		2.41		2.41	
Minimum Charge	Job		227		227	
4" x 8"						
Demolish	L.F.		.81		.81	Cost includes material and labor to
Install	L.F.	3.26	3.47		6.73	install 4" x 8" ledger board nailed to
Demolish and Install	L.F.	3.26	4.28		7.54	the face of a rafter.
Reinstall	L.F.		2.77		2.77	
Minimum Charge	Job		227		227	
Outriggers						
2" x 4"						
Demolish	L.F.		.25		.25	Cost includes material and labor to
Install	L.F.	.44	1.82		2.26	install 2" x 4" outrigger rafters for flat,
Demolish and Install	L.F.	.44	2.07		2.51	shed or gabled roofs, up to 5/12
Reinstall	L.F.		1.45		1.45	slope, per L.F.
Minimum Charge	Job		227		227	
2" x 6"						
Demolish	L.F.		.38		.38	Cost includes material and labor to
Install	L.F.	.67	1.82		2.49	install 2" x 6" outrigger rafters for flat,
Demolish and Install	L.F.	.67	2.20		2.87	shed or gabled roofs, up to 5/12
Reinstall	L.F.		1.45		1.45	slope, per L.F.
Minimum Charge	Job		227		227	
Lookout Rafter						
2" x 4"						
Demolish	L.F.		.85		.85	Cost includes material and labor to
Install	L.F.	.44	.76		1.20	install 2" x 4" lookout rafters for flat,
Demolish and Install	L.F.	.44	1.61		2.05	shed or gabled roofs, up to 5/12
Reinstall	L.F.		.61		.61	slope, per L.F.
Minimum Charge	Job		227		227	
2" x 6"						
Demolish	L.F.		.86		.86	Cost includes material and labor to
Install	L.F.	.67	.77		1.44	install 2" x 6" lookout rafters for flat,
Demolish and Install	L.F.	.67	1.63		2.30	shed or gabled roofs, up to 5/12
Reinstall	L.F.		.62		.62	slope, per L.F.
Minimum Charge	Job		227		227	
Fly Rafter						
2" x 4"						
Demolish	L.F.		.85		.85	Cost includes material and labor to
Install	L.F.	.44	.76		1.20	install 2" x 4" fly rafters for flat, shed
Demolish and Install	L.F.	.44	1.61		2.05	or gabled roofs, up to 5/12 slope,
Reinstall	L.F.		.61		.61	per L.F.
Minimum Charge	Job		227		227	
2" x 6"						
Demolish	L.F.		.86		.86	Cost includes material and labor to
Install	L.F.	.66	.77		1.43	install 2" x 6" fly rafters for flat, shed
Demolish and Install	L.F.	.66	1.63		2.29	or gabled roofs, up to 5/12 slope,
Reinstall	L.F.		.62		.62	per L.F.
Minimum Charge	Job		227		227	

For customer support on your Contractor's Pricing Guide: Residential Repair & Remodeling, call 888.606.7279.

Rough Frame / Structure

Residential Trusses

	Unit	Material	Labor	Equip.	Total	Specification
W or Fink Truss						
24' Span						
Demolish	Ea.		34.50	12.20	46.70	Includes material, labor and
Install	Ea.	84	34	12.05	130.05	equipment to install wood gang-nailed
Demolish and Install	Ea.	84	68.50	24.25	176.75	residential truss, up to 24' span.
Reinstall	Ea.		34.08	12.07	46.15	
Minimum Charge	Job		227		227	
28' Span						
Demolish	Ea.		37	13	50	Includes material, labor and
Install	Ea.	102	38.50	13.65	154.15	equipment to install wood gang-nailed
Demolish and Install	Ea.	102	75.50	26.65	204.15	residential truss, up to 28' span.
Reinstall	Ea.		38.58	13.67	52.25	
Minimum Charge	Job		227		227	
32' Span						
Demolish	Ea.		41	14.35	55.35	Includes material, labor and
Install	Ea.	121	41	14.50	176.50	equipment to install wood gang-nailed
Demolish and Install	Ea.	121	82	28.85	231.85	residential truss, up to 32' span.
Reinstall	Ea.		40.90	14.49	55.39	
Minimum Charge	Job		227		227	
36' Span						
Demolish	Ea.		44	15.50	59.50	Includes material, labor and
Install	Ea.	140	44.50	15.75	200.25	equipment to install wood gang-nailed
Demolish and Install	Ea.	140	88.50	31.25	259.75	residential truss, up to 36' span.
Reinstall	Ea.		44.45	15.75	60.20	
Minimum Charge	Job		227		227	
Gable End						
24' Span						
Demolish	Ea.		34.50	12.20	46.70	Includes material, labor and
Install	Ea.	108	36.50	12.95	157.45	equipment to install wood gang-nailed
Demolish and Install	Ea.	108	71	25.15	204.15	residential gable end truss, up to 24'
Reinstall	Ea.		36.51	12.94	49.45	span.
Clean	Ea.	1.69	44		45.69	
Paint	Ea.	4.65	50.50		55.15	
Minimum Charge	Job		227		227	
28' Span						
Demolish	Ea.		37	13	50	Includes material, labor and
Install	Ea.	138	43.50	15.40	196.90	equipment to install wood gang-nailed
Demolish and Install	Ea.	138	80.50	28.40	246.90	residential gable end truss, up to 30'
Reinstall	Ea.		43.51	15.41	58.92	span.
Clean	Ea.	1.77	60		61.77	
Paint	Ea.	4.85	68.50		73.35	
Minimum Charge	Job		227		227	
32' Span						
Demolish	Ea.		41	14.35	55.35	Includes material, labor and
Install	Ea.	160	44.50	15.75	220.25	equipment to install wood gang-nailed
Demolish and Install	Ea.	160	85.50	30.10	275.60	residential gable end truss, up to 32'
Reinstall	Ea.		44.45	15.75	60.20	span.
Clean	Ea.	1.85	78.50		80.35	
Paint	Ea.	5.05	90		95.05	
Minimum Charge	Job		227		227	
36' Span						
Demolish	Ea.		44	15.50	59.50	Includes material, labor and
Install	Ea.	204	51	18.10	273.10	equipment to install wood gang-nailed
Demolish and Install	Ea.	204	95	33.60	332.60	residential gable end truss, up to 36'
Reinstall	Ea.		51.12	18.11	69.23	span.
Clean	Ea.	1.94	99		100.94	
Paint	Ea.	5.30	113		118.30	
Minimum Charge	Job		227		227	

Rough Frame / Structure

Residential Trusses

Residential Trusses		Unit	Material	Labor	Equip.	Total	Specification
Truss Hardware							
5-1/4″ Glue-Lam Seat							
	Install	Ea.	230	2.52		232.52	Includes labor and material to install
	Minimum Charge	Job		227		227	beam hangers.
6-3/4″ Glue-Lam Seat							
	Install	Ea.	235	2.52		237.52	Includes labor and material to install
	Minimum Charge	Job		227		227	beam hangers.
2″ x 4″ Joist Hanger							
	Install	Ea.	.79	2.59		3.38	Includes labor and material to install
	Minimum Charge	Job		227		227	joist and beam hangers.
2″ x 12″ Joist Hanger							
	Install	Ea.	1.50	2.75		4.25	Cost includes labor and material to
	Minimum Charge	Job		227		227	install 2″ x 12″ joist hanger.

Commercial Trusses

Commercial Trusses		Unit	Material	Labor	Equip.	Total	Specification
Engineered Lumber, Truss / Joist							
9-1/2″							
	Demolish	S.F.		.60		.60	Includes material and labor to install
	Install	S.F.	1.84	.45		2.29	engineered lumber truss / joists per
	Demolish and Install	S.F.	1.84	1.05		2.89	S.F. of floor area based on joists 16″
	Reinstall	S.F.		.36		.36	O.C. Beams, supports and bridging
	Minimum Charge	Job		227		227	are not included.
11-7/8″							
	Demolish	S.F.		.62		.62	Includes material and labor to install
	Install	S.F.	2.02	.46		2.48	engineered lumber truss / joists per
	Demolish and Install	S.F.	2.02	1.08		3.10	S.F. of floor area based on joists 16″
	Reinstall	S.F.		.37		.37	O.C. Beams, supports and bridging
	Minimum Charge	Job		227		227	are not included.
14″							
	Demolish	S.F.		.65		.65	Includes material and labor to install
	Install	S.F.	2.10	.50		2.60	engineered lumber truss / joists per
	Demolish and Install	S.F.	2.10	1.15		3.25	S.F. of floor area based on joists 16″
	Reinstall	S.F.		.40		.40	O.C. Beams, supports and bridging
	Minimum Charge	Job		227		227	are not included.
16″							
	Demolish	S.F.		.67		.67	Includes material and labor to install
	Install	S.F.	3.54	.52		4.06	engineered lumber truss / joists per
	Demolish and Install	S.F.	3.54	1.19		4.73	S.F. of floor area based on joists 16″
	Reinstall	S.F.		.42		.42	O.C. Beams, supports and bridging
	Minimum Charge	Job		227		227	are not included.
Bowstring Truss							
100′ Clear Span							
	Demolish	SF Flr.		.62	.50	1.12	Includes material, labor and
	Install	SF Flr.	7.70	.51	.31	8.52	equipment to install bow string truss
	Demolish and Install	SF Flr.	7.70	1.13	.81	9.64	system up to 100′ clear span per S.F.
	Minimum Charge	Job		227		227	of floor area.
120′ Clear Span							
	Demolish	SF Flr.		.69	.55	1.24	Includes material, labor and
	Install	SF Flr.	8.20	.57	.34	9.11	equipment to install bow string truss
	Demolish and Install	SF Flr.	8.20	1.26	.89	10.35	system up to 120′ clear span per S.F.
	Minimum Charge	Job		227		227	of floor area.

For customer support on your Contractor's Pricing Guide: Residential Repair & Remodeling, call 888.606.7279.

Rough Frame / Structure

Roof Sheathing		Unit	Material	Labor	Equip.	Total	Specification
Plywood							
3/8"							
	Demolish	S.F.		.48		.48	Cost includes material and labor to
	Install	S.F.	.68	.60		1.28	install 3/8" CDX plywood roof
	Demolish and Install	S.F.	.68	1.08		1.76	sheathing.
	Reinstall	S.F.		.60		.60	
	Minimum Charge	Job		227		227	
1/2"							
	Demolish	S.F.		.48		.48	Cost includes material and labor to
	Install	S.F.	.75	.65		1.40	install 1/2" CDX plywood roof
	Demolish and Install	S.F.	.75	1.13		1.88	sheathing.
	Minimum Charge	Job		227		227	
5/8"							
	Demolish	S.F.		.48		.48	Cost includes material and labor to
	Install	S.F.	.84	.70		1.54	install 5/8" CDX plywood roof
	Demolish and Install	S.F.	.84	1.18		2.02	sheathing.
	Minimum Charge	Job		227		227	
3/4"							
	Demolish	S.F.		.48		.48	Cost includes material and labor to
	Install	S.F.	1	.76		1.76	install 3/4" CDX plywood roof
	Demolish and Install	S.F.	1	1.24		2.24	sheathing.
	Minimum Charge	Job		227		227	
OSB							
1/2"							
	Demolish	S.F.		.48		.48	Cost includes material and labor to
	Install	S.F.	.55	.65		1.20	install 4' x 8' x 1/2" OSB sheathing.
	Demolish and Install	S.F.	.55	1.13		1.68	
	Minimum Charge	Job		227		227	
5/8"							
	Demolish	S.F.		.48		.48	Cost includes material and labor to
	Install	S.F.	.55	.52		1.07	install 4' x 8' x 5/8" OSB sheathing.
	Demolish and Install	S.F.	.55	1		1.55	
	Minimum Charge	Job		227		227	
Plank							
	Demolish	S.F.		.48		.48	Cost includes material and labor to
	Install	S.F.	2.07	1.25		3.32	install 1" x 6" or 1" x 8" utility T&G
	Demolish and Install	S.F.	2.07	1.73		3.80	board sheathing laid diagonal.
	Reinstall	S.F.		1.25		1.25	
	Minimum Charge	Job		227		227	
Skip-type							
1" x 4"							
	Demolish	S.F.		.73		.73	Includes material and labor to install
	Install	S.F.	.64	.38		1.02	sheathing board material 1" x 4", 7"
	Demolish and Install	S.F.	.64	1.11		1.75	O.C.
	Minimum Charge	Job		227		227	
1" x 6"							
	Demolish	S.F.		.73		.73	Includes material and labor to install
	Install	S.F.	.78	.31		1.09	sheathing board material 1" x 6", 7"
	Demolish and Install	S.F.	.78	1.04		1.82	or 9" O.C.
	Minimum Charge	Job		227		227	

Rough Frame / Structure

Wood Deck	Unit	Material	Labor	Equip.	Total	Specification
Stairs						
Demolish	Ea.		440		440	Includes material and labor to install
Install	Ea.	152	785		937	wood steps. 5 - 7 risers included.
Demolish and Install	Ea.	152	1225		1377	
Reinstall	Ea.		628.99		628.99	
Clean	Flight	4.07	28.50		32.57	
Paint	Flight	21	50.50		71.50	
Minimum Charge	Job		227		227	
Bench Seating						
Demolish	L.F.		6.10		6.10	Includes material and labor to install
Install	L.F.	2.28	22.50		24.78	pressure treated bench seating.
Demolish and Install	L.F.	2.28	28.60		30.88	
Reinstall	L.F.		18.16		18.16	
Clean	L.F.	.17	1.65		1.82	
Paint	L.F.	.66	3.02		3.68	
Minimum Charge	Job		227		227	
Privacy Wall						
Demolish	SF Flr.		1.47		1.47	Includes material, labor and
Install	SF Flr.	1.58	.79	.21	2.58	equipment to install privacy wall.
Demolish and Install	SF Flr.	1.58	2.26	.21	4.05	
Reinstall	SF Flr.		.63	.17	.80	
Clean	SF Flr.	.08	.34		.42	
Paint	S.F.	.10	.29		.39	
Minimum Charge	Job		227		227	
Treated Column or Post						
4" x 4"						
Demolish	L.F.		.92		.92	Includes material and labor to install
Install	L.F.	1.49	2.33		3.82	4" x 4" treated column or post.
Demolish and Install	L.F.	1.49	3.25		4.74	
Reinstall	L.F.		1.86		1.86	
Clean	L.F.	.13	.29		.42	
Paint	L.F.	.19	.94		1.13	
Minimum Charge	Job		227		227	
4" x 6"						
Demolish	L.F.		1.33		1.33	Includes material and labor to install
Install	L.F.	2.22	3.30		5.52	4" x 6" treated column or post.
Demolish and Install	L.F.	2.22	4.63		6.85	
Reinstall	L.F.		2.64		2.64	
Clean	L.F.	.19	.41		.60	
Paint	L.F.	.19	.94		1.13	
Minimum Charge	Job		227		227	
Cedar Columns or Posts						
4" x 4"						
Demolish	L.F.		.92		.92	Includes material and labor to install
Install	L.F.	4.36	2.33		6.69	4" x 4" cedar columns or posts.
Demolish and Install	L.F.	4.36	3.25		7.61	
Reinstall	L.F.		1.86		1.86	
Clean	L.F.	.13	.29		.42	
Paint	L.F.	.19	.94		1.13	
Minimum Charge	Job		227		227	
4" x 6"						
Demolish	L.F.		1.33		1.33	Includes material and labor to install
Install	L.F.	8.45	3.30		11.75	4" x 6" cedar columns or posts.
Demolish and Install	L.F.	8.45	4.63		13.08	
Reinstall	L.F.		2.64		2.64	
Clean	L.F.	.19	.41		.60	
Paint	L.F.	.19	.94		1.13	
Minimum Charge	Job		227		227	

Rough Frame / Structure

Wood Deck	Unit	Material	Labor	Equip.	Total	Specification
Redwood Columns or Posts						
4″ x 4″						
Demolish	L.F.		.92		.92	Includes material and labor to install
Install	L.F.	7.10	2.33		9.43	4″ x 4″ redwood columns or posts.
Demolish and Install	L.F.	7.10	3.25		10.35	
Reinstall	L.F.		1.86		1.86	
Clean	L.F.	.13	.29		.42	
Paint	L.F.	.19	.94		1.13	
Minimum Charge	Job		227		227	
4″ x 6″						
Demolish	L.F.		1.33		1.33	Includes material and labor to install
Install	L.F.	13.85	3.30		17.15	4″ x 6″ redwood columns or posts.
Demolish and Install	L.F.	13.85	4.63		18.48	
Reinstall	L.F.		2.64		2.64	
Clean	L.F.	.19	.41		.60	
Paint	L.F.	.19	.94		1.13	
Minimum Charge	Job		227		227	
Treated Girder - Single						
4″ x 8″						
Demolish	L.F.		2.74		2.74	Includes material and labor to install a
Install	L.F.	4.26	1.73		5.99	treated lumber girder consisting of one
Demolish and Install	L.F.	4.26	4.47		8.73	4″ x 8″ for porch or deck construction.
Reinstall	L.F.		1.38		1.38	
Clean	L.F.	.17	.35		.52	
Minimum Charge	Job		227		227	
Treated Girder - Double						
2″ x 8″						
Demolish	L.F.		2.74		2.74	Includes material and labor to install a
Install	L.F.	2.39	1.58		3.97	treated lumber girder consisting of two
Demolish and Install	L.F.	2.39	4.32		6.71	2″ x 8″s for porch or deck
Reinstall	L.F.		1.26		1.26	construction.
Clean	L.F.	.17	.35		.52	
Minimum Charge	Job		227		227	
2″ x 10″						
Demolish	L.F.		3.42		3.42	Includes material and labor to install a
Install	L.F.	3.04	1.65		4.69	treated lumber girder consisting of two
Demolish and Install	L.F.	3.04	5.07		8.11	2″ x 10″s for porch or deck
Reinstall	L.F.		1.32		1.32	construction.
Clean	L.F.	.19	.41		.60	
Minimum Charge	Job		227		227	
2″ x 12″						
Demolish	L.F.		4.11		4.11	Includes material and labor to install a
Install	L.F.	4.26	1.73		5.99	treated lumber girder consisting of two
Demolish and Install	L.F.	4.26	5.84		10.10	2″ x 12″s for porch or deck
Reinstall	L.F.		1.38		1.38	construction.
Clean	L.F.	.21	.47		.68	
Minimum Charge	Job		227		227	
Treated Girder - Triple						
2″ x 8″						
Demolish	L.F.		4.11		4.11	Includes material and labor to install a
Install	L.F.	3.58	1.73		5.31	treated lumber girder consisting of
Demolish and Install	L.F.	3.58	5.84		9.42	three 2″ x 8″s for porch or deck
Reinstall	L.F.		1.38		1.38	construction.
Clean	L.F.	.19	.41		.60	
Minimum Charge	Job		227		227	

For customer support on your Contractor's Pricing Guide: Residential Repair & Remodeling, call 888.606.7279.

Rough Frame / Structure

Wood Deck		Unit	Material	Labor	Equip.	Total	Specification
2″ x 10″							
	Demolish	L.F.		5.15		5.15	Includes material and labor to install a
	Install	L.F.	4.54	1.82		6.36	treated lumber girder consisting of
	Demolish and Install	L.F.	4.54	6.97		11.51	three 2″ x 10″s for porch or deck
	Reinstall	L.F.		1.45		1.45	construction.
	Clean	L.F.	.21	.47		.68	
	Minimum Charge	Job		227		227	
2″ x 12″							
	Demolish	L.F.		6.10		6.10	Includes material and labor to install a
	Install	L.F.	6.40	1.91		8.31	treated lumber girder consisting of
	Demolish and Install	L.F.	6.40	8.01		14.41	three 2″ x 12″s for porch or deck
	Reinstall	L.F.		1.53		1.53	construction.
	Clean	L.F.	.24	.53		.77	
	Minimum Charge	Job		227		227	
Treated Ledger - Bolted							
2″ x 8″							
	Demolish	L.F.		.46		.46	Includes material and labor to install a
	Install	L.F.	1.33	2.33		3.66	treated lumber 2″ x 8″ ledger with
	Demolish and Install	L.F.	1.33	2.79		4.12	carriage bolts 4′ on center for porch
	Reinstall	L.F.		1.86		1.86	or deck construction.
	Clean	L.F.	.03	.24		.27	
	Minimum Charge	Job		227		227	
2″ x 10″							
	Demolish	L.F.		.46		.46	Includes material and labor to install a
	Install	L.F.	1.65	2.36		4.01	treated lumber 2″ x 10″ ledger with
	Demolish and Install	L.F.	1.65	2.82		4.47	carriage bolts 4′ on center for porch
	Reinstall	L.F.		1.89		1.89	or deck construction.
	Clean	L.F.	.04	.28		.32	
	Minimum Charge	Job		227		227	
2″ x 12″							
	Demolish	L.F.		.46		.46	Includes material and labor to install a
	Install	L.F.	2.26	2.39		4.65	treated lumber 2″ x 12″ ledger with
	Demolish and Install	L.F.	2.26	2.85		5.11	carriage bolts 4′ on center for porch
	Reinstall	L.F.		1.91		1.91	or deck construction.
	Clean	L.F.	.04	.28		.32	
	Minimum Charge	Job		227		227	
Treated Joists							
2″ x 8″							
	Demolish	L.F.		.78		.78	Includes material and labor to install
	Install	L.F.	1.19	.83		2.02	treated lumber 2″ x 8″ deck or floor
	Demolish and Install	L.F.	1.19	1.61		2.80	joists for porch or deck construction.
	Reinstall	L.F.		.66		.66	
	Clean	L.F.	.13	.29		.42	
	Minimum Charge	Job		227		227	
2″ x 10″							
	Demolish	L.F.		.81		.81	Includes material and labor to install
	Install	L.F.	1.51	1.01		2.52	treated lumber 2″ x 10″ deck or floor
	Demolish and Install	L.F.	1.51	1.82		3.33	joists for porch or deck construction.
	Reinstall	L.F.		.81		.81	
	Clean	L.F.	.17	.35		.52	
	Minimum Charge	Job		227		227	
2″ x 12″							
	Demolish	L.F.		.83		.83	Includes material and labor to install
	Install	L.F.	1.80	1.04		2.84	treated lumber 2″ x 12″ deck or floor
	Demolish and Install	L.F.	1.80	1.87		3.67	joists for porch or deck construction.
	Reinstall	L.F.		.83		.83	
	Clean	L.F.	.19	.41		.60	
	Minimum Charge	Job		227		227	

Rough Frame / Structure

Wood Deck	Unit	Material	Labor	Equip.	Total	Specification
Treated Railings and Trim						
2″ x 2″						
Demolish	L.F.		.31		.31	Includes material and labor to install
Install	L.F.	.36	1.51		1.87	treated lumber 2″ x 2″ railings and
Demolish and Install	L.F.	.36	1.82		2.18	trim for porch or deck construction.
Reinstall	L.F.		1.21		1.21	
Clean	L.F.	.01	.22		.23	
Paint	L.F.	.19	.94		1.13	
Minimum Charge	Job		227		227	
2″ x 4″						
Demolish	L.F.		.31		.31	Includes material and labor to install
Install	L.F.	.58	1.51		2.09	treated lumber 2″ x 4″ railings and
Demolish and Install	L.F.	.58	1.82		2.40	trim for porch or deck construction.
Reinstall	L.F.		1.21		1.21	
Clean	L.F.	.03	.24		.27	
Paint	L.F.	.19	.94		1.13	
Minimum Charge	Job		227		227	
2″ x 6″						
Demolish	L.F.		.31		.31	Includes material and labor to install
Install	L.F.	.86	1.51		2.37	treated lumber 2″ x 6″ railings and
Demolish and Install	L.F.	.86	1.82		2.68	trim for porch or deck construction.
Reinstall	L.F.		1.21		1.21	
Clean	L.F.	.02	.28		.30	
Paint	L.F.	.19	.94		1.13	
Minimum Charge	Job		227		227	
Cedar Railings and Trim						
1″ x 4″						
Demolish	L.F.		.31		.31	Includes material and labor to install
Install	L.F.	2.10	1.51		3.61	cedar 1″ x 4″ railings and trim for
Demolish and Install	L.F.	2.10	1.82		3.92	porch or deck construction.
Reinstall	L.F.		1.21		1.21	
Clean	L.F.	.01	.22		.23	
Paint	L.F.	.19	.94		1.13	
Minimum Charge	Job		227		227	
2″ x 4″						
Demolish	L.F.		.31		.31	Includes material and labor to install
Install	L.F.	4.25	1.51		5.76	cedar 2″ x 4″ railings and trim for
Demolish and Install	L.F.	4.25	1.82		6.07	porch or deck construction.
Reinstall	L.F.		1.21		1.21	
Clean	L.F.	.03	.24		.27	
Paint	L.F.	.19	.94		1.13	
Minimum Charge	Job		227		227	
2″ x 6″						
Demolish	L.F.		.31		.31	Includes material and labor to install
Install	L.F.	7.65	1.51		9.16	cedar 2″ x 6″ railings and trim for
Demolish and Install	L.F.	7.65	1.82		9.47	porch or deck construction.
Reinstall	L.F.		1.21		1.21	
Clean	L.F.	.02	.28		.30	
Paint	L.F.	.19	.94		1.13	
Minimum Charge	Job		227		227	
Redwood Railings and Trim						
1″ x 4″						
Demolish	L.F.		.31		.31	Includes material and labor to install
Install	L.F.	1.29	1.51		2.80	redwood 1″ x 4″ railings and trim for
Demolish and Install	L.F.	1.29	1.82		3.11	porch or deck construction.
Reinstall	L.F.		1.21		1.21	
Clean	L.F.	.01	.22		.23	
Paint	L.F.	.19	.94		1.13	
Minimum Charge	Job		227		227	

Rough Frame / Structure

Wood Deck

	Unit	Material	Labor	Equip.	Total	Specification
2″ x 6″						
Demolish	L.F.		.31		.31	Includes material and labor to install
Install	L.F.	8.25	1.51		9.76	redwood 2″ x 6″ railings and trim for
Demolish and Install	L.F.	8.25	1.82		10.07	porch or deck construction.
Reinstall	L.F.		1.21		1.21	
Clean	L.F.	.02	.28		.30	
Paint	L.F.	.19	.94		1.13	
Minimum Charge	Job		227		227	

Treated Decking

	Unit	Material	Labor	Equip.	Total	Specification
1″ x 4″						
Demolish	L.F.		.44		.44	Includes material and labor to install
Install	S.F.	2.38	1.65		4.03	treated lumber 1″ x 4″ decking for
Demolish and Install	S.F.	2.38	2.09		4.47	porch or deck construction.
Reinstall	S.F.		1.32		1.32	
Clean	S.F.	.08	.34		.42	
Paint	S.F.	.22	.76		.98	
Minimum Charge	Job		227		227	
2″ x 4″						
Demolish	L.F.		.44		.44	Includes material and labor to install
Install	S.F.	1.96	1.51		3.47	treated lumber 2″ x 4″ decking for
Demolish and Install	S.F.	1.96	1.95		3.91	porch or deck construction.
Reinstall	S.F.		1.21		1.21	
Clean	S.F.	.08	.34		.42	
Paint	S.F.	.22	.76		.98	
Minimum Charge	Job		227		227	
2″ x 6″						
Demolish	L.F.		.44		.44	Includes material and labor to install
Install	S.F.	1.85	1.42		3.27	treated lumber 2″ x 6″ decking for
Demolish and Install	S.F.	1.85	1.86		3.71	porch or deck construction.
Reinstall	S.F.		1.14		1.14	
Clean	S.F.	.08	.34		.42	
Paint	S.F.	.22	.76		.98	
Minimum Charge	Job		227		227	
5/4″ x 6″						
Demolish	L.F.		.44		.44	Includes material and labor to install
Install	S.F.	2.48	1.42		3.90	treated lumber 5/4″ x 6″ decking for
Demolish and Install	S.F.	2.48	1.86		4.34	porch or deck construction.
Reinstall	S.F.		1.14		1.14	
Clean	S.F.	.08	.34		.42	
Paint	S.F.	.22	.76		.98	
Minimum Charge	Job		227		227	

Wood / Plastic Composite Decking

	Unit	Material	Labor	Equip.	Total	Specification
5/4″ x 6″						
Demolish	L.F.		.44		.44	Includes material and labor to install
Install	L.F.	3.45	1.42		4.87	wood/plastic composite 5/4″ x 6″
Demolish and Install	L.F.	3.45	1.86		5.31	decking for porch or deck
Reinstall	L.F.		1.14		1.14	construction.
Clean	S.F.	.08	.34		.42	
Paint	S.F.	.22	.76		.98	
Minimum Charge	Job		227		227	

Square Edge Fir Decking

	Unit	Material	Labor	Equip.	Total	Specification
1″ x 4″						
Demolish	L.F.		.44		.44	Includes material and labor to install
Install	L.F.	2.40	1.65		4.05	square edge fir 1″ x 4″ decking for
Demolish and Install	L.F.	2.40	2.09		4.49	porch or deck construction.
Reinstall	L.F.		1.32		1.32	
Clean	S.F.	.08	.34		.42	
Paint	S.F.	.22	.76		.98	
Minimum Charge	Job		227		227	

Wood Deck	Unit	Material	Labor	Equip.	Total	Specification
Tongue and Groove Fir Decking						
1" x 4"						
Demolish	L.F.		.44		.44	Includes material and labor to install
Install	L.F.	1.61	2.02		3.63	treated lumber 1" x 4" decking for
Demolish and Install	L.F.	1.61	2.46		4.07	porch or deck construction.
Reinstall	L.F.		1.61		1.61	
Clean	S.F.	.08	.34		.42	
Paint	S.F.	.22	.76		.98	
Minimum Charge	Job		227		227	
Mahogany Decking						
1" x 4"						
Demolish	L.F.		.44		.44	Includes material and labor to install
Install	L.F.	2.21	1.65		3.86	treated lumber 1" x 4" decking for
Demolish and Install	L.F.	2.21	2.09		4.30	porch or deck construction.
Reinstall	L.F.		1.32		1.32	
Clean	S.F.	.08	.34		.42	
Paint	S.F.	.22	.76		.98	
Minimum Charge	Job		227		227	
PVC Decking						
5/4" x 6"						
Demolish	L.F.		.44		.44	Includes material and labor to install
Install	L.F.	4.08	1.65		5.73	PVC 5/4" x 6" decking for porch or
Demolish and Install	L.F.	4.08	2.09		6.17	deck construction.
Reinstall	L.F.		1.32		1.32	
Clean	S.F.	.08	.34		.42	
Minimum Charge	Job		227		227	
PVC Handrails and Balusters						
Straight						
Demolish	L.F.		3.06		3.06	Includes material and labor to install
Install	L.F.	28	4.73		32.73	straight PVC handrails and balusters
Demolish and Install	L.F.	28	7.79		35.79	for porch or deck construction.
Reinstall	L.F.		3.78		3.78	
Clean	L.F.	.81	2.58		3.39	
Minimum Charge	Job		227		227	
Angled						
Demolish	L.F.		3.06		3.06	Includes material and labor to install
Install	L.F.	32	6.30		38.30	angled PVC handrails and balusters
Demolish and Install	L.F.	32	9.36		41.36	for porch or deck construction.
Reinstall	L.F.		5.04		5.04	
Clean	L.F.	.81	2.58		3.39	
Minimum Charge	Job		227		227	
PVC Post Sleeve						
for 4" x 4" post						
Demolish	L.F.		.92		.92	Includes material and labor to install
Install	L.F.	14.40	4.73		19.13	PVC post sleeves over 4" x 4" porch
Demolish and Install	L.F.	14.40	5.65		20.05	or deck posts.
Reinstall	L.F.		3.78		3.78	
Clean	L.F.	.13	.29		.42	
Minimum Charge	Job		227		227	
Post Cap						
Demolish	Ea.		4.58		4.58	Includes material and labor to install
Install	Ea.	26.50	9.45		35.95	PVC post caps for porch or deck
Demolish and Install	Ea.	26.50	14.03		40.53	construction.
Reinstall	Ea.		7.57		7.57	
Clean	L.F.	.13	.29		.42	
Minimum Charge	Job		227		227	

Rough Frame / Structure

Porch

	Unit	Material	Labor	Equip.	Total	Specification
Stairs						
Demolish	Ea.		440		440	Includes material and labor to install
Install	Ea.	152	785		937	wood steps. 5 - 7 risers included.
Demolish and Install	Ea.	152	1225		1377	
Reinstall	Ea.		628.99		628.99	
Clean	Flight	4.07	28.50		32.57	
Paint	Flight	21	50.50		71.50	
Minimum Charge	Job		227		227	
Bench Seating						
Demolish	L.F.		6.10		6.10	Includes material and labor to install
Install	L.F.	2.28	22.50		24.78	pressure treated bench seating.
Demolish and Install	L.F.	2.28	28.60		30.88	
Reinstall	L.F.		18.16		18.16	
Clean	L.F.	.17	1.65		1.82	
Paint	L.F.	.66	3.02		3.68	
Minimum Charge	Job		227		227	
Privacy Wall						
Demolish	SF Flr.		1.47		1.47	Includes material, labor and
Install	SF Flr.	1.58	.79	.21	2.58	equipment to install privacy wall.
Demolish and Install	SF Flr.	1.58	2.26	.21	4.05	
Reinstall	SF Flr.		.63	.17	.80	
Clean	SF Flr.	.08	.34		.42	
Paint	S.F.	.10	.29		.39	
Minimum Charge	Job		227		227	
Wood Board Ceiling						
Demolish	S.F.		.61		.61	Includes material and labor to install
Install	S.F.	3.52	2.02		5.54	wood board on ceiling.
Demolish and Install	S.F.	3.52	2.63		6.15	
Clean	S.F.	.03	.19		.22	
Paint	S.F.	.21	.47		.68	
Minimum Charge	Job		227		227	
Re-screen						
Install	S.F.	1.77	1.82		3.59	Includes labor and material to
Minimum Charge	Job		227		227	re-screen wood frame.

For customer support on your Contractor's Pricing Guide: Residential Repair & Remodeling, call 888.606.7279.

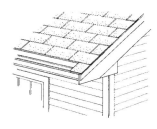
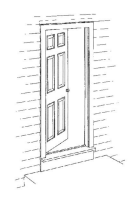

Exterior Trim

Exterior Trim	Unit	Material	Labor	Equip.	Total	Specification
Stock Lumber						
1″ x 2″						
Demolish	L.F.		.31		.31	Cost includes material and labor to
Install	L.F.	.28	1.38		1.66	install 1″ x 2″ pine trim.
Demolish and Install	L.F.	.28	1.69		1.97	
Clean	L.F.	.01	.22		.23	
Paint	L.F.	.02	.59		.61	
Minimum Charge	Job		227		227	
1″ x 3″						
Demolish	L.F.		.31		.31	Cost includes material and labor to
Install	L.F.	.30	1.57		1.87	install 1″ x 3″ pine trim.
Demolish and Install	L.F.	.30	1.88		2.18	
Clean	L.F.	.01	.22		.23	
Paint	L.F.	.02	.59		.61	
Minimum Charge	Job		227		227	
1″ x 4″						
Demolish	L.F.		.31		.31	Cost includes material and labor to
Install	L.F.	.54	1.82		2.36	install 1″ x 4″ pine trim.
Demolish and Install	L.F.	.54	2.13		2.67	
Clean	L.F.	.01	.22		.23	
Paint	L.F.	.02	.59		.61	
Minimum Charge	Job		227		227	
1″ x 6″						
Demolish	L.F.		.31		.31	Cost includes material and labor to
Install	L.F.	.80	1.82		2.62	install 1″ x 6″ pine trim.
Demolish and Install	L.F.	.80	2.13		2.93	
Clean	L.F.	.02	.22		.24	
Paint	L.F.	.09	.59		.68	
Minimum Charge	Job		227		227	
1″ x 8″						
Demolish	L.F.		.33		.33	Cost includes material and labor to
Install	L.F.	1.32	2.27		3.59	install 1″ x 8″ pine trim.
Demolish and Install	L.F.	1.32	2.60		3.92	
Reinstall	L.F.		1.82		1.82	
Clean	L.F.	.03	.24		.27	
Paint	L.F.	.09	.59		.68	
Minimum Charge	Job		227		227	

Exterior Trim

Exterior Trim	Unit	Material	Labor	Equip.	Total	Specification
Stock Lumber						
1" x 10"						
Demolish	L.F.		.51		.51	Cost includes material and labor to
Install	L.F.	1.73	2.52		4.25	install 1" x 10" pine trim.
Demolish and Install	L.F.	1.73	3.03		4.76	
Minimum Charge	Job		227		227	
1" x 12"						
Demolish	L.F.		.51		.51	Cost includes material and labor to
Install	L.F.	2.11	2.52		4.63	install 1" x 12" pine trim.
Demolish and Install	L.F.	2.11	3.03		5.14	
Minimum Charge	Job		227		227	
PVC						
1" x 4"						
Demolish	L.F.		.31		.31	Includes material and labor to install
Install	L.F.	1.61	1.82		3.43	1" x 4" PVC exterior trim.
Demolish and Install	L.F.	1.61	2.13		3.74	
Reinstall	L.F.		1.45		1.45	
Clean	L.F.	.01	.22		.23	
Paint	L.F.	.02	.59		.61	
Minimum Charge	Job		227		227	
1" x 6"						
Demolish	L.F.		.31		.31	Includes material and labor to install
Install	L.F.	2.42	1.82		4.24	1" x 6" PVC exterior trim.
Demolish and Install	L.F.	2.42	2.13		4.55	
Reinstall	L.F.		1.45		1.45	
Clean	L.F.	.02	.22		.24	
Paint	L.F.	.09	.59		.68	
Minimum Charge	Job		227		227	
1" x 8"						
Demolish	L.F.		.33		.33	Includes material and labor to install
Install	L.F.	3.22	2.02		5.24	1" x 8" PVC exterior trim.
Demolish and Install	L.F.	3.22	2.35		5.57	
Reinstall	L.F.		1.61		1.61	
Clean	L.F.	.03	.24		.27	
Paint	L.F.	.09	.59		.68	
Minimum Charge	Job		227		227	
1" x 10"						
Demolish	L.F.		.35		.35	Includes material and labor to install
Install	L.F.	4.19	2.02		6.21	1" x 10" PVC exterior trim.
Demolish and Install	L.F.	4.19	2.37		6.56	
Reinstall	L.F.		1.61		1.61	
Clean	L.F.	.04	.28		.32	
Paint	L.F.	.09	.59		.68	
Minimum Charge	Job		227		227	
1" x 12"						
Demolish	L.F.		.37		.37	Includes material and labor to install
Install	L.F.	4.80	2.27		7.07	1" x 12" PVC exterior trim.
Demolish and Install	L.F.	4.80	2.64		7.44	
Reinstall	L.F.		1.82		1.82	
Clean	L.F.	.04	.28		.32	
Paint	L.F.	.09	.59		.68	
Minimum Charge	Job		227		227	

Exterior Trim

Fascia

	Unit	Material	Labor	Equip.	Total	Specification
Aluminum						
Demolish	L.F.		.64		.64	Includes material and labor to install
Install	S.F.	2.20	2.16		4.36	aluminum fascia.
Demolish and Install	S.F.	2.20	2.80		5	
Reinstall	S.F.		1.73		1.73	
Clean	L.F.		.41		.41	
Minimum Charge	Job		227		227	
Vinyl						
Demolish	L.F.		.64		.64	Cost includes material and labor to
Install	L.F.	4.84	2.59		7.43	install 6″ vinyl fascia with 12″ vinyl
Demolish and Install	L.F.	4.84	3.23		8.07	soffit and J-channel.
Reinstall	L.F.		2.08		2.08	
Clean	L.F.		.41		.41	
Minimum Charge	Job		227		227	
Plywood						
Demolish	L.F.		.63		.63	Cost includes material and labor to
Install	L.F.	1.67	1.01		2.68	install 12″ wide plywood fascia.
Demolish and Install	L.F.	1.67	1.64		3.31	
Reinstall	L.F.		.81		.81	
Clean	L.F.	.04	.28		.32	
Paint	L.F.	.19	.94		1.13	
Minimum Charge	Job		227		227	
Cedar						
Demolish	L.F.		.63		.63	Cost includes material and labor to
Install	L.F.	3.98	1.01		4.99	install 12″ wide cedar fascia.
Demolish and Install	L.F.	3.98	1.64		5.62	
Reinstall	L.F.		.81		.81	
Clean	L.F.	.04	.28		.32	
Paint	L.F.	.19	.94		1.13	
Minimum Charge	Job		227		227	
Redwood						
1″ x 6″						
Demolish	L.F.		.49		.49	Cost includes material and labor to
Install	L.F.	1.98	1.82		3.80	install 1″ x 6″ redwood fascia board.
Demolish and Install	L.F.	1.98	2.31		4.29	
Reinstall	L.F.		1.45		1.45	
Minimum Charge	Job		227		227	
1″ x 8″						
Demolish	L.F.		.55		.55	Cost includes material and labor to
Install	L.F.	2.60	1.97		4.57	install 1″ x 8″ redwood fascia board.
Demolish and Install	L.F.	2.60	2.52		5.12	
Reinstall	L.F.		1.58		1.58	
Minimum Charge	Job		227		227	
2″ x 6″						
Demolish	L.F.		.49		.49	Cost includes material and labor to
Install	L.F.	.67	3.63		4.30	install 2″ x 6″ exterior trim board.
Demolish and Install	L.F.	.67	4.12		4.79	
Reinstall	L.F.		2.91		2.91	
Minimum Charge	Job		227		227	
2″ x 8″						
Demolish	L.F.		.55		.55	Cost includes material and labor to
Install	L.F.	.94	4.04		4.98	install 2″ x 8″ exterior trim board.
Demolish and Install	L.F.	.94	4.59		5.53	
Reinstall	L.F.		3.23		3.23	
Minimum Charge	Job		227		227	

Exterior Trim

Fascia	Unit	Material	Labor	Equip.	Total	Specification
Hem-fir Std & Better						
2" x 6"						
Demolish	L.F.		.49		.49	Cost includes material and labor to
Install	L.F.	.67	3.63		4.30	install 2" x 6" exterior trim board.
Demolish and Install	L.F.	.67	4.12		4.79	
Reinstall	L.F.		2.91		2.91	
Clean	L.F.	.02	.28		.30	
Paint	L.F.	.19	.94		1.13	
Minimum Charge	Job		227		227	
2" x 8"						
Demolish	L.F.		.55		.55	Cost includes material and labor to
Install	L.F.	.94	4.04		4.98	install 2" x 8" exterior trim board.
Demolish and Install	L.F.	.94	4.59		5.53	
Reinstall	L.F.		3.23		3.23	
Clean	L.F.	.03	.28		.31	
Paint	L.F.	.19	.94		1.13	
Minimum Charge	Job		227		227	
2" x 10"						
Demolish	L.F.		.63		.63	Cost includes material and labor to
Install	L.F.	1.47	5.05		6.52	install 2" x 10" exterior trim board.
Demolish and Install	L.F.	1.47	5.68		7.15	
Reinstall	L.F.		4.04		4.04	
Clean	L.F.	.04	.30		.34	
Paint	L.F.	.19	.94		1.13	
Minimum Charge	Job		227		227	
Barge Rafter						
2" x 6"						
Demolish	L.F.		.98		.98	Cost includes material and labor to
Install	L.F.	.67	.50		1.17	install 2" x 6" barge rafters.
Demolish and Install	L.F.	.67	1.48		2.15	
Reinstall	L.F.		.40		.40	
Clean	L.F.	.02	.28		.30	
Paint	L.F.	.19	.94		1.13	
Minimum Charge	Job		227		227	
2" x 8"						
Demolish	L.F.		.99		.99	Cost includes material and labor to
Install	L.F.	.94	.53		1.47	install 2" x 8" barge rafters.
Demolish and Install	L.F.	.94	1.52		2.46	
Reinstall	L.F.		.43		.43	
Clean	L.F.	.03	.28		.31	
Paint	L.F.	.19	.94		1.13	
Minimum Charge	Job		227		227	
2" x 10"						
Demolish	L.F.		1.01		1.01	Cost includes material and labor to
Install	L.F.	1.47	.55		2.02	install 2" x 10" barge rafters.
Demolish and Install	L.F.	1.47	1.56		3.03	
Reinstall	L.F.		.44		.44	
Clean	L.F.	.04	.30		.34	
Paint	L.F.	.19	.94		1.13	
Minimum Charge	Job		227		227	

For customer support on your Contractor's Pricing Guide: Residential Repair & Remodeling, call 888.606.7279.

Exterior Trim

Soffit

	Unit	Material	Labor	Equip.	Total	Specification
Aluminum						
12" w / Fascia						
Demolish	L.F.		.64		.64	Cost includes material and labor to
Install	L.F.	4.95	2.59		7.54	install 12" aluminum soffit with 6"
Demolish and Install	L.F.	4.95	3.23		8.18	fascia and J-channel.
Reinstall	L.F.		2.08		2.08	
Clean	L.F.		.41		.41	
Paint	L.F.	.24	.94		1.18	
Minimum Charge	Job		227		227	
18" w / Fascia						
Demolish	L.F.		.64		.64	Cost includes material and labor to
Install	L.F.	5.95	3.03		8.98	install 18" aluminum soffit with 6"
Demolish and Install	L.F.	5.95	3.67		9.62	fascia and J-channel.
Reinstall	L.F.		2.42		2.42	
Clean	L.F.		.41		.41	
Paint	L.F.	.24	.94		1.18	
Minimum Charge	Job		227		227	
24" w / Fascia						
Demolish	L.F.		.64		.64	Cost includes material and labor to
Install	L.F.	7.15	3.36		10.51	install 24" aluminum soffit with 6"
Demolish and Install	L.F.	7.15	4		11.15	fascia and J-channel.
Reinstall	L.F.		2.69		2.69	
Clean	L.F.		.60		.60	
Paint	L.F.	.24	.94		1.18	
Minimum Charge	Job		227		227	
Vinyl						
Demolish	L.F.		.64		.64	Cost includes material and labor to
Install	L.F.	4.84	2.59		7.43	install 6" vinyl fascia with 12" vinyl
Demolish and Install	L.F.	4.84	3.23		8.07	soffit and J-channel.
Reinstall	L.F.		2.08		2.08	
Clean	L.F.		.41		.41	
Paint	L.F.	.24	.94		1.18	
Minimum Charge	Job		227		227	
Plywood						
Demolish	L.F.		.70		.70	Cost includes material and labor to
Install	L.F.	1.63	1.30		2.93	install 12" wide plywood soffit.
Demolish and Install	L.F.	1.63	2		3.63	
Reinstall	L.F.		1.04		1.04	
Clean	L.F.	.04	.28		.32	
Paint	L.F.	.19	.94		1.13	
Minimum Charge	Job		227		227	
Redwood						
Demolish	L.F.		.70		.70	Cost includes material and labor to
Install	L.F.	5.20	1.30		6.50	install 12" wide redwood soffit.
Demolish and Install	L.F.	5.20	2		7.20	
Reinstall	L.F.		1.04		1.04	
Clean	L.F.	.04	.28		.32	
Paint	L.F.	.19	.94		1.13	
Minimum Charge	Job		227		227	
Cedar						
Demolish	L.F.		.70		.70	Cost includes material and labor to
Install	L.F.	3.94	1.30		5.24	install 12" wide cedar soffit.
Demolish and Install	L.F.	3.94	2		5.94	
Reinstall	L.F.		1.04		1.04	
Clean	L.F.	.04	.28		.32	
Paint	L.F.	.19	.94		1.13	
Minimum Charge	Job		227		227	

Exterior Trim

Soffit	Unit	Material	Labor	Equip.	Total	Specification
Douglas Fir						
Demolish	L.F.		.70		.70	Cost includes material and labor to
Install	L.F.	2.07	1.30		3.37	install 12″ wide Douglas fir soffit.
Demolish and Install	L.F.	2.07	2		4.07	
Reinstall	L.F.		1.04		1.04	
Clean	L.F.	.04	.28		.32	
Paint	L.F.	.19	.94		1.13	
Minimum Charge	Job		227		227	

For customer support on your Contractor's Pricing Guide: Residential Repair & Remodeling, call 888.606.7279.

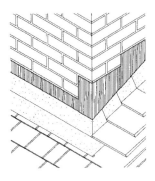

Roofing and Flashing   **Tile Roofing**

Metal Flashing		Unit	Material	Labor	Equip.	Total	Specification
Galvanized							
6″							
	Demolish	L.F.		.24		.24	Includes material and labor to install
	Install	L.F.	.73	1.06		1.79	flat galvanized steel flashing, 6″ wide.
	Demolish and Install	L.F.	.73	1.30		2.03	
	Paint	L.F.	.13	.47		.60	
	Minimum Charge	Job		212		212	
14″							
	Demolish	L.F.		.24		.24	Includes material and labor to install
	Install	L.F.	1.68	1.06		2.74	flat galvanized steel flashing, 14″
	Demolish and Install	L.F.	1.68	1.30		2.98	wide.
	Paint	L.F.	.13	.47		.60	
	Minimum Charge	Job		212		212	
20″							
	Demolish	L.F.		.24		.24	Includes material and labor to install
	Install	L.F.	2.42	1.06		3.48	flat galvanized steel flashing, 20″
	Demolish and Install	L.F.	2.42	1.30		3.72	wide.
	Paint	L.F.	.13	.47		.60	
	Minimum Charge	Job		212		212	
Aluminum							
6″							
	Demolish	L.F.		.24		.24	Includes material and labor to install
	Install	L.F.	.44	1.06		1.50	flat aluminum flashing, 6″ wide.
	Demolish and Install	L.F.	.44	1.30		1.74	
	Paint	L.F.	.13	.47		.60	
	Minimum Charge	Job		212		212	
14″							
	Demolish	L.F.		.24		.24	Includes material and labor to install
	Install	L.F.	.80	1.06		1.86	flat aluminum flashing, 14″ wide.
	Demolish and Install	L.F.	.80	1.30		2.10	
	Paint	L.F.	.13	.47		.60	
	Minimum Charge	Job		212		212	
20″							
	Demolish	L.F.		.24		.24	Includes material and labor to install
	Install	L.F.	1.12	1.06		2.18	flat aluminum flashing, 20″ wide.
	Demolish and Install	L.F.	1.12	1.30		2.42	
	Paint	L.F.	.13	.47		.60	
	Minimum Charge	Job		212		212	

Roofing

Metal Flashing	Unit	Material	Labor	Equip.	Total	Specification
Copper						
6″						
Demolish	L.F.		.24		.24	Includes material and labor to install
Install	L.F.	4.33	1.41		5.74	flat copper flashing, 6″ wide.
Demolish and Install	L.F.	4.33	1.65		5.98	
Minimum Charge	Job		252		252	
14″						
Demolish	L.F.		.24		.24	Includes material and labor to install
Install	L.F.	10.10	1.41		11.51	flat copper flashing, 14″ wide.
Demolish and Install	L.F.	10.10	1.65		11.75	
Minimum Charge	Job		252		252	
18″						
Demolish	L.F.		.24		.24	Includes material and labor to install
Install	L.F.	14.45	1.41		15.86	flat copper flashing, 18″ wide.
Demolish and Install	L.F.	14.45	1.65		16.10	
Minimum Charge	Job		252		252	
Aluminum Valley						
Demolish	L.F.		.24		.24	Includes material and labor to install
Install	L.F.	4.02	2.67		6.69	aluminum valley flashing, .024″ thick.
Demolish and Install	L.F.	4.02	2.91		6.93	
Paint	L.F.	.13	.47		.60	
Minimum Charge	Job		227		227	
Gravel Stop						
Demolish	L.F.		.39		.39	Includes material and labor to install
Install	L.F.	7.05	3.48		10.53	aluminum gravel stop, .050″ thick, 4″
Demolish and Install	L.F.	7.05	3.87		10.92	high, mill finish.
Reinstall	L.F.		2.78		2.78	
Paint	L.F.	.21	.47		.68	
Minimum Charge	Job		227		227	
Drip Edge						
Demolish	L.F.		.37		.37	Includes material and labor to install
Install	L.F.	.63	1.14		1.77	aluminum drip edge, .016″ thick, 5″
Demolish and Install	L.F.	.63	1.51		2.14	wide, mill finish.
Reinstall	L.F.		.91		.91	
Paint	L.F.	.21	.47		.68	
Minimum Charge	Job		227		227	
End Wall						
Demolish	L.F.		1.13		1.13	Includes material and labor to install
Install	L.F.	2.37	2.67		5.04	aluminum end wall flashing, .024″
Demolish and Install	L.F.	2.37	3.80		6.17	thick.
Minimum Charge	Job		227		227	
Side Wall						
Demolish	L.F.		.22		.22	Includes material and labor to install
Install	L.F.	2.31	2.67		4.98	aluminum side wall flashing, .024″
Demolish and Install	L.F.	2.31	2.89		5.20	thick.
Minimum Charge	Job		227		227	
Coping and Wall Cap						
Demolish	L.F.		1.53		1.53	Includes material and labor to install
Install	L.F.	2.54	2.92		5.46	aluminum flashing, mill finish, 0.40
Demolish and Install	L.F.	2.54	4.45		6.99	thick, 12″ wide.
Clean	L.F.	.10	1.03		1.13	
Minimum Charge	Job		212		212	

For customer support on your Contractor's Pricing Guide: Residential Repair & Remodeling, call 888.606.7279.

Roofing

Roof Shingles	Unit	Material	Labor	Equip.	Total	Specification
Underlayment						
15#						
Demolish	Sq.		12.20		12.20	Includes material and labor to install
Install	Sq.	5.85	6.65		12.50	15# felt underlayment.
Demolish and Install	Sq.	5.85	18.85		24.70	
Minimum Charge	Job		212		212	
30#						
Demolish	Sq.		12.20		12.20	Includes material and labor to install
Install	Sq.	11.35	7.30		18.65	30# felt underlayment.
Demolish and Install	Sq.	11.35	19.50		30.85	
Minimum Charge	Job		212		212	
Self Adhering						
Demolish	Sq.		30.50		30.50	Includes material and labor to install
Install	Sq.	82.50	19.25		101.75	self adhering ice barrier roofing
Demolish and Install	Sq.	82.50	49.75		132.25	underlayment.
Minimum Charge	Job		212		212	

Rolled Roofing	Unit	Material	Labor	Equip.	Total	Specification
90 lb						
Demolish	Sq.		30.50		30.50	Cost includes material and labor to
Install	Sq.	38.50	35.50		74	install 90 lb mineral surface rolled
Demolish and Install	Sq.	38.50	66		104.50	roofing.
Minimum Charge	Job		212		212	

Composition Shingles	Unit	Material	Labor	Equip.	Total	Specification
Fiberglass (3-tab)						
25 Year						
Demolish	Sq.		33.50		33.50	Cost includes material and labor to
Install	Sq.	83	47.50		130.50	install 25 year fiberglass shingles.
Demolish and Install	Sq.	83	81		164	
Minimum Charge	Job		212		212	
30 Year						
Demolish	Sq.		33.50		33.50	Cost includes material and labor to
Install	Sq.	98	53		151	install 30 year fiberglass shingles.
Demolish and Install	Sq.	98	86.50		184.50	
Minimum Charge	Job		212		212	
Emergency Repairs						
Install	Ea.	165	212		377	Includes labor and material for
Minimum Charge	Job		212		212	emergency repairs.
Architectural						
Laminated, 25 Year						
Demolish	Sq.		33.50		33.50	Cost includes material and labor to
Install	Sq.	112	59		171	install 25 year fiberglass roof shingles.
Demolish and Install	Sq.	112	92.50		204.50	
Minimum Charge	Job		212		212	
Laminated, 30 Year						
Demolish	Sq.		33.50		33.50	Cost includes material and labor to
Install	Sq.	162	75.50		237.50	install 30 year laminated roof
Demolish and Install	Sq.	162	109		271	shingles.
Minimum Charge	Job		212		212	

73

Composition Shingles	Unit	Material	Labor	Equip.	Total	Specification
Laminated, 40 Year						
Demolish	Sq.		33.50		33.50	Cost includes material and labor to
Install	Sq.	247	88		335	install 40 year laminated roof
Demolish and Install	Sq.	247	121.50		368.50	shingles.
Minimum Charge	Job		212		212	
Laminated Shake, 40 Year						
Demolish	Sq.		33.50		33.50	Cost includes material and labor to
Install	Sq.	247	88		335	install 40 year laminated roof
Demolish and Install	Sq.	247	121.50		368.50	shingles.
Minimum Charge	Job		212		212	
Emergency Repairs						
Install	Ea.	165	212		377	Includes labor and material for
Minimum Charge	Job		212		212	emergency repairs.
Roof Jack						
Demolish	Ea.		6.65		6.65	Includes material and labor to install
Install	Ea.	28.50	46		74.50	residential roof jack, w/bird screen,
Demolish and Install	Ea.	28.50	52.65		81.15	backdraft damper, 3" & 4" dia. round
Paint	Ea.	.07	4.72		4.79	duct.
Minimum Charge	Job		241		241	
Ridge Vent						
Demolish	L.F.		1.18		1.18	Includes material and labor to install a
Install	L.F.	4.27	3.25		7.52	mill finish aluminum ridge vent strip.
Demolish and Install	L.F.	4.27	4.43		8.70	
Minimum Charge	Job		227		227	
Shingle Molding						
1" x 2"						
Demolish	L.F.		.23		.23	Cost includes material and labor to
Install	L.F.	.28	1.38		1.66	install 1" x 2" pine trim.
Demolish and Install	L.F.	.28	1.61		1.89	
Paint	L.F.	.02	.59		.61	
Minimum Charge	Job		227		227	
1" x 3"						
Demolish	L.F.		.23		.23	Cost includes material and labor to
Install	L.F.	.30	1.57		1.87	install 1" x 3" pine trim.
Demolish and Install	L.F.	.30	1.80		2.10	
Paint	L.F.	.02	.59		.61	
Minimum Charge	Job		227		227	
1" x 4"						
Demolish	L.F.		.23		.23	Cost includes material and labor to
Install	L.F.	.54	1.82		2.36	install 1" x 4" pine trim.
Demolish and Install	L.F.	.54	2.05		2.59	
Paint	L.F.	.02	.59		.61	
Minimum Charge	Job		227		227	
Add for Steep Pitch						
Install	Sq.		26.50		26.50	Includes labor and material added for
Add for Additional Story						2nd story or steep roofs.
Install	Sq.		26.50		26.50	Includes labor and material added for 2nd story or steep roofs.

Roofing

Wood Shingles		Unit	Material	Labor	Equip.	Total	Specification
Cedar Wood Shingles							
16″ Red Label (#2)							
	Demolish	Sq.		81.50		81.50	Includes material and labor to install
	Install	Sq.	185	182		367	wood shingles/shakes.
	Demolish and Install	Sq.	185	263.50		448.50	
	Minimum Charge	Job		227		227	
16″ Blue Label (#1)							
	Demolish	Sq.		81.50		81.50	Includes material and labor to install
	Install	Sq.	310	140		450	wood shingles/shakes.
	Demolish and Install	Sq.	310	221.50		531.50	
	Minimum Charge	Job		227		227	
18″ Red Label (#2)							
	Demolish	Sq.		81.50		81.50	Includes material and labor to install
	Install	Sq.	310	116		426	wood shingles/shakes.
	Demolish and Install	Sq.	310	197.50		507.50	
	Minimum Charge	Job		227		227	
18″ Blue Label (#1)							
	Demolish	Sq.		81.50		81.50	Includes material and labor to install
	Install	Sq.	273	127		400	wood shingles/shakes.
	Demolish and Install	Sq.	273	208.50		481.50	
	Minimum Charge	Job		227		227	
Replace Shingle							
	Install	Ea.	71.50	11.35		82.85	Includes labor and material to replace
	Minimum Charge	Job		227		227	individual wood shingle/shake. Cost is per shingle.
Cedar Shake							
18″ Medium Handsplit							
	Demolish	Sq.		81.50		81.50	Includes material and labor to install
	Install	Sq.	197	175		372	wood shingles/shakes.
	Demolish and Install	Sq.	197	256.50		453.50	
	Minimum Charge	Job		227		227	
18″ Heavy Handsplit							
	Demolish	Sq.		81.50		81.50	Includes material and labor to install
	Install	Sq.	197	194		391	wood shingles/shakes.
	Demolish and Install	Sq.	197	275.50		472.50	
	Minimum Charge	Job		227		227	
24″ Heavy Handsplit							
	Demolish	Sq.		81.50		81.50	Includes material and labor to install
	Install	Sq.	315	155		470	wood shingles/shakes.
	Demolish and Install	Sq.	315	236.50		551.50	
	Minimum Charge	Job		227		227	
24″ Medium Handsplit							
	Demolish	Sq.		81.50		81.50	Includes material and labor to install
	Install	Sq.	315	140		455	wood shingles/shakes.
	Demolish and Install	Sq.	315	221.50		536.50	
	Minimum Charge	Job		227		227	
24″ Straight Split							
	Demolish	Sq.		81.50		81.50	Includes material and labor to install
	Install	Sq.	310	206		516	wood shingles/shakes.
	Demolish and Install	Sq.	310	287.50		597.50	
	Minimum Charge	Job		227		227	
24″ Tapersplit							
	Demolish	Sq.		81.50		81.50	Includes material and labor to install
	Install	Sq.	305	206		511	wood shingles/shakes.
	Demolish and Install	Sq.	305	287.50		592.50	
	Minimum Charge	Job		227		227	

For customer support on your Contractor's Pricing Guide: Residential Repair & Remodeling, call 888.606.7279.

Roofing

Wood Shingles	Unit	Material	Labor	Equip.	Total	Specification
18″ Straight Split						
Demolish	Sq.		81.50		81.50	Includes material and labor to install
Install	Sq.	310	206		516	wood shingles/shakes.
Demolish and Install	Sq.	310	287.50		597.50	
Minimum Charge	Job		227		227	
Replace Shingle						
Install	Ea.	71.50	11.35		82.85	Includes labor and material to replace
Minimum Charge	Job		227		227	individual wood shingle/shake. Cost is per shingle.

Slate Tile Roofing	Unit	Material	Labor	Equip.	Total	Specification
Unfading Green						
Demolish	Sq.		73.50		73.50	Includes material and labor to install
Install	Sq.	545	243		788	clear Vermont slate tile.
Demolish and Install	Sq.	545	316.50		861.50	
Reinstall	Sq.		194.19		194.19	
Minimum Charge	Job		212		212	
Unfading Purple						
Demolish	Sq.		73.50		73.50	Includes material and labor to install
Install	Sq.	480	243		723	clear Vermont slate tile.
Demolish and Install	Sq.	480	316.50		796.50	
Reinstall	Sq.		194.19		194.19	
Minimum Charge	Job		212		212	
Replace Slate Tiles						
Install	Ea.	7.70	22.50		30.20	Includes labor and material to replace
Minimum Charge	Job		212		212	individual slate tiles. Cost is per tile.
Variegated Purple						
Demolish	Sq.		73.50		73.50	Includes material and labor to install
Install	Sq.	470	243		713	clear Vermont slate tile.
Demolish and Install	Sq.	470	316.50		786.50	
Reinstall	Sq.		194.19		194.19	
Minimum Charge	Job		212		212	
Unfading Grey / Black						
Demolish	Sq.		73.50		73.50	Includes material and labor to install
Install	Sq.	525	243		768	clear Vermont slate tile.
Demolish and Install	Sq.	525	316.50		841.50	
Reinstall	Sq.		194.19		194.19	
Minimum Charge	Job		212		212	
Unfading Red						
Demolish	Sq.		73.50		73.50	Includes material and labor to install
Install	Sq.	1275	243		1518	clear Vermont slate tile.
Demolish and Install	Sq.	1275	316.50		1591.50	
Reinstall	Sq.		194.19		194.19	
Minimum Charge	Job		212		212	
Weathering Black						
Demolish	Sq.		73.50		73.50	Includes material and labor to install
Install	Sq.	540	243		783	clear Pennsylvania slate tile.
Demolish and Install	Sq.	540	316.50		856.50	
Reinstall	Sq.		194.19		194.19	
Minimum Charge	Job		212		212	
Weathering Green						
Demolish	Sq.		73.50		73.50	Includes material and labor to install
Install	Sq.	390	243		633	clear Vermont slate tile.
Demolish and Install	Sq.	390	316.50		706.50	
Reinstall	Sq.		194.19		194.19	
Minimum Charge	Job		212		212	

Roofing

Flat Clay Tile Roofing

Flat Clay Tile Roofing	Unit	Material	Labor	Equip.	Total	Specification
Glazed						
Demolish	Sq.		81.50		81.50	Includes material and labor to install
Install	Sq.	560	212		772	tile roofing. Cost based on terra cotta
Demolish and Install	Sq.	560	293.50		853.50	red flat clay tile and includes ridge
Reinstall	Sq.		169.92		169.92	and rake tiles.
Minimum Charge	Job		212		212	
Terra Cotta Red						
Demolish	Sq.		81.50		81.50	Includes material and labor to install
Install	Sq.	560	212		772	tile roofing. Cost based on terra cotta
Demolish and Install	Sq.	560	293.50		853.50	red flat clay tile and includes ridge
Reinstall	Sq.		169.92		169.92	and rake tiles.
Minimum Charge	Job		212		212	

Mission Tile Roofing

Mission Tile Roofing	Unit	Material	Labor	Equip.	Total	Specification
Glazed Red						
Demolish	Sq.		81.50		81.50	Includes material and labor to install
Install	Sq.	595	232		827	tile roofing. Cost based on glazed red
Demolish and Install	Sq.	595	313.50		908.50	tile and includes birdstop, ridge and
Reinstall	Sq.		185.37		185.37	rake tiles.
Minimum Charge	Job		212		212	
Unglazed Red						
Demolish	Sq.		81.50		81.50	Includes material and labor to install
Install	Sq.	450	232		682	tile roofing. Cost based on unglazed
Demolish and Install	Sq.	450	313.50		763.50	red tile and includes birdstop, ridge
Reinstall	Sq.		185.37		185.37	and rake tiles.
Minimum Charge	Job		212		212	
Glazed Blue						
Demolish	Sq.		81.50		81.50	Includes material and labor to install
Install	Sq.	395	232		627	tile roofing. Cost based on glazed
Demolish and Install	Sq.	395	313.50		708.50	blue tile and includes birdstop, ridge
Reinstall	Sq.		185.37		185.37	and rake tiles.
Minimum Charge	Job		212		212	
Glazed White						
Demolish	Sq.		81.50		81.50	Includes material and labor to install
Install	Sq.	430	232		662	tile roofing. Cost based on glazed
Demolish and Install	Sq.	430	313.50		743.50	white tile and includes birdstop, ridge
Reinstall	Sq.		185.37		185.37	and rake tiles.
Minimum Charge	Job		212		212	
Unglazed White						
Demolish	Sq.		81.50		81.50	Includes material and labor to install
Install	Sq.	360	232		592	tile roofing. Cost based on unglazed
Demolish and Install	Sq.	360	313.50		673.50	white tile and includes birdstop, ridge
Reinstall	Sq.		185.37		185.37	and rake tiles.
Minimum Charge	Job		212		212	
Color Blend						
Demolish	Sq.		81.50		81.50	Includes material and labor to install
Install	Sq.	360	232		592	tile roofing. Cost based on color blend
Demolish and Install	Sq.	360	313.50		673.50	tile and includes birdstop, ridge and
Reinstall	Sq.		185.37		185.37	rake tiles.
Minimum Charge	Job		212		212	

For customer support on your Contractor's Pricing Guide: Residential Repair & Remodeling, call 888.606.7279.

Roofing

Mission Tile Roofing

Mission Tile Roofing		Unit	Material	Labor	Equip.	Total	Specification
Accessories / Extras							
Add for Cap Furring Strips	Install	Sq.	99	266		365	Includes labor and material to install vertically placed wood furring strips under cap tiles.
Add for Tile Adhesive	Install	Sq.	23	15.50		38.50	Includes material and labor for adhesive material placed between tiles.
Add for Wire Attachment	Install	Sq.	15.40	185		200.40	Includes labor and material to install single wire anchors for attaching tile.
Add for Braided Runners	Install	Sq.	38.50	177		215.50	Includes labor and material to install braided wire runners for secure attachment in high wind areas.
Add for Stainless Steel Nails	Install	Sq.	25.50			25.50	Includes material only for stainless steel nails.
Add for Brass Nails	Install	Sq.	44			44	Includes material only for brass nails.
Add for Hurricane Clips	Install	Sq.	23	250		273	Includes material and labor for hurricane or wind clips for high wind areas.
Add for Copper Nails	Install	Sq.	44			44	Includes material only for copper nails.
Add for Vertical Furring Strips	Install	Sq.	19.35	49.50		68.85	Includes labor and material to install vertically placed wood furring strips under tiles.
Add for Horizontal Furring Strips	Install	Sq.	13.20	22		35.20	Includes labor and material to install horizontally placed wood furring strips under tiles.

Spanish Tile Roofing

Spanish Tile Roofing		Unit	Material	Labor	Equip.	Total	Specification
Glazed Red							
	Demolish	Sq.		81.50		81.50	Includes material and labor to install tile roofing. Cost based on glazed red tile and includes birdstop, ridge and rake tiles.
	Install	Sq.	595	177		772	
	Demolish and Install	Sq.	595	258.50		853.50	
	Reinstall	Sq.		141.60		141.60	
	Minimum Charge	Job		212		212	
Glazed White							
	Demolish	Sq.		81.50		81.50	Includes material and labor to install tile roofing. Cost based on glazed white tile and includes birdstop, ridge and rake tiles.
	Install	Sq.	385	236		621	
	Demolish and Install	Sq.	385	317.50		702.50	
	Reinstall	Sq.		188.80		188.80	
	Minimum Charge	Job		212		212	

Roofing

Spanish Tile Roofing

Spanish Tile Roofing	Unit	Material	Labor	Equip.	Total	Specification
Glazed Blue						
Demolish	Sq.		81.50		81.50	Includes material and labor to install
Install	Sq.	385	177		562	tile roofing. Cost based on glazed
Demolish and Install	Sq.	385	258.50		643.50	blue tile and includes birdstop, ridge
Reinstall	Sq.		141.60		141.60	and rake tiles.
Minimum Charge	Job		212		212	
Unglazed Red						
Demolish	Sq.		81.50		81.50	Includes material and labor to install
Install	Sq.	610	177		787	tile roofing. Cost based on unglazed
Demolish and Install	Sq.	610	258.50		868.50	red tile and includes birdstop, ridge
Reinstall	Sq.		141.60		141.60	and rake tiles.
Minimum Charge	Job		212		212	
Unglazed White						
Demolish	Sq.		81.50		81.50	Includes material and labor to install
Install	Sq.	540	177		717	tile roofing. Cost based on unglazed
Demolish and Install	Sq.	540	258.50		798.50	white tile and includes birdstop, ridge
Reinstall	Sq.		141.60		141.60	and rake tiles.
Minimum Charge	Job		212		212	
Color Blend						
Demolish	Sq.		81.50		81.50	Includes material and labor to install
Install	Sq.	660	177		837	tile roofing. Cost based on color blend
Demolish and Install	Sq.	660	258.50		918.50	tile and includes birdstop, ridge and
Reinstall	Sq.		141.60		141.60	rake tiles.
Minimum Charge	Job		212		212	
Accessories / Extras						
Add for Vertical Furring Strips						
Install	Sq.	19.35	49.50		68.85	Includes labor and material to install vertically placed wood furring strips under tiles.
Add for Wire Attachment						
Install	Sq.	8.25	142		150.25	Includes labor and material to install single wire ties.
Add for Braided Runners						
Install	Sq.	29.50	121		150.50	Includes labor and material to install braided wire runners for secure attachment in high wind areas.
Add for Stainless Steel Nails						
Install	Sq.	21			21	Includes material only for stainless steel nails.
Add for Copper Nails						
Install	Sq.	33			33	Includes material only for copper nails.
Add for Brass Nails						
Install	Sq.	33			33	Includes material only for brass nails.
Add for Hurricane Clips						
Install	Sq.	12.10	146		158.10	Includes material and labor to install hurricane or wind clips for high wind areas.
Add for Mortar Set Tiles						
Install	Sq.	50.50	212		262.50	Includes material and labor to install setting tiles by using mortar material placed between tiles.

Roofing

Spanish Tile Roofing		Unit	Material	Labor	Equip.	Total	Specification
Add for Tile Adhesive	Install	Sq.	13.20	13.80		27	Includes material and labor for adhesive placed between tiles.
Add for Cap Furring Strips	Install	Sq.	99	266		365	Includes labor and material to install vertically placed wood furring strips under cap tiles.
Add for Horizontal Furring Strips	Install	Sq.	13.20	22		35.20	Includes labor and material to install horizontally placed wood furring strips under tiles.

Concrete Tile Roofing		Unit	Material	Labor	Equip.	Total	Specification
Corrugated Red / Brown	Demolish	Sq.		81.50		81.50	Includes material and labor to install tile roofing. Cost includes birdstop, booster, ridge and rake tiles.
	Install	Sq.	113	315		428	
	Demolish and Install	Sq.	113	396.50		509.50	
	Reinstall	Sq.		251.73		251.73	
	Minimum Charge	Job		212		212	
Corrugated Gray	Demolish	Sq.		81.50		81.50	Includes material and labor to install tile roofing. Cost includes birdstop, booster, ridge and rake tiles.
	Install	Sq.	113	315		428	
	Demolish and Install	Sq.	113	396.50		509.50	
	Reinstall	Sq.		251.73		251.73	
	Minimum Charge	Job		212		212	
Corrugated Bright Red	Demolish	Sq.		81.50		81.50	Includes material and labor to install tile roofing. Cost includes birdstop, booster, ridge and rake tiles.
	Install	Sq.	113	315		428	
	Demolish and Install	Sq.	113	396.50		509.50	
	Reinstall	Sq.		251.73		251.73	
	Minimum Charge	Job		212		212	
Corrugated Black	Demolish	Sq.		81.50		81.50	Includes material and labor to install tile roofing. Cost includes birdstop, booster, ridge and rake tiles.
	Install	Sq.	113	315		428	
	Demolish and Install	Sq.	113	396.50		509.50	
	Reinstall	Sq.		251.73		251.73	
	Minimum Charge	Job		212		212	
Corrugated Green	Demolish	Sq.		81.50		81.50	Includes material and labor to install tile roofing. Cost includes birdstop, booster, ridge and rake tiles.
	Install	Sq.	113	315		428	
	Demolish and Install	Sq.	113	396.50		509.50	
	Reinstall	Sq.		251.73		251.73	
	Minimum Charge	Job		212		212	
Corrugated Blue	Demolish	Sq.		81.50		81.50	Includes material and labor to install tile roofing. Cost includes birdstop, booster, ridge and rake tiles.
	Install	Sq.	113	315		428	
	Demolish and Install	Sq.	113	396.50		509.50	
	Reinstall	Sq.		251.73		251.73	
	Minimum Charge	Job		212		212	
Flat Natural Gray	Demolish	Sq.		81.50		81.50	Includes material and labor to install tile roofing. Cost includes birdstop, booster, ridge and rake tiles.
	Install	Sq.	136	315		451	
	Demolish and Install	Sq.	136	396.50		532.50	
	Reinstall	Sq.		251.73		251.73	
	Minimum Charge	Job		212		212	

Roofing

Concrete Tile Roofing

Concrete Tile Roofing	Unit	Material	Labor	Equip.	Total	Specification
Flat Red / Brown						
Demolish	Sq.		81.50		81.50	Includes material and labor to install
Install	Sq.	136	315		451	tile roofing. Cost includes birdstop,
Demolish and Install	Sq.	136	396.50		532.50	booster, ridge and rake tiles.
Reinstall	Sq.		251.73		251.73	
Minimum Charge	Job		212		212	
Flat Bright Red						
Demolish	Sq.		81.50		81.50	Includes material and labor to install
Install	Sq.	136	315		451	tile roofing. Cost includes birdstop,
Demolish and Install	Sq.	136	396.50		532.50	booster, ridge and rake tiles.
Reinstall	Sq.		251.73		251.73	
Minimum Charge	Job		212		212	
Flat Green						
Demolish	Sq.		81.50		81.50	Includes material and labor to install
Install	Sq.	136	315		451	tile roofing. Cost includes birdstop,
Demolish and Install	Sq.	136	396.50		532.50	booster, ridge and rake tiles.
Minimum Charge	Job		212		212	
Flat Blue						
Demolish	Sq.		81.50		81.50	Includes material and labor to install
Install	Sq.	136	315		451	tile roofing. Cost includes birdstop,
Demolish and Install	Sq.	136	396.50		532.50	booster, ridge and rake tiles.
Reinstall	Sq.		251.73		251.73	
Minimum Charge	Job		212		212	
Flat Black						
Demolish	Sq.		81.50		81.50	Includes material and labor to install
Install	Sq.	136	315		451	tile roofing. Cost includes birdstop,
Demolish and Install	Sq.	136	396.50		532.50	booster, ridge and rake tiles.
Reinstall	Sq.		251.73		251.73	
Minimum Charge	Job		212		212	

Aluminum Sheet

Aluminum Sheet	Unit	Material	Labor	Equip.	Total	Specification
Corrugated						
.019" Thick Natural						
Demolish	S.F.		.87		.87	Includes material and labor to install
Install	S.F.	1.38	2.10		3.48	aluminum sheet metal roofing.
Demolish and Install	S.F.	1.38	2.97		4.35	
Reinstall	S.F.		1.68		1.68	
Minimum Charge	Job		505		505	
.016" Thick Natural						
Demolish	S.F.		.87		.87	Includes material and labor to install
Install	S.F.	1.10	2.10		3.20	aluminum sheet metal roofing.
Demolish and Install	S.F.	1.10	2.97		4.07	
Reinstall	S.F.		1.68		1.68	
Minimum Charge	Job		505		505	
.016" Colored Finish						
Demolish	S.F.		.87		.87	Includes material and labor to install
Install	S.F.	1.60	2.10		3.70	aluminum sheet metal roofing.
Demolish and Install	S.F.	1.60	2.97		4.57	
Reinstall	S.F.		1.68		1.68	
Minimum Charge	Job		505		505	
.019" Colored Finish						
Demolish	S.F.		.87		.87	Includes material and labor to install
Install	S.F.	1.71	2.10		3.81	aluminum sheet metal roofing.
Demolish and Install	S.F.	1.71	2.97		4.68	
Reinstall	S.F.		1.68		1.68	
Minimum Charge	Job		505		505	

Roofing

Aluminum Sheet	Unit	Material	Labor	Equip.	Total	Specification
Ribbed						
.019″ Thick Natural						
Demolish	S.F.		.87		.87	Includes material and labor to install
Install	S.F.	1.38	2.10		3.48	aluminum sheet metal roofing.
Demolish and Install	S.F.	1.38	2.97		4.35	
Reinstall	S.F.		1.68		1.68	
Minimum Charge	Job		505		505	
.016″ Thick Natural						
Demolish	S.F.		.87		.87	Includes material and labor to install
Install	S.F.	1.10	2.10		3.20	aluminum sheet metal roofing.
Demolish and Install	S.F.	1.10	2.97		4.07	
Reinstall	S.F.		1.68		1.68	
Minimum Charge	Job		505		505	
.050″ Colored Finish						
Demolish	S.F.		.87		.87	Includes material and labor to install
Install	S.F.	5.10	2.10		7.20	aluminum sheet metal roofing.
Demolish and Install	S.F.	5.10	2.97		8.07	
Reinstall	S.F.		1.68		1.68	
Minimum Charge	Job		505		505	
.032″ Thick Natural						
Demolish	S.F.		.87		.87	Includes material and labor to install
Install	S.F.	2.86	2.10		4.96	aluminum sheet metal roofing.
Demolish and Install	S.F.	2.86	2.97		5.83	
Reinstall	S.F.		1.68		1.68	
Minimum Charge	Job		505		505	
.040″ Thick Natural						
Demolish	S.F.		.87		.87	Includes material and labor to install
Install	S.F.	3.54	2.10		5.64	aluminum sheet metal roofing.
Demolish and Install	S.F.	3.54	2.97		6.51	
Reinstall	S.F.		1.68		1.68	
Minimum Charge	Job		505		505	
.050″ Thick Natural						
Demolish	S.F.		.87		.87	Includes material and labor to install
Install	S.F.	4.21	2.10		6.31	aluminum sheet metal roofing.
Demolish and Install	S.F.	4.21	2.97		7.18	
Reinstall	S.F.		1.68		1.68	
Minimum Charge	Job		505		505	
.016″ Colored Finish						
Demolish	S.F.		.87		.87	Includes material and labor to install
Install	S.F.	1.60	2.10		3.70	aluminum sheet metal roofing.
Demolish and Install	S.F.	1.60	2.97		4.57	
Reinstall	S.F.		1.68		1.68	
Minimum Charge	Job		505		505	
.019″ Colored Finish						
Demolish	S.F.		.87		.87	Includes material and labor to install
Install	S.F.	1.71	2.10		3.81	aluminum sheet metal roofing.
Demolish and Install	S.F.	1.71	2.97		4.68	
Reinstall	S.F.		1.68		1.68	
Minimum Charge	Job		505		505	
.032″ Colored Finish						
Demolish	S.F.		.87		.87	Includes material and labor to install
Install	S.F.	3.66	2.10		5.76	aluminum sheet metal roofing.
Demolish and Install	S.F.	3.66	2.97		6.63	
Reinstall	S.F.		1.68		1.68	
Minimum Charge	Job		505		505	

Roofing

Aluminum Sheet

Aluminum Sheet	Unit	Material	Labor	Equip.	Total	Specification
.040" Colored Finish						
Demolish	S.F.		.87		.87	Includes material and labor to install
Install	S.F.	4.32	2.10		6.42	aluminum sheet metal roofing.
Demolish and Install	S.F.	4.32	2.97		7.29	
Reinstall	S.F.		1.68		1.68	
Minimum Charge	Job		505		505	

Fiberglass Sheet

Fiberglass Sheet	Unit	Material	Labor	Equip.	Total	Specification
12 Ounce Corrugated						
Demolish	S.F.		1.55		1.55	Cost includes material and labor to
Install	S.F.	4.62	1.67		6.29	install 12 ounce corrugated fiberglass
Demolish and Install	S.F.	4.62	3.22		7.84	sheet roofing.
Reinstall	S.F.		1.34		1.34	
Minimum Charge	Job		212		212	
8 Ounce Corrugated						
Demolish	S.F.		1.55		1.55	Cost includes material and labor to
Install	S.F.	2.37	1.67		4.04	install 8 ounce corrugated fiberglass
Demolish and Install	S.F.	2.37	3.22		5.59	roofing.
Reinstall	S.F.		1.34		1.34	
Minimum Charge	Job		212		212	

Galvanized Steel

Galvanized Steel	Unit	Material	Labor	Equip.	Total	Specification
Corrugated						
24 Gauge						
Demolish	S.F.		.87		.87	Includes material and labor to install
Install	S.F.	3.19	1.76		4.95	galvanized steel sheet metal roofing.
Demolish and Install	S.F.	3.19	2.63		5.82	Cost based on 24 gauge corrugated
Reinstall	S.F.		1.41		1.41	or ribbed steel roofing.
Minimum Charge	Job		212		212	
26 Gauge						
Demolish	S.F.		.87		.87	Includes material and labor to install
Install	S.F.	2.15	1.67		3.82	galvanized steel sheet metal roofing.
Demolish and Install	S.F.	2.15	2.54		4.69	Cost based on 26 gauge corrugated
Reinstall	S.F.		1.34		1.34	or ribbed steel roofing.
Minimum Charge	Job		212		212	
28 Gauge						
Demolish	S.F.		.87		.87	Includes material and labor to install
Install	S.F.	1.86	1.59		3.45	galvanized steel sheet metal roofing.
Demolish and Install	S.F.	1.86	2.46		4.32	Cost based on 28 gauge corrugated
Reinstall	S.F.		1.27		1.27	or ribbed steel roofing.
Minimum Charge	Job		212		212	
30 Gauge						
Demolish	S.F.		.87		.87	Includes material and labor to install
Install	S.F.	1.80	1.52		3.32	galvanized steel sheet metal roofing.
Demolish and Install	S.F.	1.80	2.39		4.19	Cost based on 30 gauge corrugated
Reinstall	S.F.		1.21		1.21	or ribbed steel roofing.
Minimum Charge	Job		212		212	

Roofing

Galvanized Steel		Unit	Material	Labor	Equip.	Total	Specification
Ribbed							
24 Gauge							
	Demolish	S.F.		.87		.87	Includes material and labor to install
	Install	S.F.	3.19	1.76		4.95	galvanized steel sheet metal roofing.
	Demolish and Install	S.F.	3.19	2.63		5.82	Cost based on 24 gauge corrugated
	Reinstall	S.F.		1.41		1.41	or ribbed steel roofing.
	Minimum Charge	Job		212		212	
26 Gauge							
	Demolish	S.F.		.87		.87	Includes material and labor to install
	Install	S.F.	2.15	1.67		3.82	galvanized steel sheet metal roofing.
	Demolish and Install	S.F.	2.15	2.54		4.69	Cost based on 26 gauge corrugated
	Reinstall	S.F.		1.34		1.34	or ribbed steel roofing.
	Minimum Charge	Job		212		212	
28 Gauge							
	Demolish	S.F.		.87		.87	Includes material and labor to install
	Install	S.F.	1.86	1.59		3.45	galvanized steel sheet metal roofing.
	Demolish and Install	S.F.	1.86	2.46		4.32	Cost based on 28 gauge corrugated
	Reinstall	S.F.		1.27		1.27	or ribbed steel roofing.
	Minimum Charge	Job		212		212	
30 Gauge							
	Demolish	S.F.		.87		.87	Includes material and labor to install
	Install	S.F.	1.80	1.52		3.32	galvanized steel sheet metal roofing.
	Demolish and Install	S.F.	1.80	2.39		4.19	Cost based on 30 gauge corrugated
	Reinstall	S.F.		1.21		1.21	or ribbed steel roofing.
	Minimum Charge	Job		212		212	

Standing Seam		Unit	Material	Labor	Equip.	Total	Specification
Copper							
16 Ounce							
	Demolish	Sq.		86.50		86.50	Includes material and labor to install
	Install	Sq.	1075	390		1465	copper standing seam metal roofing.
	Demolish and Install	Sq.	1075	476.50		1551.50	Cost based on 16 ounce copper
	Reinstall	Sq.		310.40		310.40	roofing.
	Minimum Charge	Job		252		252	
18 Ounce							
	Demolish	Sq.		86.50		86.50	Includes material and labor to install
	Install	Sq.	1200	420		1620	copper standing seam metal roofing.
	Demolish and Install	Sq.	1200	506.50		1706.50	Cost based on 18 ounce copper
	Reinstall	Sq.		336.27		336.27	roofing.
	Minimum Charge	Job		252		252	
20 Ounce							
	Demolish	Sq.		86.50		86.50	Includes material and labor to install
	Install	Sq.	1425	460		1885	copper standing seam metal roofing.
	Demolish and Install	Sq.	1425	546.50		1971.50	Cost based on 20 ounce copper
	Reinstall	Sq.		366.84		366.84	roofing.
Stainless Steel							
26 Gauge							
	Demolish	Sq.		86.50		86.50	Cost includes material and labor to
	Install	Sq.	725	440		1165	install 26 gauge stainless steel
	Demolish and Install	Sq.	725	526.50		1251.50	standing seam metal roofing.
	Reinstall	Sq.		350.89		350.89	
	Minimum Charge	Job		252		252	

For customer support on your Contractor's Pricing Guide: Residential Repair & Remodeling, call 888.606.7279.

Roofing

Standing Seam

Standing Seam	Unit	Material	Labor	Equip.	Total	Specification
28 Gauge						
Demolish	Sq.		86.50		86.50	Includes material and labor to install
Install	Sq.	680	420		1100	standing seam metal roofing. Cost
Demolish and Install	Sq.	680	506.50		1186.50	based on 28 gauge stainless steel
Reinstall	Sq.		336.27		336.27	roofing.
Minimum Charge	Job		252		252	

Elastomeric Roofing

Elastomeric Roofing	Unit	Material	Labor	Equip.	Total	Specification
Neoprene						
1/16"						
Demolish	Sq.		30		30	Includes material and labor to install
Install	Sq.	229	203	46.50	478.50	neoprene membrane, fully adhered,
Demolish and Install	Sq.	229	233	46.50	508.50	1/16" thick.
Minimum Charge	Job		970	106	1076	
GRM Membrane						
1/16"						
Demolish	Sq.		30		30	Includes material and labor to install
Install	Sq.	81.50	325	35.50	442	coal tar based elastomeric roofing
Demolish and Install	Sq.	81.50	355	35.50	472	membrane, polyester fiber reinforced.
Minimum Charge	Job		970	106	1076	
Cant Strips						
3"						
Demolish	L.F.		.93		.93	Includes material and labor to install
Install	L.F.	.94	1.30		2.24	cants 3" x 3", treated timber, cut
Demolish and Install	L.F.	.94	2.23		3.17	diagonally.
Minimum Charge	Job		106		106	
4"						
Demolish	L.F.		.93		.93	Includes material and labor to install
Install	L.F.	1.98	1.30		3.28	cants 4" x 4", treated timber, cut
Demolish and Install	L.F.	1.98	2.23		4.21	diagonally.
Minimum Charge	Job		106		106	
Polyvinyl Chloride (PVC)						
45 mil (loose laid)						
Demolish	Sq.		30		30	Cost includes material and labor to
Install	Sq.	107	38	4.16	149.16	install 45 mil PVC elastomeric roofing,
Demolish and Install	Sq.	107	68	4.16	179.16	loose-laid with stone.
Minimum Charge	Job		970	106	1076	
48 mil (loose laid)						
Demolish	Sq.		30		30	Cost includes material and labor to
Install	Sq.	129	38	4.16	171.16	install 48 mil PVC elastomeric roofing,
Demolish and Install	Sq.	129	68	4.16	201.16	loose-laid with stone.
Minimum Charge	Job		970	106	1076	
60 mil (loose laid)						
Demolish	Sq.		30		30	Cost includes material and labor to
Install	Sq.	131	38	4.16	173.16	install 60 mil PVC elastomeric roofing,
Demolish and Install	Sq.	131	68	4.16	203.16	loose-laid with stone.
Minimum Charge	Job		970	106	1076	
45 mil (attached)						
Demolish	Sq.		48		48	Cost includes material and labor to
Install	Sq.	97.50	74.50	8.15	180.15	install 45 mil PVC elastomeric roofing
Demolish and Install	Sq.	97.50	122.50	8.15	228.15	fully attached to roof deck.
Minimum Charge	Job		970	106	1076	

For customer support on your Contractor's Pricing Guide: Residential Repair & Remodeling, call 888.606.7279.

Elastomeric Roofing	Unit	Material	Labor	Equip.	Total	Specification
48 mil (attached)						Cost includes material and labor to install 48 mil PVC elastomeric roofing fully attached to roof deck.
Demolish	Sq.		48		48	
Install	Sq.	169	74.50	8.15	251.65	
Demolish and Install	Sq.	169	122.50	8.15	299.65	
Minimum Charge	Job		970	106	1076	
60 mil (attached)						Cost includes material and labor to install 60 mil PVC elastomeric roofing fully attached to roof deck.
Demolish	Sq.		48		48	
Install	Sq.	172	74.50	8.15	254.65	
Demolish and Install	Sq.	172	122.50	8.15	302.65	
Minimum Charge	Job		970	106	1076	
CSPE Type						
45 mil (loose laid w / stone)						Cost includes material and labor to install 45 mil CSPE (chlorosulfonated polyethylene-hypalon) roofing material loose laid on deck and ballasted with stone.
Demolish	Sq.		30		30	
Install	Sq.	286	74.50	8.15	368.65	
Demolish and Install	Sq.	286	104.50	8.15	398.65	
Minimum Charge	Job		970	106	1076	
45 mil (attached at seams)						Cost includes material and labor to install 45 mil CSPE (chlorosulfonated polyethylene-hypalon) roofing material attached to deck at seams with batten strips.
Demolish	Sq.		36		36	
Install	Sq.	360	55.50	6.05	421.55	
Demolish and Install	Sq.	360	91.50	6.05	457.55	
Minimum Charge	Job		970	106	1076	
45 mil (fully attached)						Cost includes material and labor to install 45 mil CSPE (chlorosulfonated polyethylene-hypalon) roofing material fully attached to deck.
Demolish	Sq.		48		48	
Install	Sq.	276	55.50	6.05	337.55	
Demolish and Install	Sq.	276	103.50	6.05	385.55	
Minimum Charge	Job		970	106	1076	
EPDM Type						
45 mil (attached at seams)						Cost includes material and labor to install 45 mil EPDM (ethylene propylene diene monomer) roofing material attached to deck at seams.
Demolish	Sq.		36		36	
Install	Sq.	86.50	55.50	6.05	148.05	
Demolish and Install	Sq.	86.50	91.50	6.05	184.05	
Minimum Charge	Job		970	106	1076	
45 mil (fully attached)						Cost includes material and labor to install 45 mil EPDM (ethylene propylene diene monomer) roofing material fully attached to deck.
Demolish	Sq.		48		48	
Install	Sq.	122	74.50	8.15	204.65	
Demolish and Install	Sq.	122	122.50	8.15	252.65	
Minimum Charge	Job		970	106	1076	
45 mil (loose laid w / stone)						Cost includes material and labor to install 45 mil EPDM (ethylene propylene diene monomer) roofing material loose laid on deck and ballasted with stones (approx. 1/2 ton of stone per square).
Demolish	Sq.		30		30	
Install	Sq.	92	38	4.16	134.16	
Demolish and Install	Sq.	92	68	4.16	164.16	
Minimum Charge	Job		970	106	1076	
60 mil (attached at seams)						Cost includes material and labor to install 60 mil EPDM (ethylene propylene diene monomer) roofing material attached to deck at seams.
Demolish	Sq.		36		36	
Install	Sq.	98	55.50	6.05	159.55	
Demolish and Install	Sq.	98	91.50	6.05	195.55	
Minimum Charge	Job		970	106	1076	

Roofing

Elastomeric Roofing

Elastomeric Roofing	Unit	Material	Labor	Equip.	Total	Specification
60 mil (fully attached)						Cost includes material and labor to
Demolish	Sq.		48		48	install 60 mil EPDM (ethylene
Install	Sq.	134	74.50	8.15	216.65	propylene diene monomer) roofing
Demolish and Install	Sq.	134	122.50	8.15	264.65	material fully attached to deck.
Minimum Charge	Job		970	106	1076	
60 mil (loose laid w / stone)						Cost includes material and labor to
Demolish	Sq.		30		30	install 60 mil EPDM (ethylene
Install	Sq.	105	38	4.16	147.16	propylene diene monomer) roofing
Demolish and Install	Sq.	105	68	4.16	177.16	material loose laid on deck and
Minimum Charge	Job		970	106	1076	ballasted with stones (approx. 1/2 ton
						of stone per square).

Modified Bitumen Roofing

Modified Bitumen Roofing	Unit	Material	Labor	Equip.	Total	Specification
Fully Attached						
120 mil						
Demolish	Sq.		71.50		71.50	Cost includes material and labor to
Install	Sq.	83.50	9.45	2.16	95.11	install 120 mil modified bitumen
Demolish and Install	Sq.	83.50	80.95	2.16	166.61	roofing, fully attached to deck by
Minimum Charge	Job		106		106	torch.
150 mil						
Demolish	Sq.		71.50		71.50	Cost includes material and labor to
Install	Sq.	88	23	5.30	116.30	install 150 mil modified bitumen
Demolish and Install	Sq.	88	94.50	5.30	187.80	roofing, fully attached to deck by
Minimum Charge	Job		106		106	torch.
160 mil						
Demolish	Sq.		71.50		71.50	Cost includes material and labor to
Install	Sq.	101	24.50	5.55	131.05	install 160 mil modified bitumen
Demolish and Install	Sq.	101	96	5.55	202.55	roofing, fully attached to deck by
Minimum Charge	Job		106		106	torch.

Built-up Roofing

Built-up Roofing	Unit	Material	Labor	Equip.	Total	Specification
Emergency Repairs						
Install	Ea.	229	2000	455	2684	Minimum charge for work requiring
Minimum Charge	Job		1400	320	1720	hot asphalt.
Asphalt						
3-ply Asphalt						
Demolish	Sq.		74.50		74.50	Cost includes material and labor to
Install	Sq.	80	116	26.50	222.50	install 3-ply asphalt built-up roof
Demolish and Install	Sq.	80	190.50	26.50	297	including one 30 lb and two 15 lb
Minimum Charge	Job		1400	320	1720	layers of felt.
4-ply Asphalt						
Demolish	Sq.		88.50		88.50	Cost includes material and labor to
Install	Sq.	109	127	29	265	install 4-ply asphalt built-up roof
Demolish and Install	Sq.	109	215.50	29	353.50	including one 30 lb and three 15 lb
Minimum Charge	Job		1400	320	1720	layers of felt.

For customer support on your Contractor's Pricing Guide: Residential Repair & Remodeling, call 888.606.7279.

Built-up Roofing		Unit	Material	Labor	Equip.	Total	Specification
5-ply Asphalt							
	Demolish	Sq.		116		116	Cost includes material and labor to
	Install	Sq.	151	139	32	322	install 5-ply asphalt built-up roof
	Demolish and Install	Sq.	151	255	32	438	including one 30 lb and four 15 lb
	Minimum Charge	Job		1400	320	1720	layers of felt.
Components / Accessories							
Flood Coat							
	Install	Sq.	31	65		96	Includes labor and material to install flood coat.
R / R Gravel and Flood Coat							
	Install	Sq.	36	212		248	Includes labor and material to install flood coat and gravel.
Aluminum UV Coat							
	Install	Sq.	32.50	243	26.50	302	Includes labor and material to install aluminum-based UV coating.
Gravel Coat							
	Demolish	Sq.		37.50		37.50	Includes labor and material to install
	Install	Sq.	5	35.50		40.50	gravel coat.
	Demolish and Install	Sq.	5	73		78	
Mineral Surface Cap Sheet							
	Install	Sq.	89	7.30		96.30	Cost includes labor and material to install mineral surfaced cap sheet.
Fibrated Aluminum UV Coating							
	Install	Sq.	34.50	65		99.50	Includes labor and material to install fibered aluminum-based UV coating.
Impregnated Walk							
1"							
	Install	S.F.	3.19	2.26		5.45	Cost includes labor and material to install 1" layer of asphalt coating.
1/2"							
	Install	S.F.	2.31	1.79		4.10	Cost includes labor and material to install 1/2" layer of asphalt coating.
3/4"							
	Install	S.F.	2.86	1.88		4.74	Cost includes labor and material to install 3/4" layer of asphalt coating.

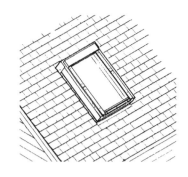

| Doors | | | Roof Window | | | Double Hung Window |

Exterior Type Door	Unit	Material	Labor	Equip.	Total	Specification
Entry						
Flush (1-3/4")						
Demolish	Ea.		42		42	Cost includes material and labor to
Install	Ea.	252	28.50		280.50	install 3' x 6' 8" pre-hung 1-3/4"
Demolish and Install	Ea.	252	70.50		322.50	door, including casing and stop,
Reinstall	Ea.		28.38		28.38	hinges, jamb, aluminum sill,
Clean	Ea.	4.07	13.75		17.82	weatherstripped.
Paint	Ea.	10.30	54		64.30	
Minimum Charge	Job		227		227	
Raised Panel						
Demolish	Ea.		42		42	Cost includes material and labor to
Install	Ea.	485	25		510	install 3' x 6' 8" pre-hung 1-3/4"
Demolish and Install	Ea.	485	67		552	door, jamb, including casing and
Reinstall	Ea.		25.22		25.22	stop, hinges, aluminum sill,
Clean	Ea.	4.07	13.75		17.82	weatherstripping.
Paint	Ea.	10.30	54		64.30	
Minimum Charge	Job		227		227	
Entry w / Lights						
Good Quality						
Demolish	Ea.		42		42	Cost includes material and labor to
Install	Ea.	350	57		407	install 3' x 6' 8" pre-hung 1-3/4"
Demolish and Install	Ea.	350	99		449	door, solid wood lower panel, with up
Reinstall	Ea.		56.75		56.75	to 9-light glass panels, casing, sill,
Clean	Ea.	4.07	13.75		17.82	weatherstripped.
Paint	Ea.	7.55	25		32.55	
Minimum Charge	Job		227		227	
Better Quality						
Demolish	Ea.		42		42	Cost includes material and labor to
Install	Ea.	360	57		417	install 3' x 6' 8" pre-hung 1-3/4" Ash
Demolish and Install	Ea.	360	99		459	door, solid wood, raised panel door
Reinstall	Ea.		56.75		56.75	without glass panels.
Clean	Ea.	4.07	13.75		17.82	
Paint	Ea.	10.30	54		64.30	
Minimum Charge	Job		227		227	
Premium Quality						
Demolish	Ea.		42		42	Cost includes material and labor to
Install	Ea.	985	26.50		1011.50	install 3' x 6' 8" pre-hung 1-3/4" Oak
Demolish and Install	Ea.	985	68.50		1053.50	door, solid wood, raised panel door
Reinstall	Ea.		26.71		26.71	without glass panels.
Clean	Ea.	4.07	13.75		17.82	
Paint	Ea.	10.30	54		64.30	
Minimum Charge	Job		227		227	

Exterior Type Door		Unit	Material	Labor	Equip.	Total	Specification
Custom Quality							Cost includes material and labor to
	Demolish	Ea.		42		42	install 3' x 6' 8" pre-hung 1-3/4"
	Install	Ea.	510	60.50		570.50	hand-made, solid wood, raised panel
	Demolish and Install	Ea.	510	102.50		612.50	door with custom stained glass.
	Reinstall	Ea.		60.53		60.53	
	Clean	Ea.	.28	13.75		14.03	
	Paint	Ea.	2.55	126		128.55	
	Minimum Charge	Job		227		227	
Miami Style							Cost includes material and labor to
	Demolish	Ea.		42		42	install 3' x 6' 8" pre-hung 1-3/4"
	Install	Ea.	232	26.50		258.50	lauan door of solid wood.
	Demolish and Install	Ea.	232	68.50		300.50	
	Reinstall	Ea.		26.71		26.71	
	Clean	Ea.	.28	13.75		14.03	
	Paint	Ea.	10.30	54		64.30	
	Minimum Charge	Job		227		227	
French Style							
1-Light							Cost includes material and labor to
	Demolish	Ea.		42		42	install 3' x 6' 8" pre-hung 1-3/4"
	Install	Ea.	660	75.50		735.50	French style door with 1 light.
	Demolish and Install	Ea.	660	117.50		777.50	
	Reinstall	Ea.		75.67		75.67	
	Clean	Ea.	2.55	20.50		23.05	
	Paint	Ea.	11.55	94.50		106.05	
	Minimum Charge	Job		227		227	
9-Light							Cost includes material and labor to
	Demolish	Ea.		42		42	install 3' x 6' 8" pre-hung 1-3/4"
	Install	Ea.	725	32.50		757.50	French style door with 9 lights.
	Demolish and Install	Ea.	725	74.50		799.50	
	Reinstall	Ea.		32.43		32.43	
	Clean	Ea.	.56	27.50		28.06	
	Paint	Ea.	11.55	94.50		106.05	
	Minimum Charge	Job		227		227	
15-Light							Cost includes material and labor to
	Demolish	Ea.		46		46	install 3' x 6' 8" pre-hung 1-3/4"
	Install	Ea.	705	65		770	French style door with 15 lights.
	Demolish and Install	Ea.	705	111		816	
	Reinstall	Ea.		64.86		64.86	
	Clean	Ea.	.56	27.50		28.06	
	Paint	Ea.	11.55	94.50		106.05	
	Minimum Charge	Job		227		227	
Jamb							
Wood							Includes material and labor to install
	Demolish	Ea.		31.50		31.50	flat pine jamb with square cut heads
	Install	Ea.	119	12.25		131.25	and rabbeted sides for 6' 8" high and
	Demolish and Install	Ea.	119	43.75		162.75	3-9/16" door including trim sets for
	Clean	Ea.	.13	8.25		8.38	both sides.
	Paint	Ea.	7.55	25		32.55	
	Minimum Charge	Job		227		227	
Metal 6' 8" High							Cost includes material and labor to
	Demolish	Ea.		25		25	install 18 gauge hollow metal door
	Install	Ea.	173	28.50		201.50	frame to fit a 4-1/2" jamb 6' 8" high
	Demolish and Install	Ea.	173	53.50		226.50	and up to 3' 6" opening.
	Reinstall	Ea.		28.38		28.38	
	Clean	Ea.	.85	6.90		7.75	
	Paint	Ea.	1.61	7.55		9.16	
	Minimum Charge	Job		227		227	

Exterior Doors and Windows

Exterior Type Door		Unit	Material	Labor	Equip.	Total	Specification
Metal 7' High							
	Demolish	Ea.		25		25	Cost includes material and labor to
	Install	Ea.	182	30.50		212.50	install 18 gauge hollow metal door
	Demolish and Install	Ea.	182	55.50		237.50	frame to fit a 4-1/2" jamb 7' high and
	Reinstall	Ea.		30.27		30.27	up to 3' 6" opening.
	Clean	Ea.	.85	6.90		7.75	
	Paint	Ea.	1.61	7.55		9.16	
	Minimum Charge	Job		227		227	
Metal 8' High							
	Demolish	Ea.		25		25	Cost includes material and labor to
	Install	Ea.	223	38		261	install 18 gauge hollow metal door
	Demolish and Install	Ea.	223	63		286	frame to fit a 4-1/2" jamb 8' high and
	Reinstall	Ea.		37.83		37.83	up to 3' 6" opening.
	Clean	Ea.	1.20	8.25		9.45	
	Paint	Ea.	1.79	8.40		10.19	
	Minimum Charge	Job		227		227	
Metal 9' High							
	Demolish	Ea.		25		25	Cost includes material and labor to
	Install	Ea.	283	45.50		328.50	install 18 gauge hollow metal door
	Demolish and Install	Ea.	283	70.50		353.50	frame to fit a 4-1/2" jamb 9' high and
	Reinstall	Ea.		45.40		45.40	up to 3' 6" opening.
	Clean	Ea.	1.20	8.25		9.45	
	Paint	Ea.	1.79	8.40		10.19	
	Minimum Charge	Job		227		227	
Casing Trim							
Single Width							
	Demolish	Ea.		8		8	Cost includes material and labor to
	Install	Opng.	26.50	38		64.50	install 11/16" x 2-1/2" pine ranch
	Demolish and Install	Opng.	26.50	46		72.50	style casing for one side of a standard
	Reinstall	Opng.		30.27		30.27	door opening.
	Clean	Opng.	.57	3.30		3.87	
	Paint	Ea.	4.73	7.55		12.28	
	Minimum Charge	Job		227		227	
Double Width							
	Demolish	Ea.		9.50		9.50	Cost includes material and labor to
	Install	Opng.	31	45.50		76.50	install 11/16" x 2-1/2" pine ranch
	Demolish and Install	Opng.	31	55		86	style door casing for one side of a
	Reinstall	Opng.		36.32		36.32	double door opening.
	Clean	Opng.	.68	6.60		7.28	
	Paint	Ea.	4.73	8.60		13.33	
	Minimum Charge	Job		227		227	

Residential Metal Door		Unit	Material	Labor	Equip.	Total	Specification
Pre-hung, Steel-clad							
2' 8" x 6' 8"							
	Demolish	Ea.		42		42	Cost includes material and labor to
	Install	Ea.	345	53.50		398.50	install 2' 6" x 6' 8" pre-hung 1-3/4"
	Demolish and Install	Ea.	345	95.50		440.50	24 gauge steel clad door with 4-1/2"
	Reinstall	Ea.		53.41		53.41	wood frame, trims, hinges, aluminum
	Clean	Ea.	4.07	13.75		17.82	sill, weatherstripped.
	Paint	Ea.	11.55	54		65.55	
	Minimum Charge	Job		227		227	

91

Exterior Doors and Windows

Residential Metal Door

	Unit	Material	Labor	Equip.	Total	Specification
3' x 6' 8"						
Demolish	Ea.		42		42	Cost includes material and labor to
Install	Ea.	290	60.50		350.50	install 3' x 6' 8" pre-hung 1-3/4" 24
Demolish and Install	Ea.	290	102.50		392.50	gauge steel clad door with 4-1/2"
Reinstall	Ea.		60.53		60.53	wood frame, trims, hinges, aluminum
Clean	Ea.	4.07	13.75		17.82	sill, weatherstripped.
Paint	Ea.	11.55	54		65.55	
Minimum Charge	Job		227		227	

Metal Door

	Unit	Material	Labor	Equip.	Total	Specification
Steel, Door Only						
2' 8" x 6' 8", 18 Gauge						
Demolish	Ea.		9		9	Cost includes material and labor to
Install	Ea.	555	28.50		583.50	install 18 gauge 2' 8" x 6' 8" steel
Demolish and Install	Ea.	555	37.50		592.50	door only, 1-3/4" thick, hinges.
Reinstall	Ea.		28.38		28.38	
Clean	Ea.	1.69	10.30		11.99	
Paint	Ea.	5.85	31.50		37.35	
Minimum Charge	Job		227		227	
3' x 6' 8", 18 Gauge						
Demolish	Ea.		9		9	Cost includes material and labor to
Install	Ea.	540	30.50		570.50	install 18 gauge 3' x 6' 8" steel door
Demolish and Install	Ea.	540	39.50		579.50	only, 1-3/4" thick, hinges.
Reinstall	Ea.		30.27		30.27	
Clean	Ea.	1.69	10.30		11.99	
Paint	Ea.	5.85	31.50		37.35	
Minimum Charge	Job		227		227	
2' 8" x 7', 18 Gauge						
Demolish	Ea.		9		9	Cost includes material and labor to
Install	Ea.	570	30.50		600.50	install 18 gauge 2' 8" x 7' steel door
Demolish and Install	Ea.	570	39.50		609.50	only, 1-3/4" thick, hinges.
Reinstall	Ea.		30.27		30.27	
Clean	Ea.	1.69	10.30		11.99	
Paint	Ea.	5.85	31.50		37.35	
Minimum Charge	Job		227		227	
3' x 7', 18 Gauge						
Demolish	Ea.		9		9	Cost includes material and labor to
Install	Ea.	585	32.50		617.50	install 18 gauge 3' x 7' steel door
Demolish and Install	Ea.	585	41.50		626.50	only, 1-3/4" thick, hinges.
Reinstall	Ea.		32.43		32.43	
Clean	Ea.	1.69	10.30		11.99	
Paint	Ea.	5.85	31.50		37.35	
Minimum Charge	Job		227		227	
3' 6" x 7', 18 Gauge						
Demolish	Ea.		9		9	Cost includes material and labor to
Install	Ea.	645	35		680	install 18 gauge 3' 6" x 7' steel door
Demolish and Install	Ea.	645	44		689	only, 1-3/4" thick, hinges.
Reinstall	Ea.		34.92		34.92	
Clean	Ea.	1.69	10.30		11.99	
Paint	Ea.	5.85	31.50		37.35	
Minimum Charge	Job		227		227	

Exterior Doors and Windows

Metal Door		Unit	Material	Labor	Equip.	Total	Specification
Casing Trim							
Single Width							
	Demolish	Ea.		8		8	Cost includes material and labor to
	Install	Opng.	26.50	38		64.50	install 11/16" x 2-1/2" pine ranch
	Demolish and Install	Opng.	26.50	46		72.50	style casing for one side of a standard
	Reinstall	Opng.		30.27		30.27	door opening.
	Clean	Opng.	.57	3.30		3.87	
	Paint	Ea.	4.73	7.55		12.28	
	Minimum Charge	Job		227		227	
Double Width							
	Demolish	Ea.		9.50		9.50	Cost includes material and labor to
	Install	Opng.	31	45.50		76.50	install 11/16" x 2-1/2" pine ranch
	Demolish and Install	Opng.	31	55		86	style door casing for one side of a
	Reinstall	Opng.		36.32		36.32	double door opening.
	Clean	Opng.	.68	6.60		7.28	
	Paint	Ea.	4.73	8.60		13.33	
	Minimum Charge	Job		227		227	

Solid Core Door		Unit	Material	Labor	Equip.	Total	Specification
Exterior Type							
Flush (1-3/4")							
	Demolish	Ea.		42		42	Cost includes material and labor to
	Install	Ea.	415	57		472	install 3' x 6' 8" pre-hung 1-3/4"
	Demolish and Install	Ea.	415	99		514	door, including casing and stop,
	Reinstall	Ea.		56.75		56.75	hinges, jamb, aluminum sill,
	Clean	Ea.	4.07	13.75		17.82	weatherstripped.
	Paint	Ea.	11.55	54		65.55	
	Minimum Charge	Job		227		227	
Raised Panel							
	Demolish	Ea.		42		42	Cost includes material and labor to
	Install	Ea.	485	25		510	install 3' x 6' 8" pre-hung 1-3/4"
	Demolish and Install	Ea.	485	67		552	door, jamb, including casing and
	Reinstall	Ea.		25.22		25.22	stop, hinges, aluminum sill,
	Clean	Ea.	4.07	13.75		17.82	weatherstripping.
	Paint	Ea.	11.55	126		137.55	
	Minimum Charge	Job		227		227	
Entry w / Lights							
	Demolish	Ea.		42		42	Cost includes material and labor to
	Install	Ea.	350	57		407	install 3' x 6' 8" pre-hung 1-3/4"
	Demolish and Install	Ea.	350	99		449	door, solid wood lower panel, with up
	Reinstall	Ea.		56.75		56.75	to 9-light glass panels, casing, sill,
	Clean	Ea.	4.07	13.75		17.82	weatherstripped.
	Paint	Ea.	2.55	126		128.55	
	Minimum Charge	Job		227		227	
Custom Entry							
	Demolish	Ea.		42		42	Cost includes material and labor to
	Install	Ea.	2150	91		2241	install prehung, fir, single glazed with
	Demolish and Install	Ea.	2150	133		2283	lead caming, 1-3/4" x 6' 8" x 3'
	Reinstall	Ea.		90.80		90.80	
	Clean	Ea.	4.07	13.75		17.82	
	Paint	Ea.	5.85	31.50		37.35	
	Minimum Charge	Job		227		227	

For customer support on your Contractor's Pricing Guide: Residential Repair & Remodeling, call 888.606.7279.

Exterior Doors and Windows

Solid Core Door		Unit	Material	Labor	Equip.	Total	Specification
Dutch Style							
	Demolish	Ea.		42		42	Includes material and labor to install
	Install	Ea.	910	75.50		985.50	dutch door, pine, 1-3/4" x 6' 8" x 2'
	Demolish and Install	Ea.	910	117.50		1027.50	8" wide.
	Reinstall	Ea.		75.67		75.67	
	Clean	Ea.	4.07	13.75		17.82	
	Paint	Ea.	2.55	126		128.55	
	Minimum Charge	Job		227		227	
Miami Style							
	Demolish	Ea.		42		42	Cost includes material and labor to
	Install	Ea.	232	26.50		258.50	install 3' x 6' 8" pre-hung 1-3/4"
	Demolish and Install	Ea.	232	68.50		300.50	lauan door of solid wood.
	Reinstall	Ea.		26.71		26.71	
	Clean	Ea.	4.07	13.75		17.82	
	Paint	Ea.	10.30	54		64.30	
	Minimum Charge	Job		227		227	
French Style							
	Demolish	Ea.		46		46	Cost includes material and labor to
	Install	Ea.	705	65		770	install 3' x 6' 8" pre-hung 1-3/4"
	Demolish and Install	Ea.	705	111		816	French style door with 15 lights.
	Reinstall	Ea.		64.86		64.86	
	Clean	Ea.	.28	13.75		14.03	
	Paint	Ea.	7.85	189		196.85	
	Minimum Charge	Job		227		227	
Double							
	Demolish	Ea.		30.50		30.50	Includes material and labor to install a
	Install	Ea.	1475	114		1589	pre-hung double entry door including
	Demolish and Install	Ea.	1475	144.50		1619.50	frame and exterior trim.
	Clean	Ea.	.15	48.50		48.65	
	Paint	Ea.	23	189		212	
	Minimum Charge	Job		114		114	
Commercial Grade							
3' x 6' 8"							
	Demolish	Ea.		42		42	Includes material and labor to install
	Install	Ea.	325	26.50		351.50	flush solid core 1-3/4" hardwood
	Demolish and Install	Ea.	325	68.50		393.50	veneer 3' x 6' 8" door only.
	Reinstall	Ea.		26.71		26.71	
	Clean	Ea.	4.07	13.75		17.82	
	Paint	Ea.	9.70	31.50		41.20	
	Minimum Charge	Job		227		227	
3' 6" x 6' 8"							
	Demolish	Ea.		42		42	Includes material and labor to install
	Install	Ea.	400	26.50		426.50	flush solid core 1-3/4" hardwood
	Demolish and Install	Ea.	400	68.50		468.50	veneer 3' 6" x 6' 8" door only.
	Reinstall	Ea.		26.71		26.71	
	Clean	Ea.	4.07	13.75		17.82	
	Paint	Ea.	9.70	31.50		41.20	
	Minimum Charge	Job		227		227	
3' x 7'							
	Demolish	Ea.		42		42	Includes material and labor to install
	Install	Ea.	350	65		415	flush solid core 1-3/4" hardwood
	Demolish and Install	Ea.	350	107		457	veneer 3' x 7' door only.
	Reinstall	Ea.		64.86		64.86	
	Clean	Ea.	4.07	13.75		17.82	
	Paint	Ea.	9.70	31.50		41.20	
	Minimum Charge	Job		227		227	

Solid Core Door		Unit	Material	Labor	Equip.	Total	Specification
3' 6" x 7'							
	Demolish	Ea.		42		42	Includes material and labor to install
	Install	Ea.	460	32.50		492.50	flush solid core 1-3/4" hardwood
	Demolish and Install	Ea.	460	74.50		534.50	veneer 3' 6" x 7' door only.
	Reinstall	Ea.		32.43		32.43	
	Clean	Ea.	4.07	13.75		17.82	
	Paint	Ea.	9.70	31.50		41.20	
	Minimum Charge	Job		227		227	
3' x 8'							
	Demolish	Ea.		42		42	Includes material and labor to install
	Install	Ea.	415	57		472	flush solid core 1-3/4" hardwood
	Demolish and Install	Ea.	415	99		514	veneer 3' x 8' door only.
	Reinstall	Ea.		56.75		56.75	
	Clean	Ea.	4.88	20.50		25.38	
	Paint	Ea.	9.70	31.50		41.20	
	Minimum Charge	Job		227		227	
3' 6" x 8'							
	Demolish	Ea.		42		42	Includes material and labor to install
	Install	Ea.	495	57		552	flush solid core 1-3/4" hardwood
	Demolish and Install	Ea.	495	99		594	veneer 3' 6" x 8' door only.
	Reinstall	Ea.		56.75		56.75	
	Clean	Ea.	4.88	20.50		25.38	
	Paint	Ea.	9.70	31.50		41.20	
	Minimum Charge	Job		227		227	
3' x 9'							
	Demolish	Ea.		42		42	Includes material and labor to install
	Install	Ea.	555	65		620	flush solid core 1-3/4" hardwood
	Demolish and Install	Ea.	555	107		662	veneer 3' x 9' door only.
	Reinstall	Ea.		64.86		64.86	
	Clean	Ea.	4.88	20.50		25.38	
	Paint	Ea.	9.70	31.50		41.20	
	Minimum Charge	Job		227		227	
Stain							
	Install	Ea.	10.55	42		52.55	Includes labor and material to stain
	Minimum Charge	Job		189		189	exterior type door and trim on both sides.
Door Only (SC)							
2' 4" x 6' 8"							
	Demolish	Ea.		9		9	Cost includes material and labor to
	Install	Ea.	215	25		240	install 2' 4" x 6' 8" lauan solid core
	Demolish and Install	Ea.	215	34		249	door slab.
	Reinstall	Ea.		25.22		25.22	
	Clean	Ea.	1.69	10.30		11.99	
	Paint	Ea.	4.92	47		51.92	
	Minimum Charge	Job		227		227	
2' 8" x 6' 8"							
	Demolish	Ea.		9		9	Cost includes material and labor to
	Install	Ea.	222	25		247	install 2' 8" x 6' 8" lauan solid core
	Demolish and Install	Ea.	222	34		256	door slab.
	Reinstall	Ea.		25.22		25.22	
	Clean	Ea.	1.69	10.30		11.99	
	Paint	Ea.	4.92	47		51.92	
	Minimum Charge	Job		227		227	

Exterior Doors and Windows

Solid Core Door	Unit	Material	Labor	Equip.	Total	Specification
3' x 6' 8"						
Demolish	Ea.		9		9	Cost includes material and labor to
Install	Ea.	230	26.50		256.50	install 3' x 6' 8" lauan solid core door
Demolish and Install	Ea.	230	35.50		265.50	slab.
Reinstall	Ea.		26.71		26.71	
Clean	Ea.	1.69	10.30		11.99	
Paint	Ea.	4.92	47		51.92	
Minimum Charge	Job		227		227	
3' 6" x 6' 8"						
Demolish	Ea.		9		9	Cost includes material and labor to
Install	Ea.	262	28.50		290.50	install 3' 6" x 6' 8" lauan solid core
Demolish and Install	Ea.	262	37.50		299.50	door slab.
Reinstall	Ea.		28.38		28.38	
Clean	Ea.	1.69	10.30		11.99	
Paint	Ea.	4.92	47		51.92	
Minimum Charge	Job		227		227	
Panel						
Demolish	Ea.		9		9	Cost includes material and labor to
Install	Ea.	224	53.50		277.50	install 3' x 6' 8" six (6) panel interior
Demolish and Install	Ea.	224	62.50		286.50	door.
Reinstall	Ea.		53.41		53.41	
Clean	Ea.	2.04	10.30		12.34	
Paint	Ea.	4.92	47		51.92	
Minimum Charge	Job		227		227	
Casing Trim						
Single Width						
Demolish	Ea.		8		8	Cost includes material and labor to
Install	Opng.	26.50	38		64.50	install 11/16" x 2-1/2" pine ranch
Demolish and Install	Opng.	26.50	46		72.50	style casing for one side of a standard
Reinstall	Opng.		30.27		30.27	door opening.
Clean	Opng.	.57	3.30		3.87	
Paint	Ea.	4.73	7.55		12.28	
Minimum Charge	Job		227		227	
Double Width						
Demolish	Ea.		9.50		9.50	Cost includes material and labor to
Install	Opng.	31	45.50		76.50	install 11/16" x 2-1/2" pine ranch
Demolish and Install	Opng.	31	55		86	style door casing for one side of a
Reinstall	Opng.		36.32		36.32	double door opening.
Clean	Opng.	.68	6.60		7.28	
Paint	Ea.	4.73	8.60		13.33	
Minimum Charge	Job		227		227	

Components	Unit	Material	Labor	Equip.	Total	Specification
Hardware						
Doorknob						
Demolish	Ea.		18		18	Includes material and labor to install
Install	Ea.	39	28.50		67.50	residential passage lockset, keyless
Demolish and Install	Ea.	39	46.50		85.50	bored type.
Reinstall	Ea.		22.70		22.70	
Clean	Ea.	.41	10.30		10.71	
Minimum Charge	Job		227		227	

Exterior Doors and Windows

Components		Unit	Material	Labor	Equip.	Total	Specification
Doorknob w / Lock							
	Demolish	Ea.		18		18	Includes material and labor to install
	Install	Ea.	40	28.50		68.50	privacy lockset.
	Demolish and Install	Ea.	40	46.50		86.50	
	Reinstall	Ea.		22.70		22.70	
	Clean	Ea.	.41	10.30		10.71	
	Minimum Charge	Job		227		227	
Lever Handle							
	Demolish	Ea.		18		18	Includes material and labor to install
	Install	Ea.	51.50	45.50		97	residential passage lockset, keyless
	Demolish and Install	Ea.	51.50	63.50		115	bored type with lever handle.
	Reinstall	Ea.		36.32		36.32	
	Clean	Ea.	.41	10.30		10.71	
	Minimum Charge	Job		227		227	
Deadbolt							
	Demolish	Ea.		15.75		15.75	Includes material and labor to install a
	Install	Ea.	70	32.50		102.50	deadbolt.
	Demolish and Install	Ea.	70	48.25		118.25	
	Reinstall	Ea.		25.94		25.94	
	Clean	Ea.	.41	10.30		10.71	
	Minimum Charge	Job		227		227	
Grip Handle Entry							
	Demolish	Ea.		14.40		14.40	Includes material and labor to install
	Install	Ea.	103	50.50		153.50	residential keyed entry lock set with
	Demolish and Install	Ea.	103	64.90		167.90	grip type handle.
	Reinstall	Ea.		40.36		40.36	
	Clean	Ea.	.41	10.30		10.71	
	Minimum Charge	Job		227		227	
Panic Device							
	Demolish	Ea.		36.50		36.50	Includes material and labor to install
	Install	Ea.	380	75.50		455.50	touch bar, low profile, exit only.
	Demolish and Install	Ea.	380	112		492	
	Reinstall	Ea.		60.53		60.53	
	Clean	Ea.	.21	18.35		18.56	
	Minimum Charge	Job		227		227	
Closer							
	Demolish	Ea.		9.45		9.45	Includes material and labor to install
	Install	Ea.	266	75.50		341.50	pneumatic heavy duty closer for
	Demolish and Install	Ea.	266	84.95		350.95	exterior type doors.
	Reinstall	Ea.		60.53		60.53	
	Clean	Ea.	.21	13.75		13.96	
	Minimum Charge	Job		227		227	
Peep-hole							
	Demolish	Ea.		8.15		8.15	Includes material and labor to install
	Install	Ea.	17.40	14.20		31.60	peep-hole for door.
	Demolish and Install	Ea.	17.40	22.35		39.75	
	Reinstall	Ea.		11.35		11.35	
	Clean	Ea.	.03	3.74		3.77	
	Minimum Charge	Job		227		227	
Exit Sign							
	Demolish	Ea.		24.50		24.50	Includes material and labor to install a
	Install	Ea.	129	60.50		189.50	wall mounted interior electric exit sign.
	Demolish and Install	Ea.	129	85		214	
	Reinstall	Ea.		48.28		48.28	
	Clean	Ea.	.03	7.60		7.63	
	Minimum Charge	Job		227		227	

For customer support on your Contractor's Pricing Guide: Residential Repair & Remodeling, call 888.606.7279.

Exterior Doors and Windows

Components	Unit	Material	Labor	Equip.	Total	Specification
Kickplate						
Demolish	Ea.		14		14	Includes material and labor to install
Install	Ea.	25	30.50		55.50	aluminum kickplate, 10" x 28".
Demolish and Install	Ea.	25	44.50		69.50	
Reinstall	Ea.		24.21		24.21	
Clean	Ea.	.41	10.30		10.71	
Minimum Charge	Job		227		227	
Threshold						
Install	L.F.	4	4.54		8.54	Includes labor and material to install
Minimum Charge	Job		114		114	oak threshold.
Weatherstripping						
Install	L.F.	.59	1.14		1.73	Includes labor and material to install
Minimum Charge	Job		114		114	rubber weatherstripping for door.
Metal Jamb						
6' 8" High						
Demolish	Ea.		25		25	Cost includes material and labor to
Install	Ea.	173	28.50		201.50	install 18 gauge hollow metal door
Demolish and Install	Ea.	173	53.50		226.50	frame to fit a 4-1/2" jamb 6' 8" high
Reinstall	Ea.		28.38		28.38	and up to 3' 6" opening.
Clean	Ea.	.85	6.90		7.75	
Paint	Ea.	1.61	7.55		9.16	
Minimum Charge	Job		227		227	
7' High						
Demolish	Ea.		25		25	Cost includes material and labor to
Install	Ea.	182	30.50		212.50	install 18 gauge hollow metal door
Demolish and Install	Ea.	182	55.50		237.50	frame to fit a 4-1/2" jamb 7' high and
Reinstall	Ea.		30.27		30.27	up to 3' 6" opening.
Clean	Ea.	.85	6.90		7.75	
Paint	Ea.	1.61	7.55		9.16	
Minimum Charge	Job		227		227	
8' High						
Demolish	Ea.		25		25	Cost includes material and labor to
Install	Ea.	223	38		261	install 18 gauge hollow metal door
Demolish and Install	Ea.	223	63		286	frame to fit a 4-1/2" jamb 8' high and
Reinstall	Ea.		37.83		37.83	up to 3' 6" opening.
Clean	Ea.	1.20	8.25		9.45	
Paint	Ea.	1.79	8.40		10.19	
Minimum Charge	Job		227		227	
9' High						
Demolish	Ea.		25		25	Cost includes material and labor to
Install	Ea.	283	45.50		328.50	install 18 gauge hollow metal door
Demolish and Install	Ea.	283	70.50		353.50	frame to fit a 4-1/2" jamb 9' high and
Reinstall	Ea.		45.40		45.40	up to 3' 6" opening.
Clean	Ea.	1.20	8.25		9.45	
Paint	Ea.	1.79	8.40		10.19	
Minimum Charge	Job		227		227	
Wood Jamb						
Demolish	Ea.		31.50		31.50	Includes material and labor to install
Install	Ea.	119	12.25		131.25	flat pine jamb with square cut heads
Demolish and Install	Ea.	119	43.75		162.75	and rabbeted sides for 6' 8" high and
Clean	Ea.	.13	8.25		8.38	3-9/16" door including trim sets for
Paint	Ea.	7.55	25		32.55	both sides.
Minimum Charge	Job		227		227	
Stain						
Install	Ea.	7.25	31.50		38.75	Includes labor and material to stain
Minimum Charge	Job		189		189	single door and trim on both sides.

Exterior Doors and Windows

Components	Unit	Material	Labor	Equip.	Total	Specification
Shave & Refit						
Install	Ea.		38		38	Includes labor to shave and rework
Minimum Charge	Job		114		114	door to fit opening at the job site.

Sliding Patio Door	Unit	Material	Labor	Equip.	Total	Specification
Aluminum						
6' Single Pane						
Demolish	Ea.		63		63	Cost includes material and labor to
Install	Ea.	500	227		727	install 6' x 6' 8" sliding anodized
Demolish and Install	Ea.	500	290		790	aluminum patio door with single pane
Reinstall	Ea.		181.60		181.60	tempered safety glass, anodized
Clean	Ea.	2.55	20.50		23.05	frame, screen, weatherstripped.
Minimum Charge	Job		227		227	
8' Single Pane						
Demolish	Ea.		63		63	Cost includes material and labor to
Install	Ea.	635	305		940	install 8' x 6' 8" sliding aluminum
Demolish and Install	Ea.	635	368		1003	patio door with single pane tempered
Reinstall	Ea.		242.13		242.13	safety glass, anodized frame, screen,
Clean	Ea.	2.55	20.50		23.05	weatherstripped.
Minimum Charge	Job		227		227	
10' Single Pane						
Demolish	Ea.		63		63	Cost includes material and labor to
Install	Ea.	710	455		1165	install 10' x 6' 8" sliding aluminum
Demolish and Install	Ea.	710	518		1228	patio door with single pane tempered
Reinstall	Ea.		363.20		363.20	safety glass, anodized frame, screen,
Clean	Ea.	2.55	20.50		23.05	weatherstripped.
Minimum Charge	Job		227		227	
6' Insulated						
Demolish	Ea.		63		63	Cost includes material and labor to
Install	Ea.	1050	227		1277	install 6' x 6' 8" sliding aluminum
Demolish and Install	Ea.	1050	290		1340	patio door with double insulated
Reinstall	Ea.		181.60		181.60	tempered glass, baked-on enamel
Clean	Ea.	2.55	20.50		23.05	finish, screen, weatherstripped.
Minimum Charge	Job		227		227	
8' Insulated						
Demolish	Ea.		63		63	Cost includes material and labor to
Install	Ea.	1250	305		1555	install 8' x 6' 8" sliding aluminum
Demolish and Install	Ea.	1250	368		1618	patio door with insulated tempered
Reinstall	Ea.		242.13		242.13	glass, baked-on enamel finish, screen,
Clean	Ea.	2.55	20.50		23.05	weatherstripped.
Minimum Charge	Job		227		227	
10' Insulated						
Demolish	Ea.		63		63	Cost includes material and labor to
Install	Ea.	1525	455		1980	install 10' x 6' 8" sliding aluminum
Demolish and Install	Ea.	1525	518		2043	patio door with double insulated
Reinstall	Ea.		363.20		363.20	tempered glass, baked-on enamel
Clean	Ea.	2.55	20.50		23.05	finish, screen, weatherstripped.
Minimum Charge	Job		227		227	

Sliding Patio Door	Unit	Material	Labor	Equip.	Total	Specification
Vinyl						
6' Single Pane						
Demolish	Ea.		63		63	Cost includes material and labor to
Install	Ea.	460	227		687	install 6' x 6' 8" sliding vinyl patio
Demolish and Install	Ea.	460	290		750	door with single pane tempered safety
Reinstall	Ea.		181.60		181.60	glass, anodized frame, screen,
Clean	Ea.	2.55	20.50		23.05	weatherstripped.
Minimum Charge	Job		227		227	
8' Single Pane						
Demolish	Ea.		63		63	Cost includes material and labor to
Install	Ea.	645	305		950	install 8' x 6' 8" sliding vinyl patio
Demolish and Install	Ea.	645	368		1013	door with single pane tempered safety
Reinstall	Ea.		242.13		242.13	glass, anodized frame, screen,
Clean	Ea.	2.55	20.50		23.05	weatherstripped.
Minimum Charge	Job		227		227	
10' Single Pane						
Demolish	Ea.		63		63	Cost includes material and labor to
Install	Ea.	700	455		1155	install 10' x 6' 8" sliding vinyl patio
Demolish and Install	Ea.	700	518		1218	door with single pane tempered safety
Reinstall	Ea.		363.20		363.20	glass, anodized frame, screen,
Clean	Ea.	2.55	20.50		23.05	weatherstripped.
Minimum Charge	Job		227		227	
6' Insulated						
Demolish	Ea.		63		63	Cost includes material and labor to
Install	Ea.	1275	227		1502	install 6' x 6' 8" sliding vinyl patio
Demolish and Install	Ea.	1275	290		1565	door with double insulated tempered
Reinstall	Ea.		181.60		181.60	glass, anodized frame, screen,
Clean	Ea.	2.55	20.50		23.05	weatherstripped.
Minimum Charge	Job		227		227	
8' Insulated						
Demolish	Ea.		63		63	Cost includes material and labor to
Install	Ea.	1050	305		1355	install 8' x 6' 8" sliding vinyl patio
Demolish and Install	Ea.	1050	368		1418	door with double insulated tempered
Reinstall	Ea.		242.13		242.13	glass, anodized frame, screen,
Clean	Ea.	2.55	20.50		23.05	weatherstripped.
Minimum Charge	Job		227		227	
10' Insulated						
Demolish	Ea.		63		63	Cost includes material and labor to
Install	Ea.	1850	227		2077	install 10' x 6' 8" sliding vinyl patio
Demolish and Install	Ea.	1850	290		2140	door with double insulated tempered
Reinstall	Ea.		181.60		181.60	glass, anodized frame, screen,
Clean	Ea.	2.55	20.50		23.05	weatherstripped.
Minimum Charge	Job		227		227	
Wood Framed						
6' Premium Insulated						
Demolish	Ea.		63		63	Cost includes material and labor to
Install	Ea.	1600	227		1827	install 6' x 6' 8" sliding wood patio
Demolish and Install	Ea.	1600	290		1890	door with pine frame, double insulated
Reinstall	Ea.		181.60		181.60	glass, screen, lock and dead bolt,
Clean	Ea.	2.55	20.50		23.05	weatherstripped. Add for casing and
Minimum Charge	Job		227		227	finishing.

Exterior Doors and Windows

Sliding Patio Door	Unit	Material	Labor	Equip.	Total	Specification
8' Premium Insulated						Cost includes material and labor to
Demolish	Ea.		63		63	install 8' x 6' 8" sliding wood patio
Install	Ea.	2025	305		2330	door with pine frame, double insulated
Demolish and Install	Ea.	2025	368		2393	glass, screen, lock and dead bolt,
Reinstall	Ea.		242.13		242.13	weatherstripped. Add for casing and
Clean	Ea.	2.55	20.50		23.05	finishing.
Minimum Charge	Job		227		227	
9' Premium Insulated						Cost includes material and labor to
Demolish	Ea.		63		63	install 9' x 6' 8" sliding wood patio
Install	Ea.	2025	305		2330	door with pine frame, double insulated
Demolish and Install	Ea.	2025	368		2393	glass, screen, lock and dead bolt,
Reinstall	Ea.		242.13		242.13	weatherstripped. Add for casing and
Clean	Ea.	2.55	20.50		23.05	finishing.
Minimum Charge	Job		227		227	
Stain						Includes labor and material to stain
Install	Ea.	10.55	42		52.55	exterior type door and trim on both
Minimum Charge	Job		189		189	sides.
Casing Trim						Cost includes material and labor to
Demolish	Ea.		9.50		9.50	install 11/16" x 2-1/2" pine ranch
Install	Opng.	31	45.50		76.50	style door casing for one side of a
Demolish and Install	Opng.	31	55		86	double door opening.
Reinstall	Opng.		36.32		36.32	
Clean	Opng.	.68	6.60		7.28	
Paint	Ea.	4.73	8.60		13.33	
Minimum Charge	Job		227		227	
Sliding Screen						Includes material and labor to install
Demolish	Ea.		9		9	sliding screen door, 6' 8" x 3'.
Install	Ea.	143	22.50		165.50	
Demolish and Install	Ea.	143	31.50		174.50	
Reinstall	Ea.		18.16		18.16	
Clean	Ea.	.85	6.90		7.75	
Minimum Charge	Job		227		227	

Screen Door	Unit	Material	Labor	Equip.	Total	Specification
Aluminum						Includes material and labor to install
Demolish	Ea.		42		42	aluminum frame screen door with
Install	Ea.	171	65		236	fiberglass screen cloth, closer, hinges,
Demolish and Install	Ea.	171	107		278	latch.
Reinstall	Ea.		51.89		51.89	
Clean	Ea.	4.07	13.75		17.82	
Paint	Ea.	4.54	15.10		19.64	
Minimum Charge	Job		227		227	
Wood						Includes material and labor to install
Demolish	Ea.		42		42	wood screen door with aluminum cloth
Install	Ea.	415	75.50		490.50	screen. Frame and hardware not
Demolish and Install	Ea.	415	117.50		532.50	included.
Reinstall	Ea.		60.53		60.53	
Clean	Ea.	4.07	13.75		17.82	
Paint	Ea.	4.54	15.10		19.64	
Minimum Charge	Job		227		227	
Shave & Refit						Includes labor to shave and rework
Install	Ea.		38		38	door to fit opening at the job site.
Minimum Charge	Job		227		227	

Exterior Doors and Windows

Storm Door		Unit	Material	Labor	Equip.	Total	Specification
Good Grade							
	Demolish	Ea.		42		42	Includes material and labor to install
	Install	Ea.	234	65		299	aluminum frame storm door with
	Demolish and Install	Ea.	234	107		341	fiberglass screen cloth, upper glass
	Reinstall	Ea.		51.89		51.89	panel, closer, hinges, latch.
	Clean	Ea.	4.07	13.75		17.82	
	Paint	Ea.	4.54	15.10		19.64	
	Minimum Charge	Job		227		227	
Better Grade							
	Demolish	Ea.		42		42	Includes material and labor to install
	Install	Ea.	297	65		362	aluminum frame storm door with
	Demolish and Install	Ea.	297	107		404	fiberglass screen cloth, upper and
	Reinstall	Ea.		51.89		51.89	lower glass panel, closer, latch.
	Clean	Ea.	4.07	13.75		17.82	
	Paint	Ea.	4.54	15.10		19.64	
	Minimum Charge	Job		227		227	
Premium Grade							
	Demolish	Ea.		42		42	Includes material and labor to install
	Install	Ea.	305	65		370	aluminum frame storm door with
	Demolish and Install	Ea.	305	107		412	fiberglass screen cloth, upper and
	Reinstall	Ea.		51.89		51.89	lower glass panel, closer, latch.
	Clean	Ea.	4.07	13.75		17.82	
	Paint	Ea.	4.54	15.10		19.64	
	Minimum Charge	Job		227		227	
Insulated							
	Demolish	Ea.		42		42	Includes material and labor to install
	Install	Ea.	286	32.50		318.50	aluminum frame storm door with
	Demolish and Install	Ea.	286	74.50		360.50	fiberglass screen cloth, upper and
	Reinstall	Ea.		25.94		25.94	lower glass panel, closer, latch.
	Clean	Ea.	4.07	13.75		17.82	
	Paint	Ea.	4.54	15.10		19.64	
	Minimum Charge	Job		227		227	

Garage Door		Unit	Material	Labor	Equip.	Total	Specification
9' One Piece Metal							
	Demolish	Ea.		106		106	Includes material and labor to install
	Install	Ea.	660	114		774	one piece steel garage door 9' x 7'
	Demolish and Install	Ea.	660	220		880	primed including hardware.
	Reinstall	Ea.		90.80		90.80	
	Clean	Ea.	5.10	41.50		46.60	
	Paint	Ea.	9.80	94.50		104.30	
	Minimum Charge	Job		227		227	
16' One Piece Metal							
	Demolish	Ea.		126		126	Includes material and labor to install
	Install	Ea.	880	151		1031	one piece steel garage door 16' x 7'
	Demolish and Install	Ea.	880	277		1157	primed including hardware.
	Reinstall	Ea.		121.07		121.07	
	Clean	Ea.	10.20	55		65.20	
	Paint	Ea.	19.60	108		127.60	
	Minimum Charge	Job		227		227	

Exterior Doors and Windows

Garage Door		Unit	Material	Labor	Equip.	Total	Specification
9' Metal Sectional							
	Demolish	Ea.		106		106	Includes material and labor to install
	Install	Ea.	920	114		1034	sectional metal garage door 9' x 7'
	Demolish and Install	Ea.	920	220		1140	primed including hardware.
	Reinstall	Ea.		90.80		90.80	
	Clean	Ea.	5.20	47		52.20	
	Paint	Ea.	19.60	108		127.60	
	Minimum Charge	Job		227		227	
16' Metal Sectional							
	Demolish	Ea.		126		126	Includes material and labor to install
	Install	Ea.	1125	151		1276	sectional metal garage door 16' x 7'
	Demolish and Install	Ea.	1125	277		1402	primed including hardware.
	Reinstall	Ea.		121.07		121.07	
	Clean	Ea.	10.20	55		65.20	
	Paint	Ea.	19.60	108		127.60	
	Minimum Charge	Job		227		227	
Metal / Steel Roll Up							
8' x 8' Steel							
	Demolish	Ea.		144		144	Cost includes material and labor to
	Install	Ea.	1300	655		1955	install 8' x 8' steel overhead roll-up,
	Demolish and Install	Ea.	1300	799		2099	chain hoist operated service door.
	Reinstall	Ea.		523.20		523.20	
	Clean	Ea.	5.20	47		52.20	
	Paint	Ea.	21	126		147	
	Minimum Charge	Job		1050		1050	
10' x 10' Steel							
	Demolish	Ea.		168		168	Cost includes material and labor to
	Install	Ea.	2150	745		2895	install 10' x 10' steel overhead roll-up,
	Demolish and Install	Ea.	2150	913		3063	chain hoist operated service door.
	Reinstall	Ea.		597.94		597.94	
	Clean	Ea.	8.15	66		74.15	
	Paint	Ea.	32.50	126		158.50	
	Minimum Charge	Job		1050		1050	
12' x 12' Steel							
	Demolish	Ea.		202		202	Cost includes material and labor to
	Install	Ea.	2175	870		3045	install 12' x 12' steel overhead roll-up,
	Demolish and Install	Ea.	2175	1072		3247	chain hoist operated service door.
	Reinstall	Ea.		697.60		697.60	
	Clean	Ea.	11.70	73.50		85.20	
	Paint	Ea.	47.50	151		198.50	
	Minimum Charge	Job		1050		1050	
14' x 14' Steel							
	Demolish	Ea.		252		252	Cost includes material and labor to
	Install	Ea.	3475	1300		4775	install 14' x 14' steel overhead roll-up,
	Demolish and Install	Ea.	3475	1552		5027	chain hoist operated service door.
	Reinstall	Ea.		1046.40		1046.40	
	Clean	Ea.	15.95	94.50		110.45	
	Paint	Ea.	55	189		244	
	Minimum Charge	Job		1050		1050	
18' x 18' Steel							
	Demolish	Ea.		280		280	Cost includes material and labor to
	Install	Opng.	2825	1750		4575	install 18' x 18' steel overhead roll-up,
	Demolish and Install	Opng.	2825	2030		4855	chain hoist operated service door.
	Reinstall	Opng.		1395.20		1395.20	
	Clean	Ea.	15.95	55		70.95	
	Paint	Ea.	106	380		486	
	Minimum Charge	Job		1050		1050	
Automatic Motor							
	Install	Ea.	570	45.50		615.50	Includes labor and material to install
	Minimum Charge	Job		227		227	electric opener motor, 1/3 HP.

103

Exterior Doors and Windows

Garage Door		Unit	Material	Labor	Equip.	Total	Specification
9' One Piece Wood							
	Demolish	Ea.		106		106	Includes material and labor to install
	Install	Ea.	770	114		884	one piece wood garage door 9' x 7'
	Demolish and Install	Ea.	770	220		990	primed including hardware.
	Reinstall	Ea.		90.80		90.80	
	Clean	Ea.	5.10	41.50		46.60	
	Paint	Ea.	9.80	94.50		104.30	
	Minimum Charge	Job		227		227	
9' Wood Sectional							
	Demolish	Ea.		106		106	Includes material and labor to install
	Install	Ea.	1075	114		1189	sectional wood garage door 9' x 7'
	Demolish and Install	Ea.	1075	220		1295	unfinished including hardware.
	Reinstall	Ea.		90.80		90.80	
	Clean	Ea.	5.10	41.50		46.60	
	Paint	Ea.	9.80	94.50		104.30	
	Minimum Charge	Job		227		227	
16' Wood Sectional							
	Demolish	Ea.		126		126	Includes material and labor to install
	Install	Ea.	1725	151		1876	sectional fiberglass garage door 16' x
	Demolish and Install	Ea.	1725	277		2002	7' unfinished including hardware.
	Reinstall	Ea.		121.07		121.07	
	Clean	Ea.	10.20	55		65.20	
	Paint	Ea.	19.60	108		127.60	
	Minimum Charge	Job		227		227	
Electric Opener							
	Demolish	Ea.		73.50		73.50	Includes material and labor to install
	Install	Ea.	415	85		500	electric opener, 1/3 HP economy.
	Demolish and Install	Ea.	415	158.50		573.50	
	Reinstall	Ea.		68.14		68.14	
	Clean	Ea.	.21	18.35		18.56	
	Minimum Charge	Job		227		227	
Springs							
	Install	Pr.	178	114		292	Includes labor and material to replace
	Minimum Charge	Job		227		227	springs for garage door.

Commercial Metal Door		Unit	Material	Labor	Equip.	Total	Specification
Pre-hung Steel Door							
2' 8" x 6' 8"							
	Demolish	Ea.		63		63	Cost includes material and labor to
	Install	Ea.	730	28.50		758.50	install 2' 8" x 6' 8" pre-hung 1-3/4"
	Demolish and Install	Ea.	730	91.50		821.50	18 gauge steel door with steel frame,
	Reinstall	Ea.		28.38		28.38	hinges, aluminum sill,
	Clean	Ea.	4.07	13.75		17.82	weather-stripped.
	Paint	Ea.	5.85	31.50		37.35	
	Minimum Charge	Job		227		227	
3' x 6' 8"							
	Demolish	Ea.		63		63	Cost includes material and labor to
	Install	Ea.	690	30.50		720.50	install 3' x 6' 8" pre-hung 1-3/4" 18
	Demolish and Install	Ea.	690	93.50		783.50	gauge steel door with steel frame,
	Reinstall	Ea.		30.27		30.27	hinges, aluminum sill,
	Clean	Ea.	4.07	13.75		17.82	weather-stripped.
	Paint	Ea.	5.85	31.50		37.35	
	Minimum Charge	Job		227		227	

Commercial Metal Door	Unit	Material	Labor	Equip.	Total	Specification
3' 6" x 6' 8"						
Demolish	Ea.		63		63	Cost includes material and labor to
Install	Ea.	785	35		820	install 3' 6" x 6' 8" pre-hung 1-3/4"
Demolish and Install	Ea.	785	98		883	18 gauge steel door with steel frame,
Reinstall	Ea.		34.92		34.92	hinges, aluminum sill,
Clean	Ea.	4.07	13.75		17.82	weather-stripped.
Paint	Ea.	5.85	31.50		37.35	
Minimum Charge	Job		227		227	
4' x 6' 8"						
Demolish	Ea.		63		63	Cost includes material and labor to
Install	Ea.	855	45.50		900.50	install 4' x 6' 8" pre-hung 1-3/4" 18
Demolish and Install	Ea.	855	108.50		963.50	gauge steel door with steel frame,
Reinstall	Ea.		45.40		45.40	hinges, aluminum sill,
Clean	Ea.	4.07	13.75		17.82	weather-stripped.
Paint	Ea.	5.85	31.50		37.35	
Minimum Charge	Job		227		227	
Metal Jamb						
6' 8" High						
Demolish	Ea.		25		25	Cost includes material and labor to
Install	Ea.	173	28.50		201.50	install 18 gauge hollow metal door
Demolish and Install	Ea.	173	53.50		226.50	frame to fit a 4-1/2" jamb 6' 8" high
Reinstall	Ea.		28.38		28.38	and up to 3' 6" opening.
Clean	Ea.	.85	6.90		7.75	
Paint	Ea.	1.61	7.55		9.16	
Minimum Charge	Job		227		227	
7' High						
Demolish	Ea.		25		25	Cost includes material and labor to
Install	Ea.	182	30.50		212.50	install 18 gauge hollow metal door
Demolish and Install	Ea.	182	55.50		237.50	frame to fit a 4-1/2" jamb 7' high and
Reinstall	Ea.		30.27		30.27	up to 3' 6" opening.
Clean	Ea.	.85	6.90		7.75	
Paint	Ea.	1.61	7.55		9.16	
Minimum Charge	Job		227		227	
8' High						
Demolish	Ea.		25		25	Cost includes material and labor to
Install	Ea.	223	38		261	install 18 gauge hollow metal door
Demolish and Install	Ea.	223	63		286	frame to fit a 4-1/2" jamb 8' high and
Reinstall	Ea.		37.83		37.83	up to 3' 6" opening.
Clean	Ea.	1.20	8.25		9.45	
Paint	Ea.	1.79	8.40		10.19	
Minimum Charge	Job		227		227	
9' High						
Demolish	Ea.		25		25	Cost includes material and labor to
Install	Ea.	283	45.50		328.50	install 18 gauge hollow metal door
Demolish and Install	Ea.	283	70.50		353.50	frame to fit a 4-1/2" jamb 9' high and
Reinstall	Ea.		45.40		45.40	up to 3' 6" opening.
Clean	Ea.	1.20	8.25		9.45	
Paint	Ea.	1.79	8.40		10.19	
Minimum Charge	Job		227		227	
Hardware						
Doorknob w / Lock						
Demolish	Ea.		18		18	Includes material and labor to install
Install	Ea.	197	38		235	keyed entry lock set, bored type
Demolish and Install	Ea.	197	56		253	including knob trim, strike and strike
Reinstall	Ea.		30.27		30.27	box for standard commercial
Clean	Ea.	.41	10.30		10.71	application.
Minimum Charge	Job		227		227	

Exterior Doors and Windows

Commercial Metal Door		Unit	Material	Labor	Equip.	Total	Specification
Lever Handle							
	Demolish	Ea.		18		18	Includes material and labor to install
	Install	Ea.	155	45.50		200.50	passage lockset, keyless bored type
	Demolish and Install	Ea.	155	63.50		218.50	with lever handle, non-locking
	Reinstall	Ea.		36.32		36.32	standard commercial latchset.
	Clean	Ea.	.41	10.30		10.71	
	Minimum Charge	Job		227		227	
Deadbolt							
	Demolish	Ea.		15.75		15.75	Includes material and labor to install a
	Install	Ea.	70	32.50		102.50	deadbolt.
	Demolish and Install	Ea.	70	48.25		118.25	
	Reinstall	Ea.		25.94		25.94	
	Clean	Ea.	.41	10.30		10.71	
	Minimum Charge	Job		227		227	
Panic Device							
	Demolish	Ea.		36.50		36.50	Includes material and labor to install
	Install	Ea.	380	75.50		455.50	touch bar, low profile, exit only.
	Demolish and Install	Ea.	380	112		492	
	Reinstall	Ea.		60.53		60.53	
	Clean	Ea.	.21	18.35		18.56	
	Minimum Charge	Job		227		227	
Closer							
	Demolish	Ea.		9.45		9.45	Includes material and labor to install
	Install	Ea.	310	75.50		385.50	pneumatic heavy duty model for
	Demolish and Install	Ea.	310	84.95		394.95	exterior type doors.
	Reinstall	Ea.		60.53		60.53	
	Clean	Ea.	.21	13.75		13.96	
	Minimum Charge	Job		227		227	
Peep-Hole							
	Demolish	Ea.		8.15		8.15	Includes material and labor to install
	Install	Ea.	17.40	14.20		31.60	peep-hole for door.
	Demolish and Install	Ea.	17.40	22.35		39.75	
	Reinstall	Ea.		11.35		11.35	
	Clean	Ea.	.03	3.74		3.77	
	Minimum Charge	Job		227		227	
Exit Sign							
	Demolish	Ea.		24.50		24.50	Includes material and labor to install a
	Install	Ea.	129	60.50		189.50	wall mounted interior electric exit sign.
	Demolish and Install	Ea.	129	85		214	
	Reinstall	Ea.		48.28		48.28	
	Clean	Ea.	.03	7.60		7.63	
	Minimum Charge	Job		227		227	
Kickplate							
	Demolish	Ea.		14		14	Includes material and labor to install
	Install	Ea.	88.50	30.50		119	stainless steel kickplate, 10" x 36".
	Demolish and Install	Ea.	88.50	44.50		133	
	Reinstall	Ea.		24.21		24.21	
	Clean	Ea.	.41	10.30		10.71	
	Minimum Charge	Job		227		227	

Aluminum Window		Unit	Material	Labor	Equip.	Total	Specification
Single Hung Two Light							
2' x 2'							
	Demolish	Ea.		23		23	Cost includes material and labor to
	Install	Ea.	191	95		286	install 2' x 2' single hung, two light,
	Demolish and Install	Ea.	191	118		309	double glazed aluminum window
	Reinstall	Ea.		76.10		76.10	including screen.
	Clean	Ea.	.04	8.65		8.69	
	Minimum Charge	Job		262		262	

Exterior Doors and Windows

Aluminum Window	Unit	Material	Labor	Equip.	Total	Specification
2' x 2' 6"						
Demolish	Ea.		23		23	Cost includes material and labor to
Install	Ea.	84.50	95		179.50	install 2' x 2' 6" single hung, two
Demolish and Install	Ea.	84.50	118		202.50	light, double glazed aluminum
Reinstall	Ea.		76.10		76.10	window including screen.
Clean	Ea.	.04	8.65		8.69	
Minimum Charge	Job		262		262	
3' x 1' 6"						
Demolish	Ea.		23		23	Cost includes material and labor to
Install	Ea.	77	105		182	install 3' x 1' 6" single hung, two
Demolish and Install	Ea.	77	128		205	light, double glazed aluminum
Reinstall	Ea.		83.71		83.71	window including screen.
Clean	Ea.	.03	7.60		7.63	
Minimum Charge	Job		262		262	
3' x 2'						
Demolish	Ea.		23		23	Cost includes material and labor to
Install	Ea.	93.50	105		198.50	install 3' x 2' single hung, two light,
Demolish and Install	Ea.	93.50	128		221.50	double glazed aluminum window
Reinstall	Ea.		83.71		83.71	including screen.
Clean	Ea.	.03	7.60		7.63	
Minimum Charge	Job		262		262	
3' x 2' 6"						
Demolish	Ea.		23		23	Cost includes material and labor to
Install	Ea.	101	105		206	install 3' x 2' 6" single hung, two
Demolish and Install	Ea.	101	128		229	light, double glazed aluminum
Reinstall	Ea.		83.71		83.71	window including screen.
Clean	Ea.	.04	8.65		8.69	
Minimum Charge	Job		262		262	
3' x 3'						
Demolish	Ea.		23		23	Cost includes material and labor to
Install	Ea.	107	105		212	install 3' x 3' single hung, two light,
Demolish and Install	Ea.	107	128		235	double glazed aluminum window
Reinstall	Ea.		83.71		83.71	including screen.
Clean	Ea.	.04	8.65		8.69	
Minimum Charge	Job		262		262	
3' x 3' 6"						
Demolish	Ea.		23		23	Cost includes material and labor to
Install	Ea.	264	105		369	install 3' x 3' 6" single hung, two
Demolish and Install	Ea.	264	128		392	light, double glazed aluminum
Reinstall	Ea.		83.71		83.71	window including screen.
Clean	Ea.	.07	10.10		10.17	
Minimum Charge	Job		262		262	
3' x 4'						
Demolish	Ea.		23		23	Cost includes material and labor to
Install	Ea.	277	105		382	install 3' x 4' single hung, two light,
Demolish and Install	Ea.	277	128		405	double glazed aluminum window
Reinstall	Ea.		83.71		83.71	including screen.
Clean	Ea.	.07	10.10		10.17	
Minimum Charge	Job		262		262	
3' x 5'						
Demolish	Ea.		33.50		33.50	Cost includes material and labor to
Install	Ea.	138	105		243	install 3' x 5' single hung, two light,
Demolish and Install	Ea.	138	138.50		276.50	double glazed aluminum window
Reinstall	Ea.		83.71		83.71	including screen.
Clean	Ea.	.08	12.15		12.23	
Minimum Charge	Job		262		262	

For customer support on your Contractor's Pricing Guide: Residential Repair & Remodeling, call 888.606.7279.

Exterior Doors and Windows

Aluminum Window	Unit	Material	Labor	Equip.	Total	Specification
3' x 6'						
Demolish	Ea.		33.50		33.50	Cost includes material and labor to
Install	Ea.	380	116		496	install 3' x 6' single hung, two light,
Demolish and Install	Ea.	380	149.50		529.50	double glazed aluminum window
Reinstall	Ea.		93.01		93.01	including screen.
Clean	Ea.	.09	14.30		14.39	
Minimum Charge	Job		262		262	
Double Hung Two Light						
2' x 2'						
Demolish	Ea.		23		23	Cost includes material and labor to
Install	Ea.	129	95		224	install 2' x 2' double hung, two light,
Demolish and Install	Ea.	129	118		247	double glazed aluminum window
Reinstall	Ea.		76.10		76.10	including screen.
Clean	Ea.	.03	7.60		7.63	
Minimum Charge	Job		262		262	
2' x 2' 6"						
Demolish	Ea.		23		23	Cost includes material and labor to
Install	Ea.	142	95		237	install 2' x 2' 6" double hung, two
Demolish and Install	Ea.	142	118		260	light, double glazed aluminum
Reinstall	Ea.		76.10		76.10	window including screen.
Clean	Ea.	.03	7.60		7.63	
Minimum Charge	Job		262		262	
3' x 1' 6"						
Demolish	Ea.		23		23	Cost includes material and labor to
Install	Ea.	297	105		402	install 3' x 1' 6" double hung, two
Demolish and Install	Ea.	297	128		425	light, double glazed aluminum
Reinstall	Ea.		83.71		83.71	window including screen.
Clean	Ea.	.03	7.60		7.63	
Minimum Charge	Job		262		262	
3' x 2'						
Demolish	Ea.		23		23	Cost includes material and labor to
Install	Ea.	345	105		450	install 3' x 2' double hung, two light,
Demolish and Install	Ea.	345	128		473	double glazed aluminum window
Reinstall	Ea.		83.71		83.71	including screen.
Clean	Ea.	.03	7.60		7.63	
Minimum Charge	Job		262		262	
3' x 2' 6"						
Demolish	Ea.		23		23	Cost includes material and labor to
Install	Ea.	420	105		525	install 3' x 2' 6" double hung, two
Demolish and Install	Ea.	420	128		548	light, double glazed aluminum
Reinstall	Ea.		83.71		83.71	window including screen.
Clean	Ea.	.04	8.65		8.69	
Minimum Charge	Job		262		262	
3' x 3'						
Demolish	Ea.		23		23	Cost includes material and labor to
Install	Ea.	470	105		575	install 3' x 3' double hung, two light,
Demolish and Install	Ea.	470	128		598	double glazed aluminum window
Reinstall	Ea.		83.71		83.71	including screen.
Clean	Ea.	.04	8.65		8.69	
Minimum Charge	Job		262		262	
3' x 3' 6"						
Demolish	Ea.		23		23	Cost includes material and labor to
Install	Ea.	490	105		595	install 3' x 3' 6" double hung, two
Demolish and Install	Ea.	490	128		618	light, double glazed aluminum
Reinstall	Ea.		83.71		83.71	window including screen.
Clean	Ea.	.07	10.10		10.17	
Minimum Charge	Job		262		262	

Exterior Doors and Windows

Aluminum Window		Unit	Material	Labor	Equip.	Total	Specification
3' x 4'							
	Demolish	Ea.		23		23	Cost includes material and labor to
	Install	Ea.	530	105		635	install 3' x 4' double hung, two light,
	Demolish and Install	Ea.	530	128		658	double glazed aluminum window
	Reinstall	Ea.		83.71		83.71	including screen.
	Clean	Ea.	.07	10.10		10.17	
	Minimum Charge	Job		262		262	
3' x 5'							
	Demolish	Ea.		33.50		33.50	Cost includes material and labor to
	Install	Ea.	585	105		690	install 3' x 5' double hung, two light,
	Demolish and Install	Ea.	585	138.50		723.50	double glazed aluminum window
	Reinstall	Ea.		83.71		83.71	including screen.
	Clean	Ea.	.08	12.15		12.23	
	Minimum Charge	Job		262		262	
3' x 6'							
	Demolish	Ea.		33.50		33.50	Cost includes material and labor to
	Install	Ea.	615	116		731	install 3' x 6' double hung, two light,
	Demolish and Install	Ea.	615	149.50		764.50	double glazed aluminum window
	Reinstall	Ea.		93.01		93.01	including screen.
	Clean	Ea.	.09	14.30		14.39	
	Minimum Charge	Job		262		262	
Sliding Sash							
1' 6" x 3'							
	Demolish	Ea.		23		23	Cost includes material and labor to
	Install	Ea.	240	95		335	install 1' 6" x 3' sliding sash, two
	Demolish and Install	Ea.	240	118		358	light, double glazed aluminum
	Reinstall	Ea.		76.10		76.10	window including screen.
	Clean	Ea.	.03	7.60		7.63	
	Minimum Charge	Job		262		262	
2' x 2'							
	Demolish	Ea.		23		23	Cost includes material and labor to
	Install	Ea.	66	105		171	install 2' x 2' sliding sash, two light,
	Demolish and Install	Ea.	66	128		194	double glazed aluminum window
	Reinstall	Ea.		83.71		83.71	including screen.
	Clean	Ea.	.03	7.60		7.63	
	Minimum Charge	Job		262		262	
3' x 2'							
	Demolish	Ea.		23		23	Cost includes material and labor to
	Install	Ea.	253	105		358	install 3' x 2' sliding sash, two light,
	Demolish and Install	Ea.	253	128		381	double glazed aluminum window
	Reinstall	Ea.		83.71		83.71	including screen.
	Clean	Ea.	.03	7.60		7.63	
	Minimum Charge	Job		262		262	
3' x 2' 6"							
	Demolish	Ea.		23		23	Cost includes material and labor to
	Install	Ea.	90	105		195	install 3' x 2' 6" sliding sash, two
	Demolish and Install	Ea.	90	128		218	light, double glazed aluminum
	Reinstall	Ea.		83.71		83.71	window including screen.
	Clean	Ea.	.04	8.65		8.69	
	Minimum Charge	Job		262		262	
3' x 3'							
	Demolish	Ea.		23		23	Cost includes material and labor to
	Install	Ea.	305	105		410	install 3' x 3' sliding sash, two light,
	Demolish and Install	Ea.	305	128		433	double glazed aluminum window
	Reinstall	Ea.		83.71		83.71	including screen.
	Clean	Ea.	.04	8.65		8.69	
	Minimum Charge	Job		262		262	

For customer support on your Contractor's Pricing Guide: Residential Repair & Remodeling, call 888.606.7279.

Aluminum Window	Unit	Material	Labor	Equip.	Total	Specification
3' x 3' 6"						
Demolish	Ea.		23		23	Cost includes material and labor to
Install	Ea.	109	105		214	install 3' x 3' 6" sliding sash, two
Demolish and Install	Ea.	109	128		237	light, double glazed aluminum
Reinstall	Ea.		83.71		83.71	window including screen.
Clean	Ea.	.07	10.10		10.17	
Minimum Charge	Job		262		262	
3' x 4'						
Demolish	Ea.		23		23	Cost includes material and labor to
Install	Ea.	116	116		232	install 3' x 4' sliding sash, two light,
Demolish and Install	Ea.	116	139		255	double glazed aluminum window
Reinstall	Ea.		93.01		93.01	including screen.
Clean	Ea.	.07	10.10		10.17	
Minimum Charge	Job		262		262	
3' x 5'						
Demolish	Ea.		33.50		33.50	Cost includes material and labor to
Install	Ea.	149	116		265	install 3' x 5' sliding sash, two light,
Demolish and Install	Ea.	149	149.50		298.50	double glazed aluminum window
Reinstall	Ea.		93.01		93.01	including screen.
Clean	Ea.	.08	12.15		12.23	
Minimum Charge	Job		262		262	
4' x 5'						
Demolish	Ea.		33.50		33.50	Cost includes material and labor to
Install	Ea.	151	149		300	install 4' x 5' sliding sash, two light,
Demolish and Install	Ea.	151	182.50		333.50	double glazed aluminum window
Reinstall	Ea.		119.59		119.59	including screen.
Clean	Ea.	.13	17.35		17.48	
Minimum Charge	Job		262		262	
5' x 5'						
Demolish	Ea.		33.50		33.50	Cost includes material and labor to
Install	Ea.	360	149		509	install 5' x 5' sliding sash, two light,
Demolish and Install	Ea.	360	182.50		542.50	double glazed aluminum window
Reinstall	Ea.		119.59		119.59	including screen.
Clean	Ea.	.13	17.35		17.48	
Minimum Charge	Job		262		262	
Awning Sash						
26" x 26", 2 Light						
Demolish	Ea.		23		23	Cost includes material and labor to
Install	Ea.	320	116		436	install 26" x 26" two light awning,
Demolish and Install	Ea.	320	139		459	double glazed aluminum window
Reinstall	Ea.		93.01		93.01	including screen.
Clean	Ea.	.04	8.65		8.69	
Minimum Charge	Job		262		262	
26" x 37", 2 Light						
Demolish	Ea.		23		23	Cost includes material and labor to
Install	Ea.	360	116		476	install 26" x 37" two light awning,
Demolish and Install	Ea.	360	139		499	double glazed aluminum window
Reinstall	Ea.		93.01		93.01	including screen.
Clean	Ea.	.04	8.65		8.69	
Minimum Charge	Job		262		262	
38" x 26", 3 Light						
Demolish	Ea.		23		23	Cost includes material and labor to
Install	Ea.	395	116		511	install 38" x 26" three light awning,
Demolish and Install	Ea.	395	139		534	double glazed aluminum window
Reinstall	Ea.		93.01		93.01	including screen.
Clean	Ea.	.07	10.10		10.17	
Minimum Charge	Job		262		262	

Exterior Doors and Windows

Aluminum Window	Unit	Material	Labor	Equip.	Total	Specification
38" x 37", 3 Light						
Demolish	Ea.		23		23	Cost includes material and labor to
Install	Ea.	400	131		531	install 38" x 37" three light awning,
Demolish and Install	Ea.	400	154		554	double glazed aluminum window
Reinstall	Ea.		104.64		104.64	including screen.
Clean	Ea.	.07	10.10		10.17	
Minimum Charge	Job		262		262	
38" x 53", 3 Light						
Demolish	Ea.		33.50		33.50	Cost includes material and labor to
Install	Ea.	670	131		801	install 38" x 53" three light awning,
Demolish and Install	Ea.	670	164.50		834.50	double glazed aluminum window
Reinstall	Ea.		104.64		104.64	including screen.
Clean	Ea.	.08	12.15		12.23	
Minimum Charge	Job		262		262	
50" x 37", 4 Light						
Demolish	Ea.		33.50		33.50	Cost includes material and labor to
Install	Ea.	660	131		791	install 50" x 37" four light awning,
Demolish and Install	Ea.	660	164.50		824.50	double glazed aluminum window
Reinstall	Ea.		104.64		104.64	including screen.
Clean	Ea.	.08	12.15		12.23	
Minimum Charge	Job		262		262	
50" x 53", 4 Light						
Demolish	Ea.		33.50		33.50	Cost includes material and labor to
Install	Ea.	1100	149		1249	install 50" x 53" four light awning,
Demolish and Install	Ea.	1100	182.50		1282.50	double glazed aluminum window
Reinstall	Ea.		119.59		119.59	including screen.
Clean	Ea.	.09	14.30		14.39	
Minimum Charge	Job		262		262	
63" x 37", 5 Light						
Demolish	Ea.		33.50		33.50	Cost includes material and labor to
Install	Ea.	860	149		1009	install 63" x 37" five light awning,
Demolish and Install	Ea.	860	182.50		1042.50	double glazed aluminum window
Reinstall	Ea.		119.59		119.59	including screen.
Clean	Ea.	.09	14.30		14.39	
Minimum Charge	Job		262		262	
63" x 53", 5 Light						
Demolish	Ea.		33.50		33.50	Cost includes material and labor to
Install	Ea.	1200	149		1349	install 63" x 53" five light awning,
Demolish and Install	Ea.	1200	182.50		1382.50	double glazed aluminum window
Reinstall	Ea.		119.59		119.59	including screen.
Clean	Ea.	.13	17.35		17.48	
Minimum Charge	Job		262		262	
Mullion Bars						
Install	Ea.	13.55	42		55.55	Includes labor and material to install
Minimum Charge	Job		262		262	mullions for sliding aluminum window.
Bay						
4' 8" x 6' 4"						
Demolish	Ea.		24.50		24.50	Cost includes material and labor to
Install	Ea.	1650	114		1764	install 4' 8" x 6' 4" angle bay double
Demolish and Install	Ea.	1650	138.50		1788.50	hung unit, pine frame with aluminum
Reinstall	Ea.		90.80		90.80	cladding, picture center sash,
Clean	Ea.	.18	20		20.18	hardware, screens.
Minimum Charge	Job		262		262	

111

Exterior Doors and Windows

Aluminum Window		Unit	Material	Labor	Equip.	Total	Specification
4' 8" x 7' 2"							
	Demolish	Ea.		24.50		24.50	Cost includes material and labor to
	Install	Ea.	1675	114		1789	install 4' 8" x 7' 2" angle bay double
	Demolish and Install	Ea.	1675	138.50		1813.50	hung unit, pine frame with aluminum
	Reinstall	Ea.		90.80		90.80	cladding, picture center sash,
	Clean	Ea.	.18	20		20.18	hardware, screens.
	Minimum Charge	Job		262		262	
4' 8" x 7' 10"							
	Demolish	Ea.		24.50		24.50	Cost includes material and labor to
	Install	Ea.	1700	114		1814	install 4' 8" x 7' 10" angle bay
	Demolish and Install	Ea.	1700	138.50		1838.50	double hung unit, pine frame with
	Reinstall	Ea.		90.80		90.80	aluminum cladding, picture center
	Clean	Ea.	.18	20		20.18	sash, hardware, screens.
	Minimum Charge	Job		262		262	
4' 8" x 8' 6"							
	Demolish	Ea.		28		28	Cost includes material and labor to
	Install	Ea.	1650	130		1780	install 4' 8" x 8' 6" angle bay double
	Demolish and Install	Ea.	1650	158		1808	hung unit, pine frame with aluminum
	Reinstall	Ea.		103.77		103.77	cladding, picture center sash,
	Clean	Ea.	.02	27		27.02	hardware, screens.
	Minimum Charge	Job		262		262	
4' 8" x 8' 10"							
	Demolish	Ea.		28		28	Cost includes material and labor to
	Install	Ea.	1700	130		1830	install 4' 8" x 8' 10" angle bay
	Demolish and Install	Ea.	1700	158		1858	double hung unit, pine frame with
	Reinstall	Ea.		103.77		103.77	aluminum cladding, picture center
	Clean	Ea.	.02	27		27.02	sash, hardware, screens.
	Minimum Charge	Job		262		262	
4' 8" x 9' 2"							
	Demolish	Ea.		28		28	Cost includes material and labor to
	Install	Ea.	1750	227		1977	install 4' 8" x 9' 2" angle bay double
	Demolish and Install	Ea.	1750	255		2005	hung unit, pine frame with aluminum
	Reinstall	Ea.		181.60		181.60	cladding, picture center sash,
	Clean	Ea.	.02	27		27.02	hardware, screens.
	Minimum Charge	Job		262		262	
4' 8" x 9' 10"							
	Demolish	Ea.		28		28	Cost includes material and labor to
	Install	Ea.	1850	227		2077	install 4' 8" x 9' 10" angle bay
	Demolish and Install	Ea.	1850	255		2105	double hung unit, pine frame with
	Reinstall	Ea.		181.60		181.60	aluminum cladding, picture center
	Clean	Ea.	.02	27		27.02	sash, hardware, screens.
	Minimum Charge	Job		262		262	
Casing							
	Demolish	Ea.		10.50		10.50	Cost includes material and labor to
	Install	Opng.	34.50	35		69.50	install 11/16" x 2-1/2" pine ranch
	Demolish and Install	Opng.	34.50	45.50		80	style window casing.
	Reinstall	Opng.		27.94		27.94	
	Clean	Opng.	.57	3.30		3.87	
	Paint	Ea.	4.73	7.55		12.28	
	Minimum Charge	Job		262		262	
Storm Window							
2' x 3' 6"							
	Demolish	Ea.		13.60		13.60	Includes material and labor to install
	Install	Ea.	95.50	30.50		126	residential aluminum storm window
	Demolish and Install	Ea.	95.50	44.10		139.60	with mill finish, 2' x 3' 6" high.
	Reinstall	Ea.		24.21		24.21	
	Clean	Ea.	.03	7.60		7.63	
	Minimum Charge	Job		262		262	

112

Exterior Doors and Windows

Aluminum Window

	Unit	Material	Labor	Equip.	Total	Specification
2' 6" x 5'						
Demolish	Ea.		13.60		13.60	Includes material and labor to install
Install	Ea.	101	32.50		133.50	residential aluminum storm window
Demolish and Install	Ea.	101	46.10		147.10	with mill finish, 2' 6" x 5' high.
Reinstall	Ea.		25.94		25.94	
Clean	Ea.	.07	10.10		10.17	
Minimum Charge	Job		262		262	
4' x 6'						
Demolish	Ea.		26		26	Includes material and labor to install
Install	Ea.	123	36.50		159.50	residential aluminum storm window, 4'
Demolish and Install	Ea.	123	62.50		185.50	x 6' high.
Reinstall	Ea.		29.06		29.06	
Clean	Ea.	.12	15.20		15.32	
Minimum Charge	Job		262		262	
Hardware						
Weatherstripping						
Install	L.F.	.69	3.78		4.47	Includes labor and material to install
Minimum Charge	Job		151		151	vinyl V strip weatherstripping for double-hung window.
Trim Set						
Demolish	Ea.		10.50		10.50	Cost includes material and labor to
Install	Opng.	34.50	35		69.50	install 11/16" x 2-1/2" pine ranch
Demolish and Install	Opng.	34.50	45.50		80	style window casing.
Reinstall	Opng.		27.94		27.94	
Clean	Opng.	.57	3.30		3.87	
Paint	Ea.	4.73	7.55		12.28	
Minimum Charge	Job		227		227	

Vinyl Window

	Unit	Material	Labor	Equip.	Total	Specification
Double Hung						
2' x 2' 6"						
Demolish	Ea.		16.65		16.65	Cost includes material and labor to
Install	Ea.	221	60.50		281.50	install 2' x 2' 6" double hung, two
Demolish and Install	Ea.	221	77.15		298.15	light, vinyl window with screen.
Reinstall	Ea.		48.43		48.43	
Clean	Ea.	.03	7.60		7.63	
Minimum Charge	Job		151		151	
2' x 3' 6"						
Demolish	Ea.		16.65		16.65	Cost includes material and labor to
Install	Ea.	235	65		300	install 2' x 3' 6" double hung, two
Demolish and Install	Ea.	235	81.65		316.65	light, vinyl window with screen.
Reinstall	Ea.		51.89		51.89	
Clean	Ea.	.04	8.65		8.69	
Minimum Charge	Job		151		151	
2' 6" x 4' 6"						
Demolish	Ea.		16.65		16.65	Cost includes material and labor to
Install	Ea.	278	70		348	install 2' 6" x 4' 6" double hung, two
Demolish and Install	Ea.	278	86.65		364.65	light, vinyl window with screen.
Reinstall	Ea.		55.88		55.88	
Clean	Ea.	.07	10.10		10.17	
Minimum Charge	Job		151		151	
3' x 4'						
Demolish	Ea.		16.65		16.65	Includes material and labor to install
Install	Ea.	248	91		339	medium size double hung, 3' x 4',
Demolish and Install	Ea.	248	107.65		355.65	two light, vinyl window with screen.
Reinstall	Ea.		72.64		72.64	
Clean	Ea.	.07	10.10		10.17	
Minimum Charge	Job		151		151	

113

Exterior Doors and Windows

Vinyl Window	Unit	Material	Labor	Equip.	Total	Specification
3' x 4' 6"						
Demolish	Ea.		20.50		20.50	Cost includes material and labor to
Install	Ea.	320	101		421	install 3' x 4' 6" double hung, two
Demolish and Install	Ea.	320	121.50		441.50	light, vinyl window with screen.
Reinstall	Ea.		80.71		80.71	
Clean	Ea.	.08	12.15		12.23	
Minimum Charge	Job		151		151	
4' x 4' 6"						
Demolish	Ea.		20.50		20.50	Cost includes material and labor to
Install	Ea.	340	114		454	install 4' x 4' 6" double hung, two
Demolish and Install	Ea.	340	134.50		474.50	light, vinyl window with screen.
Reinstall	Ea.		90.80		90.80	
Clean	Ea.	.09	14.30		14.39	
Minimum Charge	Job		151		151	
4' x 6'						
Demolish	Ea.		20.50		20.50	Includes material and labor to install
Install	Ea.	420	130		550	large double hung, 4' x 6', two light,
Demolish and Install	Ea.	420	150.50		570.50	vinyl window with screen.
Reinstall	Ea.		103.77		103.77	
Clean	Ea.	.13	17.35		17.48	
Minimum Charge	Job		151		151	
Trim Set						
Demolish	Ea.		10.50		10.50	Cost includes material and labor to
Install	Opng.	34.50	35		69.50	install 11/16" x 2-1/2" pine ranch
Demolish and Install	Opng.	34.50	45.50		80	style window casing.
Reinstall	Opng.		27.94		27.94	
Clean	Opng.	.57	3.30		3.87	
Paint	Ea.	4.73	7.55		12.28	
Minimum Charge	Job		262		262	

Wood Window	Unit	Material	Labor	Equip.	Total	Specification
Double Hung						
2' 2" x 3' 4"						
Demolish	Ea.		16.65		16.65	Cost includes material and labor to
Install	Ea.	233	60.50		293.50	install 2' 2" x 3' 4" double hung, two
Demolish and Install	Ea.	233	77.15		310.15	light, double glazed wood window
Reinstall	Ea.		48.43		48.43	with screen.
Clean	Ea.	.04	8.65		8.69	
Paint	Ea.	2.96	15.75		18.71	
Minimum Charge	Job		227		227	
2' 2" x 4' 4"						
Demolish	Ea.		16.65		16.65	Cost includes material and labor to
Install	Ea.	254	65		319	install 2' 2" x 4' 4" double hung, two
Demolish and Install	Ea.	254	81.65		335.65	light, double glazed wood window
Reinstall	Ea.		51.89		51.89	with screen.
Clean	Ea.	.04	8.65		8.69	
Paint	Ea.	2.96	15.75		18.71	
Minimum Charge	Job		227		227	
2' 6" x 3' 4"						
Demolish	Ea.		16.65		16.65	Cost includes material and labor to
Install	Ea.	243	70		313	install 2' 6" x 3' 4" double hung, two
Demolish and Install	Ea.	243	86.65		329.65	light, double glazed wood window
Reinstall	Ea.		55.88		55.88	with screen.
Clean	Ea.	.07	10.10		10.17	
Paint	Ea.	3.78	18.90		22.68	
Minimum Charge	Job		227		227	

For customer support on your Contractor's Pricing Guide: Residential Repair & Remodeling, call 888.606.7279.

Exterior Doors and Windows

Wood Window	Unit	Material	Labor	Equip.	Total	Specification
2' 6" x 4'						
Demolish	Ea.		16.65		16.65	Cost includes material and labor to
Install	Ea.	253	75.50		328.50	install 2' 6" x 4' double hung, two
Demolish and Install	Ea.	253	92.15		345.15	light, double glazed wood window
Reinstall	Ea.		60.53		60.53	with screen.
Clean	Ea.	.07	10.10		10.17	
Paint	Ea.	3.78	18.90		22.68	
Minimum Charge	Job		227		227	
2' 6" x 4' 8"						
Demolish	Ea.		16.65		16.65	Cost includes material and labor to
Install	Ea.	278	75.50		353.50	install 2' 6" x 4' 8" double hung, two
Demolish and Install	Ea.	278	92.15		370.15	light, double glazed wood window
Reinstall	Ea.		60.53		60.53	with screen.
Clean	Ea.	.07	10.10		10.17	
Paint	Ea.	3.78	18.90		22.68	
Minimum Charge	Job		227		227	
2' 10" x 3' 4"						
Demolish	Ea.		16.65		16.65	Cost includes material and labor to
Install	Ea.	249	91		340	install 2' 10" x 3' 4" double hung, two
Demolish and Install	Ea.	249	107.65		356.65	light, double glazed wood window
Reinstall	Ea.		72.64		72.64	with screen.
Clean	Ea.	.07	10.10		10.17	
Paint	Ea.	3.78	18.90		22.68	
Minimum Charge	Job		227		227	
2' 10" x 4'						
Demolish	Ea.		16.65		16.65	Cost includes material and labor to
Install	Ea.	273	91		364	install 2' 10" x 4' double hung, two
Demolish and Install	Ea.	273	107.65		380.65	light, double glazed wood window
Reinstall	Ea.		72.64		72.64	with screen.
Clean	Ea.	.07	10.10		10.17	
Paint	Ea.	3.78	18.90		22.68	
Minimum Charge	Job		227		227	
3' 7" x 3' 4"						
Demolish	Ea.		16.65		16.65	Cost includes material and labor to
Install	Ea.	282	101		383	install 3' 7" x 3' 4" double hung, two
Demolish and Install	Ea.	282	117.65		399.65	light, double glazed wood window
Reinstall	Ea.		80.71		80.71	with screen.
Clean	Ea.	.07	10.10		10.17	
Paint	Ea.	3.78	18.90		22.68	
Minimum Charge	Job		227		227	
3' 7" x 5' 4"						
Demolish	Ea.		20.50		20.50	Cost includes material and labor to
Install	Ea.	315	101		416	install 3' 7" x 5' 4" double hung, two
Demolish and Install	Ea.	315	121.50		436.50	light, double glazed wood window
Reinstall	Ea.		80.71		80.71	with screen.
Clean	Ea.	.13	17.35		17.48	
Paint	Ea.	6.35	31.50		37.85	
Minimum Charge	Job		227		227	
4' 3" x 5' 4"						
Demolish	Ea.		20.50		20.50	Cost includes material and labor to
Install	Ea.	565	114		679	install 4' 3" x 5' 4" double hung, two
Demolish and Install	Ea.	565	134.50		699.50	light, double glazed wood window
Reinstall	Ea.		90.80		90.80	with screen.
Clean	Ea.	.13	17.35		17.48	
Paint	Ea.	4.60	38		42.60	
Minimum Charge	Job		227		227	

For customer support on your Contractor's Pricing Guide: Residential Repair & Remodeling, call 888.606.7279.

Exterior Doors and Windows

Wood Window

	Unit	Material	Labor	Equip.	Total	Specification
Casement						
2' x 3' 4"						
Demolish	Ea.		16.65		16.65	Cost includes material and labor to
Install	Ea.	290	45.50		335.50	install 2' x 3' 4" casement window
Demolish and Install	Ea.	290	62.15		352.15	with 3/4" insulated glass, screens,
Reinstall	Ea.		36.32		36.32	weatherstripping, hardware.
Clean	Ea.	.04	8.65		8.69	
Paint	Ea.	2.96	15.75		18.71	
Minimum Charge	Job		455		455	
2' x 4'						
Demolish	Ea.		16.65		16.65	Cost includes material and labor to
Install	Ea.	310	50.50		360.50	install 2' x 4' casement window with
Demolish and Install	Ea.	310	67.15		377.15	3/4" insulated glass, screens,
Reinstall	Ea.		40.36		40.36	weatherstripping, hardware.
Clean	Ea.	.04	8.65		8.69	
Paint	Ea.	2.96	15.75		18.71	
Minimum Charge	Job		455		455	
2' x 5'						
Demolish	Ea.		16.65		16.65	Cost includes material and labor to
Install	Ea.	340	53.50		393.50	install 2' x 5' casement window with
Demolish and Install	Ea.	340	70.15		410.15	3/4" insulated glass, screens,
Reinstall	Ea.		42.73		42.73	weatherstripping, hardware.
Clean	Ea.	.07	10.10		10.17	
Paint	Ea.	3.78	18.90		22.68	
Minimum Charge	Job		455		455	
2' x 6'						
Demolish	Ea.		16.65		16.65	Cost includes material and labor to
Install	Ea.	335	57		392	install 2' x 6' casement window with
Demolish and Install	Ea.	335	73.65		408.65	3/4" insulated glass, screens,
Reinstall	Ea.		45.40		45.40	weatherstripping, hardware.
Clean	Ea.	.07	10.10		10.17	
Paint	Ea.	3.78	18.90		22.68	
Minimum Charge	Job		455		455	
4' x 3' 4"						
Demolish	Ea.		20.50		20.50	Cost includes material and labor to
Install	Ea.	650	60.50		710.50	install 4' x 3' 4" casement windows
Demolish and Install	Ea.	650	81		731	with 3/4" insulated glass, screens,
Reinstall	Ea.		48.43		48.43	weatherstripping, hardware.
Clean	Ea.	.07	10.10		10.17	
Paint	Ea.	2.30	18.90		21.20	
Minimum Charge	Job		455		455	
4' x 4'						
Demolish	Ea.		20.50		20.50	Cost includes material and labor to
Install	Ea.	725	60.50		785.50	install 4' x 4' casement windows with
Demolish and Install	Ea.	725	81		806	3/4" insulated glass, screens,
Reinstall	Ea.		48.43		48.43	weatherstripping, hardware.
Clean	Ea.	.09	14.30		14.39	
Paint	Ea.	2.72	23.50		26.22	
Minimum Charge	Job		455		455	
4' x 5'						
Demolish	Ea.		20.50		20.50	Cost includes material and labor to
Install	Ea.	820	65		885	install 4' x 5' casement windows with
Demolish and Install	Ea.	820	85.50		905.50	3/4" insulated glass, screens,
Reinstall	Ea.		51.89		51.89	weatherstripping, hardware.
Clean	Ea.	.13	17.35		17.48	
Paint	Ea.	3.55	31.50		35.05	
Minimum Charge	Job		455		455	

116

Exterior Doors and Windows

Wood Window		Unit	Material	Labor	Equip.	Total	Specification
4' x 6'							
	Demolish	Ea.		20.50		20.50	Cost includes material and labor to
	Install	Ea.	930	75.50		1005.50	install 4' x 6' casement windows with
	Demolish and Install	Ea.	930	96		1026	3/4" insulated glass, screens,
	Reinstall	Ea.		60.53		60.53	weatherstripping, hardware.
	Clean	Ea.	.13	17.35		17.48	
	Paint	Ea.	4.18	38		42.18	
	Minimum Charge	Job		455		455	
Casement Bow							
4' 8" x 8' 1"							
	Demolish	Ea.		28		28	Includes material and labor to install
	Install	Ea.	1750	170		1920	casement bow window unit with 4
	Demolish and Install	Ea.	1750	198		1948	lights, pine frame, 1/2" insulated
	Reinstall	Ea.		136.20		136.20	glass, weatherstripping, hardware,
	Clean	Ea.	.18	20		20.18	screens.
	Paint	Ea.	7.75	75.50		83.25	
	Minimum Charge	Job		455		455	
4' 8" x 10'							
	Demolish	Ea.		28		28	Includes material and labor to install
	Install	Ea.	2125	227		2352	casement bow window unit with 5
	Demolish and Install	Ea.	2125	255		2380	lights, pine frame, 1/2" insulated
	Reinstall	Ea.		181.60		181.60	glass, weatherstripping, hardware,
	Clean	Ea.	.02	27		27.02	screens.
	Paint	Ea.	9.65	94.50		104.15	
	Minimum Charge	Job		455		455	
5' 4" x 8' 1"							
	Demolish	Ea.		28		28	Includes material and labor to install
	Install	Ea.	1875	170		2045	casement bow window unit with 4
	Demolish and Install	Ea.	1875	198		2073	lights, pine frame, 1/2" insulated
	Reinstall	Ea.		136.20		136.20	glass, weatherstripping, hardware,
	Clean	Ea.	.02	27		27.02	screens.
	Paint	Ea.	8.80	75.50		84.30	
	Minimum Charge	Job		455		455	
5' 4" x 10'							
	Demolish	Ea.		28		28	Includes material and labor to install
	Install	Ea.	2125	227		2352	casement bow window unit with 5
	Demolish and Install	Ea.	2125	255		2380	lights, pine frame, 1/2" insulated
	Reinstall	Ea.		181.60		181.60	glass, weatherstripping, hardware,
	Clean	Ea.	.02	27		27.02	screens.
	Paint	Ea.	11.05	94.50		105.55	
	Minimum Charge	Job		455		455	
6' x 8' 1"							
	Demolish	Ea.		28		28	Includes material and labor to install
	Install	Ea.	2125	227		2352	casement bow window unit with 5
	Demolish and Install	Ea.	2125	255		2380	lights, pine frame, 1/2" insulated
	Reinstall	Ea.		181.60		181.60	glass, weatherstripping, hardware,
	Clean	Ea.	.02	27		27.02	screens.
	Paint	Ea.	10	94.50		104.50	
	Minimum Charge	Job		455		455	
6' x 10'							
	Demolish	Ea.		28		28	Includes material and labor to install
	Install	Ea.	2225	227		2452	casement bow window unit with 5
	Demolish and Install	Ea.	2225	255		2480	lights, pine frame, 1/2" insulated
	Reinstall	Ea.		181.60		181.60	glass, weatherstripping, hardware,
	Clean	Ea.	.02	27		27.02	screens.
	Paint	Ea.	12.55	126		138.55	
	Minimum Charge	Job		455		455	

Exterior Doors and Windows

Wood Window		Unit	Material	Labor	Equip.	Total	Specification
Casing							
	Demolish	Ea.		10.50		10.50	Cost includes material and labor to
	Install	Opng.	34.50	35		69.50	install 11/16" x 2-1/2" pine ranch
	Demolish and Install	Opng.	34.50	45.50		80	style window casing.
	Reinstall	Opng.		27.94		27.94	
	Clean	Opng.	.57	3.30		3.87	
	Paint	Ea.	4.73	7.55		12.28	
	Minimum Charge	Job		262		262	
Trim Set							
	Demolish	Ea.		8		8	Cost includes material and labor to
	Install	Opng.	34.50	35		69.50	install 11/16" x 2-1/2" pine ranch
	Demolish and Install	Opng.	34.50	43		77.50	style window casing.
	Reinstall	Opng.		27.94		27.94	
	Clean	Opng.	.57	3.30		3.87	
	Paint	Ea.	4.73	7.55		12.28	
	Minimum Charge	Job		227		227	
Wood Picture Window							
3' 6" x 4'							
	Demolish	Ea.		20.50		20.50	Includes material and labor to install a
	Install	Ea.	480	75.50		555.50	wood framed picture window
	Demolish and Install	Ea.	480	96		576	including exterior trim.
	Clean	Ea.	.13	17.35		17.48	
	Paint	Ea.	3.51	29		32.51	
	Minimum Charge	Job		455		455	
4' x 4' 6"							
	Demolish	Ea.		20.50		20.50	Includes material and labor to install a
	Install	Ea.	585	82.50		667.50	wood framed picture window
	Demolish and Install	Ea.	585	103		688	including exterior trim.
	Clean	Ea.	.13	17.35		17.48	
	Paint	Ea.	3.51	29		32.51	
	Minimum Charge	Job		455		455	
5' x 4'							
	Demolish	Ea.		20.50		20.50	Includes material and labor to install a
	Install	Ea.	665	82.50		747.50	wood framed picture window
	Demolish and Install	Ea.	665	103		768	including exterior trim.
	Clean	Ea.	.13	17.35		17.48	
	Paint	Ea.	7.50	47		54.50	
	Minimum Charge	Job		455		455	
6' x 4' 6"							
	Demolish	Ea.		20.50		20.50	Includes material and labor to install a
	Install	Ea.	705	91		796	wood framed picture window
	Demolish and Install	Ea.	705	111.50		816.50	including exterior trim.
	Clean	Ea.	.13	17.35		17.48	
	Paint	Ea.	7.50	47		54.50	
	Minimum Charge	Job		455		455	

Window Screens		Unit	Material	Labor	Equip.	Total	Specification
Small							
	Demolish	Ea.		3.82		3.82	Includes material and labor to install
	Install	Ea.	21.50	13.75		35.25	small window screen, 4 S.F.
	Demolish and Install	Ea.	21.50	17.57		39.07	
	Reinstall	Ea.		11.01		11.01	
	Clean	Ea.	.41	6.90		7.31	
	Minimum Charge	Job		114		114	

For customer support on your Contractor's Pricing Guide: Residential Repair & Remodeling, call 888.606.7279.

Exterior Doors and Windows

Window Screens	Unit	Material	Labor	Equip.	Total	Specification
Standard Size						
Demolish	Ea.		3.82		3.82	Includes material and labor to install
Install	Ea.	51.50	13.75		65.25	medium window screen, 12 S.F.
Demolish and Install	Ea.	51.50	17.57		69.07	
Reinstall	Ea.		11.01		11.01	
Clean	Ea.	.41	6.90		7.31	
Minimum Charge	Job		114		114	
Large						
Demolish	Ea.		3.82		3.82	Includes material and labor to install
Install	Ea.	69	13.75		82.75	large window screen, 16 S.F.
Demolish and Install	Ea.	69	17.57		86.57	
Reinstall	Ea.		11.01		11.01	
Clean	Ea.	.41	6.90		7.31	
Minimum Charge	Job		114		114	

Window Glass	Unit	Material	Labor	Equip.	Total	Specification
Plate, Clear, 1/4"						
Demolish	S.F.		1.83		1.83	Includes material and labor to install
Install	S.F.	9.90	7.35		17.25	clear glass.
Demolish and Install	S.F.	9.90	9.18		19.08	
Reinstall	S.F.		5.87		5.87	
Clean	S.F.	.03	.19		.22	
Minimum Charge	Job		220		220	
Insulated						
Demolish	S.F.		1.83		1.83	Includes material and labor to install
Install	S.F.	15.90	12.60		28.50	insulated glass.
Demolish and Install	S.F.	15.90	14.43		30.33	
Reinstall	S.F.		10.07		10.07	
Clean	S.F.	.03	.19		.22	
Minimum Charge	Job		220		220	
Tinted						
Demolish	S.F.		1.83		1.83	Includes material and labor to install
Install	S.F.	8.90	6.80		15.70	tinted glass.
Demolish and Install	S.F.	8.90	8.63		17.53	
Reinstall	S.F.		5.42		5.42	
Clean	S.F.	.03	.19		.22	
Minimum Charge	Job		220		220	
Obscure / Florentine						
Demolish	S.F.		1.83		1.83	Includes material and labor to install
Install	S.F.	12.75	6.30		19.05	obscure glass.
Demolish and Install	S.F.	12.75	8.13		20.88	
Reinstall	S.F.		5.03		5.03	
Clean	S.F.	.03	.19		.22	
Minimum Charge	Job		220		220	
Laminated Safety						
Demolish	S.F.		1.83		1.83	Includes material and labor to install
Install	S.F.	13.75	9.80		23.55	clear laminated glass.
Demolish and Install	S.F.	13.75	11.63		25.38	
Reinstall	S.F.		7.83		7.83	
Clean	S.F.	.03	.19		.22	
Minimum Charge	Job		220		220	

For customer support on your Contractor's Pricing Guide: Residential Repair & Remodeling, call 888.606.7279.

Exterior Doors and Windows

Window Glass		Unit	Material	Labor	Equip.	Total	Specification
Wired							
	Demolish	S.F.		1.83		1.83	Includes material and labor to install
	Install	S.F.	26	6.50		32.50	rough obscure wire glass.
	Demolish and Install	S.F.	26	8.33		34.33	
	Reinstall	S.F.		5.22		5.22	
	Clean	S.F.	.03	.19		.22	
	Minimum Charge	Job		220		220	
Tempered							
	Demolish	S.F.		1.83		1.83	Includes material and labor to install
	Install	S.F.	9.90	7.35		17.25	clear glass.
	Demolish and Install	S.F.	9.90	9.18		19.08	
	Reinstall	S.F.		5.87		5.87	
	Clean	S.F.	.03	.19		.22	
	Minimum Charge	Job		220		220	
Stained							
	Demolish	S.F.		1.83		1.83	Includes material and labor to install
	Install	S.F.	67.50	7.35		74.85	stained glass.
	Demolish and Install	S.F.	67.50	9.18		76.68	
	Reinstall	S.F.		5.87		5.87	
	Clean	S.F.	.03	.19		.22	
	Minimum Charge	Job		220		220	
Etched							
	Demolish	S.F.		1.83		1.83	Includes material and labor to install
	Install	S.F.	56	7.35		63.35	etched glass.
	Demolish and Install	S.F.	56	9.18		65.18	
	Reinstall	S.F.		5.87		5.87	
	Clean	S.F.	.03	.19		.22	
	Minimum Charge	Job		220		220	
Wall Mirror							
	Demolish	S.F.		1.83		1.83	Includes material and labor to install
	Install	S.F.	10.45	7.05		17.50	distortion-free float glass mirror, 1/4"
	Demolish and Install	S.F.	10.45	8.88		19.33	thick, cut and polished edges.
	Reinstall	S.F.		5.64		5.64	
	Clean	S.F.	.03	.19		.22	
	Minimum Charge	Job		220		220	
Bullet Resistant							
	Demolish	S.F.		1.83		1.83	Includes material and labor to install
	Install	S.F.	132	88		220	bullet resistant clear laminated glass.
	Demolish and Install	S.F.	132	89.83		221.83	
	Reinstall	S.F.		70.46		70.46	
	Clean	S.F.	.03	.19		.22	
	Minimum Charge	Job		220		220	

Storefront System		Unit	Material	Labor	Equip.	Total	Specification
Clear Glass							
Stub Wall to 8'							
	Demolish	S.F.		2.86		2.86	Includes material and labor to install
	Install	S.F.	25.50	12.10		37.60	commercial grade 4-1/2" section,
	Demolish and Install	S.F.	25.50	14.96		40.46	anodized aluminum with clear glass in
	Reinstall	S.F.		9.69		9.69	upper section and safety glass in
	Clean	S.F.		.40		.40	lower section.
	Minimum Charge	Job		262		262	

Exterior Doors and Windows

Storefront System	Unit	Material	Labor	Equip.	Total	Specification
Floor to 10'						
Demolish	S.F.		2.86		2.86	Includes material and labor to install
Install	S.F.	26	12.10		38.10	commercial grade 4-1/2" section,
Demolish and Install	S.F.	26	14.96		40.96	anodized aluminum with clear glass in
Reinstall	S.F.		9.69		9.69	upper section and safety glass in
Clean	S.F.		.40		.40	lower section.
Minimum Charge	Job		262		262	
Polished Plate Glass						
Stub Wall to 9'						
Demolish	S.F.		2.86		2.86	Cost includes material and labor to
Install	S.F.	16.95	12.10		29.05	install 4' x 9' polished plate glass
Demolish and Install	S.F.	16.95	14.96		31.91	window built on 8" stub wall.
Reinstall	S.F.		9.69		9.69	
Clean	S.F.		.40		.40	
Minimum Charge	Job		262		262	
Stub Wall to 13'						
Demolish	S.F.		2.86		2.86	Cost includes material and labor to
Install	S.F.	18.25	12.10		30.35	install 4' x 13' polished plate glass
Demolish and Install	S.F.	18.25	14.96		33.21	window built on 8" stub wall.
Reinstall	S.F.		9.69		9.69	
Clean	S.F.		.40		.40	
Minimum Charge	Job		262		262	
Floor to 9'						
Demolish	S.F.		2.86		2.86	Cost includes material and labor to
Install	S.F.	18.85	12.10		30.95	install 4' x 9' polished plate glass
Demolish and Install	S.F.	18.85	14.96		33.81	window built on floor.
Reinstall	S.F.		9.69		9.69	
Clean	S.F.		.40		.40	
Minimum Charge	Job		262		262	
Floor to 13'						
Demolish	S.F.		2.86		2.86	Cost includes material and labor to
Install	S.F.	20	12.10		32.10	install 4' x 13' polished plate glass
Demolish and Install	S.F.	20	14.96		34.96	window built on floor.
Reinstall	S.F.		9.69		9.69	
Clean	S.F.		.40		.40	
Minimum Charge	Job		262		262	
Clear Plate Glass						
Floor to 9'						
Demolish	S.F.		2.86		2.86	Cost includes material and labor to
Install	S.F.	27	12.10		39.10	install 4' x 9' clear plate glass window
Demolish and Install	S.F.	27	14.96		41.96	built on floor.
Reinstall	S.F.		9.69		9.69	
Clean	S.F.		.40		.40	
Minimum Charge	Job		262		262	
Floor to 11'						
Demolish	S.F.		2.86		2.86	Cost includes material and labor to
Install	S.F.	26.50	12.10		38.60	install 4' x 11' clear plate glass
Demolish and Install	S.F.	26.50	14.96		41.46	window built on floor.
Reinstall	S.F.		9.69		9.69	
Clean	S.F.		.40		.40	
Minimum Charge	Job		262		262	
Floor to 13'						
Demolish	S.F.		2.86		2.86	Cost includes material and labor to
Install	S.F.	26.50	12.10		38.60	install 4' x 13' clear plate glass
Demolish and Install	S.F.	26.50	14.96		41.46	window built on floor.
Reinstall	S.F.		9.69		9.69	
Clean	S.F.		.40		.40	
Minimum Charge	Job		262		262	

For customer support on your Contractor's Pricing Guide: Residential Repair & Remodeling, call 888.606.7279.

Exterior Doors and Windows

Storefront System		Unit	Material	Labor	Equip.	Total	Specification
Entrance							
Narrow Stile 3' x 7"							
	Demolish	Ea.		25		25	Cost includes material and labor to
	Install	Ea.	1025	305		1330	install 3' x 7' narrow stile door, center
	Demolish and Install	Ea.	1025	330		1355	pivot concealed closer.
	Reinstall	Ea.		242.13		242.13	
	Clean	Ea.	5.10	41.50		46.60	
	Minimum Charge	Job		262		262	
Tempered Glass							
	Demolish	Ea.		25		25	Cost includes material and labor to
	Install	Ea.	3525	565		4090	install 3' x 7' tempered glass door,
	Demolish and Install	Ea.	3525	590		4115	center pivot concealed closer.
	Reinstall	Ea.		452.50		452.50	
	Clean	Ea.	5.10	41.50		46.60	
	Minimum Charge	Job		262		262	
Heavy Section							
	Demolish	Ea.		25		25	Includes material and labor to install
	Install	Ea.	1375	525		1900	wide stile door with 3' 6" x 7'
	Demolish and Install	Ea.	1375	550		1925	opening.
	Reinstall	Ea.		418.56		418.56	
	Clean	Ea.	1.02	13.75		14.77	
	Minimum Charge	Job		880		880	

Curtain Wall System		Unit	Material	Labor	Equip.	Total	Specification
Regular Weight							
Bronze Anodized							
	Demolish	S.F.		2.86		2.86	Includes material and labor to install
	Install	S.F.	42	12.10		54.10	structural curtain wall system with
	Demolish and Install	S.F.	42	14.96		56.96	bronze anodized frame of regular
	Clean	S.F.		.40		.40	weight, float glass.
	Minimum Charge	Job		227		227	
Black Anodized							
	Demolish	S.F.		2.86		2.86	Includes material and labor to install
	Install	S.F.	45	12.10		57.10	structural curtain wall system with
	Demolish and Install	S.F.	45	14.96		59.96	black anodized frame of regular
	Clean	S.F.		.40		.40	weight, float glass.
	Minimum Charge	Job		227		227	
Heavy Weight							
Bronze Anodized							
	Demolish	S.F.		2.86		2.86	Includes material and labor to install
	Install	S.F.	42	12.10		54.10	structural curtain wall system with
	Demolish and Install	S.F.	42	14.96		56.96	bronze anodized frame of heavy
	Clean	S.F.		.40		.40	weight, float glass.
	Minimum Charge	Job		227		227	
Black Anodized							
	Demolish	S.F.		2.86		2.86	Includes material and labor to install
	Install	S.F.	43.50	12.10		55.60	structural curtain wall system with
	Demolish and Install	S.F.	43.50	14.96		58.46	black anodized frame of heavy
	Clean	S.F.		.40		.40	weight, float glass.
	Minimum Charge	Job		227		227	
Glass							
1" Insulated							
	Install	S.F.	18.85	11.75		30.60	Cost includes labor and material to
	Minimum Charge	Job		220		220	install 1" insulated glass.

For customer support on your Contractor's Pricing Guide: Residential Repair & Remodeling, call 888.606.7279.

Exterior Doors and Windows

Skylight

	Unit	Material	Labor	Equip.	Total	Specification
Dome						
22" x 22"						
Demolish	Ea.		89		89	Cost includes material and labor to
Install	Ea.	238	139		377	install 22" x 22" fixed dome skylight.
Demolish and Install	Ea.	238	228		466	
Reinstall	Ea.		111.25		111.25	
Clean	Ea.	.46	11.80		12.26	
Minimum Charge	Job		835		835	
22" x 46"						
Demolish	Ea.		89		89	Cost includes material and labor to
Install	Ea.	296	167		463	install 22" x 46" fixed dome skylight.
Demolish and Install	Ea.	296	256		552	
Reinstall	Ea.		133.50		133.50	
Clean	Ea.		18.35		18.35	
Minimum Charge	Job		835		835	
30" x 30"						
Demolish	Ea.		89		89	Cost includes material and labor to
Install	Ea.	315	139		454	install 30" x 30" fixed dome skylight.
Demolish and Install	Ea.	315	228		543	
Reinstall	Ea.		111.25		111.25	
Clean	Ea.		18.35		18.35	
Minimum Charge	Job		835		835	
30" x 46"						
Demolish	Ea.		89		89	Cost includes material and labor to
Install	Ea.	420	167		587	install 30" x 46" fixed dome skylight.
Demolish and Install	Ea.	420	256		676	
Reinstall	Ea.		133.50		133.50	
Clean	Ea.		18.35		18.35	
Minimum Charge	Job		835		835	
Fixed						
22" x 27"						
Demolish	Ea.		89		89	Includes material and labor to install
Install	Ea.	305	139		444	double glazed skylight, 22" x 27",
Demolish and Install	Ea.	305	228		533	fixed.
Reinstall	Ea.		111.25		111.25	
Clean	Ea.	.46	11.80		12.26	
Minimum Charge	Job		835		835	
22" x 46"						
Demolish	Ea.		89		89	Includes material and labor to install
Install	Ea.	320	167		487	double glazed skylight, 22" x 46",
Demolish and Install	Ea.	320	256		576	fixed.
Reinstall	Ea.		133.50		133.50	
Clean	Ea.		18.35		18.35	
Minimum Charge	Job		835		835	
44" x 46"						
Demolish	Ea.		89		89	Includes material and labor to install
Install	Ea.	475	167		642	double glazed skylight, 44" x 46",
Demolish and Install	Ea.	475	256		731	fixed.
Reinstall	Ea.		133.50		133.50	
Clean	Ea.		20.50		20.50	
Minimum Charge	Job		835		835	
Operable						
22" x 27"						
Demolish	Ea.		89		89	Includes material and labor to install
Install	Ea.	445	139		584	double glazed skylight, 22" x 27",
Demolish and Install	Ea.	445	228		673	operable.
Reinstall	Ea.		111.25		111.25	
Clean	Ea.	.46	11.80		12.26	
Minimum Charge	Job		835		835	

Exterior Doors and Windows

Skylight

	Unit	Material	Labor	Equip.	Total	Specification
22" x 46"						
Demolish	Ea.		89		89	Includes material and labor to install
Install	Ea.	510	167		677	double glazed skylight, 22" x 46",
Demolish and Install	Ea.	510	256		766	operable.
Reinstall	Ea.		133.50		133.50	
Clean	Ea.		18.35		18.35	
Minimum Charge	Job		835		835	
44" x 46"						
Demolish	Ea.		89		89	Includes material and labor to install
Install	Ea.	995	167		1162	double glazed skylight, 44" x 46",
Demolish and Install	Ea.	995	256		1251	operable.
Reinstall	Ea.		133.50		133.50	
Clean	Ea.		20.50		20.50	
Minimum Charge	Job		835		835	

Industrial Steel Window

	Unit	Material	Labor	Equip.	Total	Specification
100% Fixed						
Demolish	S.F.		2.35		2.35	Includes material and labor to install
Install	S.F.	32.50	5.25		37.75	commercial grade steel windows,
Demolish and Install	S.F.	32.50	7.60		40.10	including glazing.
Reinstall	S.F.		4.19		4.19	
Clean	Ea.	.19	1.26		1.45	
Paint	S.F.	.17	.84		1.01	
Minimum Charge	Job		262		262	
50% Vented						
Demolish	S.F.		2.35		2.35	Includes material and labor to install
Install	S.F.	60.50	5.25		65.75	commercial grade steel windows,
Demolish and Install	S.F.	60.50	7.60		68.10	including glazing.
Reinstall	S.F.		4.19		4.19	
Clean	Ea.	.19	1.26		1.45	
Paint	S.F.	.17	.84		1.01	
Minimum Charge	Job		262		262	

Brick Veneer

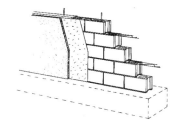

Concrete Masonry Units

Masonry Wall	Unit	Material	Labor	Equip.	Total	Specification
Unreinforced CMU						
4″ x 8″ x 16″						
Demolish	S.F.		.13	.31	.44	Includes material and labor to install
Install	S.F.	2.02	4.36		6.38	normal weight 4″ x 8″ x 16″ block.
Demolish and Install	S.F.	2.02	4.49	.31	6.82	
Clean	S.F.	.09	.51		.60	
Paint	S.F.	.17	.86		1.03	
Minimum Charge	Job		216		216	
6″ x 8″ x 16″						
Demolish	S.F.		.18	.42	.60	Includes material and labor to install
Install	S.F.	2.64	4.56		7.20	normal weight 6″ x 8″ x 16″ block.
Demolish and Install	S.F.	2.64	4.74	.42	7.80	
Clean	S.F.	.09	.51		.60	
Paint	S.F.	.17	.86		1.03	
Minimum Charge	Job		216		216	
8″ x 8″ x 16″						
Demolish	S.F.		.35	.84	1.19	Includes material and labor to install
Install	S.F.	2.82	5		7.82	normal weight 8″ x 8″ x 16″ block.
Demolish and Install	S.F.	2.82	5.35	.84	9.01	
Clean	S.F.	.09	.51		.60	
Paint	S.F.	.17	.86		1.03	
Minimum Charge	Job		216		216	
12″ x 8″ x 16″						
Demolish	S.F.		.35	.84	1.19	Includes material and labor to install
Install	S.F.	4.59	7.60		12.19	normal weight 12″ x 8″ x 16″ block.
Demolish and Install	S.F.	4.59	7.95	.84	13.38	
Clean	S.F.	.09	.51		.60	
Paint	S.F.	.17	.86		1.03	
Minimum Charge	Job		216		216	
Reinforced CMU						
4″ x 8″ x 16″						
Demolish	S.F.		2.04	1.14	3.18	Includes material and labor to install
Install	S.F.	2.20	4.46		6.66	normal weight 4″ x 8″ x 16″ block,
Demolish and Install	S.F.	2.20	6.50	1.14	9.84	mortar and 1/2″ re-bar installed
Clean	S.F.	.09	.51		.60	vertically every 24″ O.C.
Paint	S.F.	.17	.86		1.03	
Minimum Charge	Job		395		395	
6″ x 8″ x 16″						
Demolish	S.F.		2.14	.58	2.72	Includes material and labor to install
Install	S.F.	2.82	4.66		7.48	normal weight 6″ x 8″ x 16″ block,
Demolish and Install	S.F.	2.82	6.80	.58	10.20	mortar and 1/2″ re-bar installed
Clean	S.F.	.09	.51		.60	vertically every 24″ O.C.
Paint	S.F.	.17	.86		1.03	
Minimum Charge	Job		395		395	

Masonry

Masonry Wall		Unit	Material	Labor	Equip.	Total	Specification
8″ x 8″ x 16″							
	Demolish	S.F.		2.26	.61	2.87	Includes material and labor to install
	Install	S.F.	3.04	5.10		8.14	normal weight 8″ x 8″ x 16″ block,
	Demolish and Install	S.F.	3.04	7.36	.61	11.01	mortar and 1/2″ re-bar installed
	Clean	S.F.	.09	.51		.60	vertically every 24″ O.C.
	Paint	S.F.	.17	.86		1.03	
	Minimum Charge	Job		395		395	
12″ x 8″ x 16″							
	Demolish	S.F.		2.34	.63	2.97	Includes material and labor to install
	Install	S.F.	4.77	7.85		12.62	normal weight 12″ x 8″ x 16″ block,
	Demolish and Install	S.F.	4.77	10.19	.63	15.59	mortar and 1/2″ re-bar installed
	Clean	S.F.	.09	.51		.60	vertically every 24″ O.C.
	Paint	S.F.	.17	.86		1.03	
	Minimum Charge	Job		395		395	

Structural Brick
4″ Single Row

		Unit	Material	Labor	Equip.	Total	Specification
	Demolish	S.F.		2.93		2.93	Cost includes material and labor to
	Install	S.F.	4.25	9.35		13.60	install 4″ red brick wall, running bond
	Demolish and Install	S.F.	4.25	12.28		16.53	with 3/8″ concave joints, mortar and
	Clean	S.F.	.09	.43		.52	ties.
	Paint	S.F.	.17	.86		1.03	
	Minimum Charge	Job		395		395	

8″ Double Row

		Unit	Material	Labor	Equip.	Total	Specification
	Demolish	S.F.		2.93		2.93	Cost includes material and labor to
	Install	S.F.	9.40	14.85		24.25	install 8″ red brick wall, running bond,
	Demolish and Install	S.F.	9.40	17.78		27.18	with 3/8″ concave joints, mortar and
	Clean	S.F.	.09	.43		.52	wall ties.
	Paint	S.F.	.17	.86		1.03	
	Minimum Charge	Job		395		395	

12″ Triple Row

		Unit	Material	Labor	Equip.	Total	Specification
	Demolish	S.F.		4.07		4.07	Cost includes material and labor to
	Install	S.F.	14.15	21		35.15	install 12″ red brick wall, running
	Demolish and Install	S.F.	14.15	25.07		39.22	bond, with 3/8″ concave joints,
	Clean	S.F.	.09	.43		.52	mortar and wall ties.
	Paint	S.F.	.17	.86		1.03	
	Minimum Charge	Job		395		395	

Brick Veneer
4″ Red Common

		Unit	Material	Labor	Equip.	Total	Specification
	Demolish	S.F.		2.93		2.93	Cost includes material and labor to
	Install	S.F.	4.42	8.70		13.12	install 4″ common red face brick,
	Demolish and Install	S.F.	4.42	11.63		16.05	running bond with 3/8″ concave
	Clean	S.F.	.09	.43		.52	joints, mortar and wall ties.
	Paint	S.F.	.17	.86		1.03	
	Minimum Charge	Job		216		216	

4″ Used

		Unit	Material	Labor	Equip.	Total	Specification
	Demolish	S.F.		2.93		2.93	Cost includes material and labor to
	Install	S.F.	4.25	9.35		13.60	install 4″ red brick wall, running bond
	Demolish and Install	S.F.	4.25	12.28		16.53	with 3/8″ concave joints, mortar and
	Clean	S.F.	.09	.43		.52	ties.
	Paint	S.F.	.17	.86		1.03	
	Minimum Charge	Job		216		216	

Masonry

Masonry Wall		Unit	Material	Labor	Equip.	Total	Specification
6" Jumbo							
	Demolish	S.F.		2.93		2.93	Cost includes material and labor to
	Install	S.F.	5.35	4.61		9.96	install 6" red jumbo brick veneer,
	Demolish and Install	S.F.	5.35	7.54		12.89	running bond with 3/8" concave
	Clean	S.F.	.09	.43		.52	joints, mortar and ties.
	Paint	S.F.	.17	.86		1.03	
	Minimum Charge	Job		216		216	
Norman (11-1/2")							
	Demolish	S.F.		2.93		2.93	Cost includes material and labor to
	Install	S.F.	6.65	6.25		12.90	install 4" norman red brick wall,
	Demolish and Install	S.F.	6.65	9.18		15.83	running bond with 3/8" concave
	Clean	S.F.	.09	.43		.52	joints, mortar and ties.
	Paint	S.F.	.17	.86		1.03	
	Minimum Charge	Job		216		216	
Glazed							
	Demolish	S.F.		2.93		2.93	Includes material, labor and
	Install	S.F.	19.75	9.55		29.30	equipment to install glazed brick.
	Demolish and Install	S.F.	19.75	12.48		32.23	
	Clean	S.F.	.09	.43		.52	
	Paint	S.F.	.17	.86		1.03	
	Minimum Charge	Job		216		216	
Slumpstone							
6" x 4"							
	Demolish	S.F.		.13	.31	.44	Cost includes material and labor to
	Install	S.F.	6.90	4.92		11.82	install 6" x 4" x 16" concrete slump
	Demolish and Install	S.F.	6.90	5.05	.31	12.26	block.
	Clean	S.F.	.09	.51		.60	
	Paint	S.F.	.17	.86		1.03	
	Minimum Charge	Job		216		216	
6" x 6"							
	Demolish	S.F.		.18	.42	.60	Cost includes material and labor to
	Install	S.F.	6.40	5.10		11.50	install 6" x 6" x 16" concrete slump
	Demolish and Install	S.F.	6.40	5.28	.42	12.10	block.
	Clean	S.F.	.09	.51		.60	
	Paint	S.F.	.17	.86		1.03	
	Minimum Charge	Job		216		216	
8" x 6"							
	Demolish	S.F.		.18	.42	.60	Cost includes material and labor to
	Install	S.F.	9.65	5.25		14.90	install 8" x 6" x 16" concrete slump
	Demolish and Install	S.F.	9.65	5.43	.42	15.50	block.
	Clean	S.F.	.09	.51		.60	
	Paint	S.F.	.17	.86		1.03	
	Minimum Charge	Job		216		216	
12" x 4"							
	Demolish	S.F.		.13	.31	.44	Cost includes material and labor to
	Install	S.F.	14.40	6.05		20.45	install 12" x 4" x 16" concrete slump
	Demolish and Install	S.F.	14.40	6.18	.31	20.89	block.
	Clean	S.F.	.09	.51		.60	
	Paint	S.F.	.17	.86		1.03	
	Minimum Charge	Job		216		216	
12" x 6"							
	Demolish	S.F.		.18	.42	.60	Cost includes material and labor to
	Install	S.F.	15.45	6.55		22	install 12" x 6" x 16" concrete slump
	Demolish and Install	S.F.	15.45	6.73	.42	22.60	block.
	Clean	S.F.	.09	.51		.60	
	Paint	S.F.	.17	.86		1.03	
	Minimum Charge	Job		216		216	

For customer support on your Contractor's Pricing Guide: Residential Repair & Remodeling, call 888.606.7279.

Masonry Wall		Unit	Material	Labor	Equip.	Total	Specification
Stucco							
Three Coat							
	Demolish	S.Y.		2.53		2.53	Includes material and labor to install
	Install	S.Y.	4.02	9.65	.77	14.44	stucco on exterior masonry walls.
	Demolish and Install	S.Y.	4.02	12.18	.77	16.97	
	Clean	S.Y.		4.58		4.58	
	Paint	S.F.	.22	.62		.84	
	Minimum Charge	Job		206		206	
Three Coat w / Wire Mesh							
	Demolish	S.Y.		4.58		4.58	Includes material and labor to install
	Install	S.Y.	7.05	37.50	2.45	47	stucco on wire mesh over wood
	Demolish and Install	S.Y.	7.05	42.08	2.45	51.58	framing.
	Clean	S.Y.		4.58		4.58	
	Paint	S.F.	.22	.62		.84	
	Minimum Charge	Job		206		206	
Fill Cracks							
	Install	S.F.	.45	3.29		3.74	Includes labor and material to fill in
	Minimum Charge	Job		206		206	hairline cracks up to 1/8" with filler,
							including sand and fill per S.F.
Furring							
1" x 2"							
	Demolish	L.F.		.15		.15	Cost includes material and labor to
	Install	L.F.	.29	.92		1.21	install 1" x 2" furring on masonry wall.
	Demolish and Install	L.F.	.29	1.07		1.36	
	Minimum Charge	Job		227		227	
2" x 2"							
	Demolish	L.F.		.16		.16	Cost includes material and labor to
	Install	L.F.	.30	1.01		1.31	install 2" x 2" furring on masonry wall.
	Demolish and Install	L.F.	.30	1.17		1.47	
	Minimum Charge	Job		227		227	
2" x 4"							
	Demolish	L.F.		.17		.17	Cost includes material and labor to
	Install	L.F.	.53	1.15		1.68	install 2" x 4" furring on masonry wall.
	Demolish and Install	L.F.	.53	1.32		1.85	
	Reinstall	L.F.		.92		.92	
	Minimum Charge	Job		227		227	
Maintenance / Cleaning							
Repoint							
	Install	S.F.	.64	5.40		6.04	Includes labor and material to cut and
	Minimum Charge	Hr.		54		54	repoint brick, hard mortar, running
							bond.
Acid Etch							
	Install	S.F.	.06	.77		.83	Includes labor and material to acid
	Minimum Charge	Job		165	45	210	wash smooth brick.
Pressure Wash							
	Clean	S.F.		.88	.21	1.09	Includes labor and material to high
	Minimum Charge	Job		165	45	210	pressure water wash masonry,
							average.
Waterproof							
	Install	S.F.	.40	.85		1.25	Includes labor and material to install
	Minimum Charge	Job		212		212	masonry waterproofing.

Chimney

Chimney		Unit	Material	Labor	Equip.	Total	Specification
Demo & Haul Exterior							
	Demolish	C.F.		6		6	Includes minimum labor and
	Minimum Charge	Job		183		183	equipment to remove masonry and haul debris to truck or dumpster.
Demo & Haul Interior							
	Demolish	C.F.		8.25		8.25	Includes minimum labor and
	Minimum Charge	Job		183		183	equipment to remove masonry and haul debris to truck or dumpster.
Install New (w / flue)							
Brick							
	Install	V.L.F.	43	49		92	Includes material and labor to install
	Minimum Charge	Job		216		216	brick chimney with flue. Foundation not included.
Natural Stone							
	Install	L.F.	75.50	78.50		154	Includes labor and material to install
	Minimum Charge	Job		216		216	natural stone chimney.
Install From Roof Up							
Brick							
	Install	V.L.F.	43	49		92	Includes material and labor to install
	Minimum Charge	Job		216		216	brick chimney with flue. Foundation not included.
Natural Stone							
	Install	L.F.	75.50	65.50		141	Includes labor and material to install
	Minimum Charge	Job		216		216	natural stone chimney.
Install Base Pad							
	Install	Ea.	279	246	1.45	526.45	Includes labor and material to replace
	Minimum Charge	Job		1050	35	1085	spread footing, 5' square x 16" deep.
Metal Cap							
	Install	Ea.	59.50	27		86.50	Includes labor and materials to install
	Minimum Charge	Job		227		227	bird and squirrel screens in a chimney flue.

Fireplace

Fireplace		Unit	Material	Labor	Equip.	Total	Specification
Common Brick							
30"							
	Demolish	Ea.		1300		1300	Includes material and labor to install a
	Install	Ea.	2100	3925		6025	complete unit which includes common
	Demolish and Install	Ea.	2100	5225		7325	brick, foundation, damper, flue lining
	Minimum Charge	Job		216		216	stack to 15' high and a 30" fire box.
36"							
	Demolish	Ea.		1300		1300	Includes material and labor to install a
	Install	Ea.	2225	4375		6600	complete unit which includes common
	Demolish and Install	Ea.	2225	5675		7900	brick, foundation, damper, flue lining
	Minimum Charge	Job		216		216	stack to 15' high and a 36" fire box.
42"							
	Demolish	Ea.		1300		1300	Includes material and labor to install a
	Install	Ea.	2400	4925		7325	complete unit which includes common
	Demolish and Install	Ea.	2400	6225		8625	brick, foundation, damper, flue lining
	Minimum Charge	Job		216		216	stack to 15' high and a 42" fire box.

Masonry

Fireplace	Unit	Material	Labor	Equip.	Total	Specification
48"						
Demolish	Ea.		1300		1300	Includes material and labor to install a
Install	Ea.	2525	5600		8125	complete unit which includes common
Demolish and Install	Ea.	2525	6900		9425	brick, foundation, damper, flue lining
Minimum Charge	Job		216		216	stack to 15' high and a 48" fire box.
Raised Hearth						
Brick Hearth						
Demolish	S.F.		5.75		5.75	Includes material and labor to install a
Install	S.F.	19.45	33		52.45	brick fireplace hearth based on 30" x
Demolish and Install	S.F.	19.45	38.75		58.20	29" opening.
Clean	S.F.	.09	.43		.52	
Minimum Charge	Job		216		216	
Marble Hearth / Facing						
Demolish	Ea.		36.50		36.50	Includes material and labor to install
Install	Ea.	385	132		517	marble fireplace hearth.
Demolish and Install	Ea.	385	168.50		553.50	
Reinstall	Ea.		105.95		105.95	
Clean	Ea.	44.50	6.60		51.10	
Minimum Charge	Job		216		216	
Breast (face) Brick						
Demolish	S.F.		5.75		5.75	Includes material and labor to install
Install	S.F.	4.32	9.10		13.42	standard red brick, running bond.
Demolish and Install	S.F.	4.32	14.85		19.17	
Clean	S.F.		.88	.21	1.09	
Minimum Charge	Hr.		54		54	
Fire Box						
Demolish	Ea.		293		293	Includes material and labor to install
Install	Ea.	178	395		573	fireplace box only (110 brick).
Demolish and Install	Ea.	178	688		866	
Clean	Ea.	.15	15.20		15.35	
Minimum Charge	Hr.		54		54	
Natural Stone						
Demolish	S.F.		5.75		5.75	Includes material and labor to install
Install	S.F.	19	16.70		35.70	high priced stone, split or rock face.
Demolish and Install	S.F.	19	22.45		41.45	
Clean	S.F.	.10	.98		1.08	
Minimum Charge	Job		395		395	
Slumpstone Veneer						
Demolish	S.F.		.18	.42	.60	Cost includes material and labor to
Install	S.F.	9.65	5.25		14.90	install 8" x 6" x 16" concrete slump
Demolish and Install	S.F.	9.65	5.43	.42	15.50	block.
Clean	S.F.	.09	.51		.60	
Paint	S.F.	.17	.86		1.03	
Minimum Charge	Job		216		216	
Prefabricated						
Firebox						
Demolish	Ea.		52.50		52.50	Includes material and labor to install
Install	Ea.	2400	415		2815	up to 43" radiant heat, zero
Demolish and Install	Ea.	2400	467.50		2867.50	clearance, prefabricated fireplace.
Reinstall	Ea.		330.18		330.18	Flues, doors or blowers not included.
Clean	Ea.	2.04	20.50		22.54	
Minimum Charge	Job		227		227	

Fireplace		Unit	Material	Labor	Equip.	Total	Specification
Flue							
	Demolish	V.L.F.		6.10		6.10	Includes material and labor to install
	Install	V.L.F.	25	17.45		42.45	prefabricated chimney vent, 8"
	Demolish and Install	V.L.F.	25	23.55		48.55	exposed metal flue.
	Reinstall	V.L.F.		13.97		13.97	
	Clean	L.F.	.81	4.13		4.94	
	Minimum Charge	Job		330		330	
Blower							
	Demolish	Ea.		4.89		4.89	Includes material and labor to install
	Install	Ea.	154	31.50		185.50	blower for metal fireplace flue.
	Demolish and Install	Ea.	154	36.39		190.39	
	Reinstall	Ea.		25.22		25.22	
	Clean	Ea.	4.07	41.50		45.57	
	Minimum Charge	Job		330		330	
Accessories							
Wood Mantle							
	Demolish	L.F.		6.10		6.10	Includes material and labor to install
	Install	L.F.	11.90	12.95		24.85	wood fireplace mantel.
	Demolish and Install	L.F.	11.90	19.05		30.95	
	Reinstall	L.F.		12.97		12.97	
	Clean	S.F.		.41		.41	
	Paint	L.F.	.13	10.25		10.38	
	Minimum Charge	Job		227		227	
Colonial Style							
	Demolish	Ea.		56		56	Includes material and labor to install
	Install	Opng.	1650	227		1877	wood fireplace mantel.
	Demolish and Install	Opng.	1650	283		1933	
	Reinstall	Opng.		227		227	
	Clean	Ea.		.67		.67	
	Paint	Ea.	19.05	141		160.05	
	Minimum Charge	Job		227		227	
Modern Design							
	Demolish	Ea.		56		56	Includes material and labor to install
	Install	Opng.	650	91		741	wood fireplace mantel.
	Demolish and Install	Opng.	650	147		797	
	Reinstall	Opng.		90.80		90.80	
	Clean	Ea.		.67		.67	
	Paint	Ea.	19.05	141		160.05	
	Minimum Charge	Job		227		227	
Glass Door Unit							
	Demolish	Ea.		7.65		7.65	Includes material and labor to install a
	Install	Ea.	330	9		339	fireplace glass door unit.
	Demolish and Install	Ea.	330	16.65		346.65	
	Reinstall	Ea.		9		9	
	Clean	Ea.	.48	12.70		13.18	
	Minimum Charge	Job		227		227	
Fire Screen							
	Demolish	Ea.		3.82		3.82	Includes material and labor to install a
	Install	Ea.	130	4.50		134.50	fireplace fire screen.
	Demolish and Install	Ea.	130	8.32		138.32	
	Reinstall	Ea.		4.50		4.50	
	Clean	Ea.	.24	11		11.24	
	Minimum Charge	Job		227		227	

For customer support on your Contractor's Pricing Guide: Residential Repair & Remodeling, call 888.606.7279.

Siding

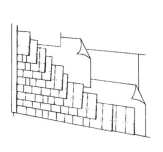

Shingles

Board & Batten

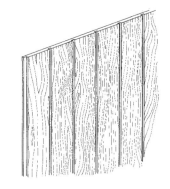

Texture 1-11

Exterior Siding		Unit	Material	Labor	Equip.	Total	Specification
Trim							
1″ x 2″							
	Demolish	L.F.		.31		.31	Cost includes material and labor to
	Install	L.F.	.28	1.38		1.66	install 1″ x 2″ pine trim.
	Demolish and Install	L.F.	.28	1.69		1.97	
	Clean	L.F.	.01	.22		.23	
	Paint	L.F.	.02	.59		.61	
	Minimum Charge	Job		227		227	
1″ x 3″							
	Demolish	L.F.		.31		.31	Cost includes material and labor to
	Install	L.F.	.30	1.57		1.87	install 1″ x 3″ pine trim.
	Demolish and Install	L.F.	.30	1.88		2.18	
	Clean	L.F.	.01	.22		.23	
	Paint	L.F.	.02	.59		.61	
	Minimum Charge	Job		227		227	
1″ x 4″							
	Demolish	L.F.		.31		.31	Cost includes material and labor to
	Install	L.F.	.54	1.82		2.36	install 1″ x 4″ pine trim.
	Demolish and Install	L.F.	.54	2.13		2.67	
	Clean	L.F.	.01	.22		.23	
	Paint	L.F.	.02	.59		.61	
	Minimum Charge	Job		227		227	
1″ x 6″							
	Demolish	L.F.		.31		.31	Cost includes material and labor to
	Install	L.F.	.80	1.82		2.62	install 1″ x 6″ pine trim.
	Demolish and Install	L.F.	.80	2.13		2.93	
	Clean	L.F.	.02	.22		.24	
	Paint	L.F.	.09	.59		.68	
	Minimum Charge	Job		227		227	
1″ x 8″							
	Demolish	L.F.		.33		.33	Cost includes material and labor to
	Install	L.F.	1.32	2.27		3.59	install 1″ x 8″ pine trim.
	Demolish and Install	L.F.	1.32	2.60		3.92	
	Clean	L.F.	.03	.24		.27	
	Paint	L.F.	.09	.59		.68	
	Minimum Charge	Job		227		227	
Aluminum							
8″ Non-Insulated							
	Demolish	S.F.		.83		.83	Includes material and labor to install
	Install	S.F.	2.72	1.80		4.52	horizontal, colored clapboard, 8″ to
	Demolish and Install	S.F.	2.72	2.63		5.35	10″ wide, plain.
	Reinstall	S.F.		1.80		1.80	
	Clean	S.F.		.33		.33	
	Minimum Charge	Job		227		227	

Siding

Exterior Siding	Unit	Material	Labor	Equip.	Total	Specification
8″ Insulated						
Demolish	S.F.		.83		.83	Includes material and labor to install
Install	S.F.	3.14	1.80		4.94	horizontal, colored insulated
Demolish and Install	S.F.	3.14	2.63		5.77	clapboard, 8″ wide, plain.
Reinstall	S.F.		1.80		1.80	
Clean	S.F.		.33		.33	
Minimum Charge	Job		227		227	
12″ Non-Insulated						
Demolish	S.F.		.83		.83	Includes material and labor to install
Install	S.F.	2.92	1.55		4.47	horizontal, colored clapboard, plain.
Demolish and Install	S.F.	2.92	2.38		5.30	
Reinstall	S.F.		1.55		1.55	
Clean	S.F.		.33		.33	
Minimum Charge	Job		227		227	
12″ Insulated						
Demolish	S.F.		.83		.83	Includes material and labor to install
Install	S.F.	3.14	1.55		4.69	horizontal, colored insulated
Demolish and Install	S.F.	3.14	2.38		5.52	clapboard, plain.
Reinstall	S.F.		1.55		1.55	
Clean	S.F.		.33		.33	
Minimum Charge	Job		227		227	
Clapboard 8″ to 10″						
Demolish	S.F.		.83		.83	Includes material and labor to install
Install	S.F.	2.72	1.80		4.52	horizontal, colored clapboard, 8″ to
Demolish and Install	S.F.	2.72	2.63		5.35	10″ wide, plain.
Reinstall	S.F.		1.80		1.80	
Clean	S.F.		.33		.33	
Minimum Charge	Job		227		227	
Corner Strips						
Demolish	L.F.		.43		.43	Includes material and labor to install
Install	V.L.F.	3.96	1.80		5.76	corners for horizontal lapped
Demolish and Install	V.L.F.	3.96	2.23		6.19	aluminum siding.
Reinstall	V.L.F.		1.80		1.80	
Clean	L.F.		.22		.22	
Minimum Charge	Job		227		227	
Vinyl						
Non-Insulated						
Demolish	S.F.		.65		.65	Cost includes material and labor to
Install	S.F.	1.78	1.65		3.43	install 10″ non-insulated solid vinyl
Demolish and Install	S.F.	1.78	2.30		4.08	siding.
Reinstall	S.F.		1.32		1.32	
Clean	S.F.		.33		.33	
Minimum Charge	Job		227		227	
Insulated						
Demolish	S.F.		.65		.65	Cost includes material and labor to
Install	S.F.	3.42	2.27		5.69	install insulated solid vinyl siding
Demolish and Install	S.F.	3.42	2.92		6.34	panels.
Reinstall	S.F.		1.82		1.82	
Clean	S.F.		.33		.33	
Minimum Charge	Job		227		227	
Corner Strips						
Demolish	L.F.		.41		.41	Cost includes material and labor to
Install	L.F.	2.29	1.30		3.59	install corner posts for vinyl siding
Demolish and Install	L.F.	2.29	1.71		4	panels.
Reinstall	L.F.		1.04		1.04	
Clean	L.F.		.22		.22	
Minimum Charge	Job		227		227	

Siding

Exterior Siding	Unit	Material	Labor	Equip.	Total	Specification
Fiber Cement Siding						
4' x 8' Panel						
Demolish	S.F.		.68		.68	Cost includes material and labor to
Install	S.F.	1.17	.99		2.16	install 4' x 8' fiber cement 7/16"
Demolish and Install	S.F.	1.17	1.67		2.84	thick.
Clean	S.F.		.33		.33	
Paint	S.F.	.33	1.27		1.60	
Minimum Charge	Job		227		227	
Lap						
Demolish	S.F.		.67		.67	Cost includes material and labor to
Install	S.F.	1.15	1.80		2.95	install 8" or 12" fiber cement lap
Demolish and Install	S.F.	1.15	2.47		3.62	siding 7/16" thick with wood grain
Clean	S.F.		.33		.33	finish.
Paint	S.F.	.33	1.27		1.60	
Minimum Charge	Job		227		227	
Plywood Panel						
Douglas Fir (T-1-11)						
Demolish	S.F.		.51		.51	Includes material and labor to install
Install	S.F.	1.61	1.37		2.98	Douglas fir 4' x 8' plywood exterior
Demolish and Install	S.F.	1.61	1.88		3.49	siding plain or patterns.
Reinstall	S.F.		1.37		1.37	
Clean	S.F.		.33		.33	
Paint	S.F.	.33	1.27		1.60	
Minimum Charge	Job		227		227	
Redwood						
Demolish	S.F.		.51		.51	Includes material and labor to install
Install	S.F.	2.11	1.37		3.48	redwood 4' x 8' plywood exterior
Demolish and Install	S.F.	2.11	1.88		3.99	siding plain or patterns.
Reinstall	S.F.		1.37		1.37	
Clean	S.F.		.33		.33	
Paint	S.F.	.33	1.27		1.60	
Minimum Charge	Job		227		227	
Cedar						
Demolish	S.F.		.51		.51	Cost includes material and labor to
Install	S.F.	2.97	1.37		4.34	install 4' x 8' cedar plywood exterior
Demolish and Install	S.F.	2.97	1.88		4.85	siding plain or patterns.
Reinstall	S.F.		1.37		1.37	
Clean	S.F.		.33		.33	
Paint	S.F.	.33	1.27		1.60	
Minimum Charge	Job		227		227	
Tongue and Groove						
Fir, Tongue and Groove						
Demolish	S.F.		.92		.92	Includes material and labor to install
Install	S.F.	5.50	1.38		6.88	fir tongue and groove exterior siding
Demolish and Install	S.F.	5.50	2.30		7.80	1" x 8".
Reinstall	S.F.		1.10		1.10	
Clean	S.F.		.33		.33	
Paint	S.F.	.33	1.27		1.60	
Minimum Charge	Job		227		227	
Pine, Tongue and Groove						
Demolish	S.F.		.92		.92	Includes labor and material to install
Install	S.F.	2.41	1.38		3.79	Pine tongue and groove exterior
Demolish and Install	S.F.	2.41	2.30		4.71	siding 1" x 8".
Reinstall	S.F.		1.38		1.38	
Clean	S.F.		.33		.33	
Paint	S.F.	.33	1.27		1.60	
Minimum Charge	Job		227		227	

Siding

Exterior Siding	Unit	Material	Labor	Equip.	Total	Specification
Cedar, Tongue and Groove						
Demolish	S.F.		.92		.92	Includes labor and material to install
Install	S.F.	7.20	2.06		9.26	cedar tongue and groove exterior
Demolish and Install	S.F.	7.20	2.98		10.18	siding 1" x 8".
Reinstall	S.F.		1.65		1.65	
Clean	S.F.		.33		.33	
Paint	S.F.	.33	1.27		1.60	
Minimum Charge	Job		227		227	
Redwood, Tongue and Groove						
Demolish	S.F.		.92		.92	Includes labor and material to install
Install	S.F.	3.58	2.06		5.64	redwood tongue and groove exterior
Demolish and Install	S.F.	3.58	2.98		6.56	siding 1" x 4".
Reinstall	S.F.		2.06		2.06	
Clean	S.F.		.33		.33	
Paint	S.F.	.33	1.27		1.60	
Minimum Charge	Job		227		227	

Board and Batten
Fir

	Unit	Material	Labor	Equip.	Total	Specification
Demolish	S.F.		.92		.92	Cost includes material and labor to
Install	S.F.	6.20	1.38		7.58	install 1" x 8" fir boards with 1" x 3"
Demolish and Install	S.F.	6.20	2.30		8.50	battens.
Reinstall	S.F.		1.10		1.10	
Clean	S.F.		.33		.33	
Paint	S.F.	.33	1.27		1.60	
Minimum Charge	Job		227		227	

Pine

	Unit	Material	Labor	Equip.	Total	Specification
Demolish	S.F.		.92		.92	Cost includes material and labor to
Install	S.F.	3.50	1.21		4.71	install 1" x 10" pine boards with 1" x
Demolish and Install	S.F.	3.50	2.13		5.63	3" battens.
Reinstall	S.F.		.97		.97	
Clean	S.F.		.33		.33	
Paint	S.F.	.33	1.27		1.60	
Minimum Charge	Job		227		227	

Cedar

	Unit	Material	Labor	Equip.	Total	Specification
Demolish	S.F.		.92		.92	Cost includes material and labor to
Install	S.F.	5.65	1.21		6.86	install 1" x 10" cedar boards with 1" x
Demolish and Install	S.F.	5.65	2.13		7.78	3" battens.
Reinstall	S.F.		.97		.97	
Clean	S.F.		.33		.33	
Paint	S.F.	.33	1.27		1.60	
Minimum Charge	Job		227		227	

Redwood

	Unit	Material	Labor	Equip.	Total	Specification
Demolish	S.F.		.92		.92	Cost includes material and labor to
Install	S.F.	5.90	1.21		7.11	install 1" x 10" redwood boards with
Demolish and Install	S.F.	5.90	2.13		8.03	1" x 3" battens.
Reinstall	S.F.		.97		.97	
Clean	S.F.		.33		.33	
Paint	S.F.	.33	1.27		1.60	
Minimum Charge	Job		227		227	

Wood Shingle

	Unit	Material	Labor	Equip.	Total	Specification
Demolish	Sq.		84.50		84.50	Includes material and labor to install
Install	Sq.	200	202		402	wood shingles/shakes.
Demolish and Install	Sq.	200	286.50		486.50	
Clean	Sq.		34.50		34.50	
Paint	Sq.	22.50	113		135.50	
Minimum Charge	Job		227		227	

Siding

Exterior Siding	Unit	Material	Labor	Equip.	Total	Specification
Synthetic Stucco System						
Demolish	S.F.		3.06		3.06	Cost includes material and labor to
Install	S.F.	2.98	7.20	.58	10.76	install 2 coats of adhesive mixture,
Demolish and Install	S.F.	2.98	10.26	.58	13.82	glass fiber mesh, 1" insulation board,
Clean	S.F.		.30		.30	and textured finish coat.
Paint	S.F.	.22	.62		.84	
Minimum Charge	Job		965	77	1042	

For customer support on your Contractor's Pricing Guide: Residential Repair & Remodeling, call 888.606.7279.

Site Work

Front-end Loader

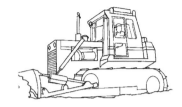

Bulldozer

Grading		Unit	Material	Labor	Equip.	Total	Specification
Hand (fine)							
	Install	S.F.		.66		.66	Includes labor and material to install
	Minimum Charge	Job		165		165	fine grade for slab on grade, hand grade.
Machine (fine)							
	Install	S.Y.		.59	.58	1.17	Includes labor and material to install
	Minimum Charge	Job		400	395	795	fine grade for roadway, base or leveling course.

Asphalt Paving		Unit	Material	Labor	Equip.	Total	Specification
Seal Coat							
	Install	S.F.	.11	.12		.23	Includes labor and material to install
	Minimum Charge	Job		165		165	sealcoat for a small area.
Wearing Course							
	Install	S.F.	.19	.37	.29	.85	Cost includes labor and material to
	Minimum Charge	Job		165		165	install 1" thick wearing course.

Masonry Pavers		Unit	Material	Labor	Equip.	Total	Specification
Natural Concrete							
Sand Base							
	Demolish	S.F.		3.67		3.67	Includes material and labor to install
	Install	S.F.	3.33	7.20		10.53	concrete pavers on sand base.
	Demolish and Install	S.F.	3.33	10.87		14.20	
	Reinstall	S.F.		5.76		5.76	
	Clean	S.F.		1.32		1.32	
	Minimum Charge	Job		216		216	
Mortar Base							
	Demolish	S.F.		6.15	2.23	8.38	Includes material and labor to install
	Install	S.F.	3.50	6		9.50	concrete pavers on mortar base.
	Demolish and Install	S.F.	3.50	12.15	2.23	17.88	
	Reinstall	S.F.		4.80		4.80	
	Clean	S.F.		1.20		1.20	
	Minimum Charge	Job		216		216	

139

Site Work

Masonry Pavers	Unit	Material	Labor	Equip.	Total	Specification
Adobe						
Sand Base						
Demolish	S.F.		3.67		3.67	Includes material and labor to install
Install	S.F.	1.97	4.80		6.77	adobe pavers on sand base.
Demolish and Install	S.F.	1.97	8.47		10.44	
Reinstall	S.F.		3.84		3.84	
Clean	S.F.		1.32		1.32	
Minimum Charge	Job		216		216	
Mortar Base						
Demolish	S.F.		6.15	2.23	8.38	Includes material and labor to install
Install	S.F.	3.04	6		9.04	adobe pavers on mortar base.
Demolish and Install	S.F.	3.04	12.15	2.23	17.42	
Reinstall	S.F.		4.80		4.80	
Clean	S.F.		1.20		1.20	
Minimum Charge	Job		216		216	
Standard Brick						
Sand Base						
Demolish	S.F.		3.67		3.67	Includes material and labor to install
Install	S.F.	3.14	7.85		10.99	standard brick pavers on sand base.
Demolish and Install	S.F.	3.14	11.52		14.66	
Reinstall	S.F.		6.30		6.30	
Clean	S.F.		1.32		1.32	
Minimum Charge	Job		216		216	
Mortar Base						
Demolish	S.F.		6.15	2.23	8.38	Includes material and labor to install
Install	S.F.	3.30	2.88		6.18	standard brick pavers on mortar base.
Demolish and Install	S.F.	3.30	9.03	2.23	14.56	
Reinstall	S.F.		2.30		2.30	
Clean	S.F.		1.20		1.20	
Minimum Charge	Job		216		216	
Deluxe Brick						
Sand Base						
Demolish	S.F.		3.67		3.67	Includes material and labor to install
Install	S.F.	3.85	6		9.85	deluxe brick pavers on sand base.
Demolish and Install	S.F.	3.85	9.67		13.52	
Reinstall	S.F.		4.80		4.80	
Clean	S.F.		1.32		1.32	
Minimum Charge	Job		216		216	
Mortar Base						
Demolish	S.F.		6.15	2.23	8.38	Includes material and labor to install
Install	S.F.	4.40	10.80		15.20	deluxe brick pavers on mortar base.
Demolish and Install	S.F.	4.40	16.95	2.23	23.58	
Reinstall	S.F.		8.64		8.64	
Clean	S.F.		1.20		1.20	
Minimum Charge	Job		216		216	

Fencing	Unit	Material	Labor	Equip.	Total	Specification
Block w / Footing						
4"						
Demolish	S.F.		.13	.31	.44	Cost includes material and labor to
Install	S.F.	1.91	4.66		6.57	install 4" non-reinforced concrete
Demolish and Install	S.F.	1.91	4.79	.31	7.01	block.
Clean	S.F.	.26	1.10	.26	1.62	
Paint	S.F.	.33	.72		1.05	
Minimum Charge	Job		216		216	

Fencing

Fencing	Unit	Material	Labor	Equip.	Total	Specification
4" Reinforced						
Demolish	S.F.		2.04	1.14	3.18	Cost includes material and labor to
Install	S.F.	2.10	4.72		6.82	install 4" reinforced concrete block.
Demolish and Install	S.F.	2.10	6.76	1.14	10	
Clean	S.F.	.26	1.10	.26	1.62	
Paint	S.F.	.33	.72		1.05	
Minimum Charge	Job		216		216	
6"						
Demolish	S.F.		.18	.42	.60	Cost includes material and labor to
Install	S.F.	2.53	5		7.53	install 6" non-reinforced concrete
Demolish and Install	S.F.	2.53	5.18	.42	8.13	block.
Clean	S.F.	.26	1.10	.26	1.62	
Paint	S.F.	.33	.72		1.05	
Minimum Charge	Job		216		216	
6" Reinforced						
Demolish	S.F.		2.14	.58	2.72	Cost includes material and labor to
Install	S.F.	2.71	5.10		7.81	install 6" reinforced concrete block.
Demolish and Install	S.F.	2.71	7.24	.58	10.53	
Clean	S.F.	.26	1.10	.26	1.62	
Paint	S.F.	.33	.72		1.05	
Minimum Charge	Job		216		216	
Cap for Block Wall						
Demolish	L.F.		2.29		2.29	Includes labor and equipment to install
Install	L.F.	19.20	10.50		29.70	precast concrete coping.
Demolish and Install	L.F.	19.20	12.79		31.99	
Clean	L.F.	.02	3.44		3.46	
Paint	S.F.	.33	.72		1.05	
Minimum Charge	Job		216		216	
Block Pilaster						
Demolish	Ea.		157		157	Includes material and labor to install
Install	Ea.	62.50	157		219.50	block pilaster for fence.
Demolish and Install	Ea.	62.50	314		376.50	
Reinstall	Ea.		157.38		157.38	
Paint	Ea.	4.87	16.80		21.67	
Minimum Charge	Job		216		216	
Block w / o Footing						
4"						
Demolish	S.F.		.13	.31	.44	Cost includes material and labor to
Install	S.F.	2.04	3.58		5.62	install 4" non-reinforced hollow
Demolish and Install	S.F.	2.04	3.71	.31	6.06	lightweight block.
Clean	S.F.	.26	1.10	.26	1.62	
Paint	S.F.	.33	.72		1.05	
Minimum Charge	Job		216		216	
4" Reinforced						
Demolish	S.F.		2.04	1.14	3.18	Cost includes material and labor to
Install	S.F.	2.21	4.61		6.82	install 4" reinforced hollow lightweight
Demolish and Install	S.F.	2.21	6.65	1.14	10	block.
Clean	S.F.	.26	1.10	.26	1.62	
Paint	S.F.	.33	.72		1.05	
Minimum Charge	Job		216		216	
6"						
Demolish	S.F.		.18	.42	.60	Cost includes material and labor to
Install	S.F.	2.89	3.93		6.82	install 6" non-reinforced hollow
Demolish and Install	S.F.	2.89	4.11	.42	7.42	lightweight block.
Clean	S.F.	.26	1.10	.26	1.62	
Paint	S.F.	.33	.72		1.05	
Minimum Charge	Job		216		216	

Fencing	Unit	Material	Labor	Equip.	Total	Specification
6" Reinforced						
Demolish	S.F.		2.14	.58	2.72	Cost includes material and labor to
Install	S.F.	3.04	4.95		7.99	install 6" reinforced hollow lightweight
Demolish and Install	S.F.	3.04	7.09	.58	10.71	block.
Clean	S.F.	.26	1.10	.26	1.62	
Paint	S.F.	.33	.72		1.05	
Minimum Charge	Job		216		216	
Cap for Block Wall						
Demolish	L.F.		2.29		2.29	Includes labor and equipment to install
Install	L.F.	19.20	10.50		29.70	precast concrete coping.
Demolish and Install	L.F.	19.20	12.79		31.99	
Clean	L.F.	.02	3.44		3.46	
Paint	S.F.	.33	.72		1.05	
Minimum Charge	Job		216		216	
Block Pilaster						
Demolish	Ea.		157		157	Includes material and labor to install
Install	Ea.	62.50	157		219.50	block pilaster for fence.
Demolish and Install	Ea.	62.50	314		376.50	
Reinstall	Ea.		157.38		157.38	
Paint	Ea.	4.87	16.80		21.67	
Minimum Charge	Job		216		216	
Chain Link						
3' High w / o Top Rail						
Demolish	L.F.		2.90	1.05	3.95	Includes material and labor to install
Install	L.F.	6.25	2.01	.54	8.80	residential galvanized fence 3' high
Demolish and Install	L.F.	6.25	4.91	1.59	12.75	with 1-3/8" steel post 10' O.C. set
Reinstall	L.F.		1.60	.43	2.03	without concrete.
Clean	L.F.	.12	1.20		1.32	
Minimum Charge	Job		510		510	
4' High w / o Top Rail						
Demolish	L.F.		2.90	1.05	3.95	Includes material and labor to install
Install	L.F.	6.60	2.48	.66	9.74	residential galvanized fence 4' high
Demolish and Install	L.F.	6.60	5.38	1.71	13.69	with 1-3/8" steel post 10' O.C. set
Reinstall	L.F.		1.98	.53	2.51	without concrete.
Clean	L.F.	.17	1.32		1.49	
Minimum Charge	Job		510		510	
5' High w / o Top Rail						
Demolish	L.F.		2.90	1.05	3.95	Includes material and labor to install
Install	L.F.	8.25	3.24	.87	12.36	residential galvanized fence 5' high
Demolish and Install	L.F.	8.25	6.14	1.92	16.31	with 1-3/8" steel post 10' O.C. set
Reinstall	L.F.		2.59	.69	3.28	without concrete.
Clean	L.F.	.21	1.47		1.68	
Minimum Charge	Job		510		510	
3' High w / Top Rail						
Demolish	L.F.		3.08	1.12	4.20	Includes material and labor to install
Install	L.F.	2.38	2.11	.56	5.05	residential galvanized fence 3' high
Demolish and Install	L.F.	2.38	5.19	1.68	9.25	with 1-3/8" steel post 10' O.C. set
Reinstall	L.F.		1.68	.45	2.13	without concrete.
Clean	L.F.	.12	1.22		1.34	
Minimum Charge	Job		510		510	
4' High w / Top Rail						
Demolish	L.F.		3.08	1.12	4.20	Includes material and labor to install
Install	L.F.	8.35	2.63	.71	11.69	residential galvanized fence 4' high
Demolish and Install	L.F.	8.35	5.71	1.83	15.89	with 1-5/8" steel post 10' O.C. set
Reinstall	L.F.		2.11	.56	2.67	without concrete and 1-5/8" top rail
Clean	L.F.	.17	1.35		1.52	with sleeves.
Minimum Charge	Job		510		510	

Fencing		Unit	Material	Labor	Equip.	Total	Specification
5' High w / Top Rail							Includes material and labor to install
	Demolish	L.F.		3.08	1.12	4.20	residential galvanized fence 5' high
	Install	L.F.	9	3.51	.94	13.45	with 1-5/8" steel post 10' O.C. set
	Demolish and Install	L.F.	9	6.59	2.06	17.65	without concrete and 1-5/8" top rail
	Reinstall	L.F.		2.81	.75	3.56	with sleeves.
	Clean	L.F.	.21	1.50		1.71	
	Minimum Charge	Job		510		510	
6' High w / Top Rail							Includes material and labor to install
	Demolish	L.F.		3.08	1.12	4.20	residential galvanized fence 6' high
	Install	L.F.	10.60	5.25	1.41	17.26	with 2" steel post 10' O.C. set without
	Demolish and Install	L.F.	10.60	8.33	2.53	21.46	concrete and 1-3/8" top rail with
	Reinstall	L.F.		4.21	1.13	5.34	sleeves.
	Clean	L.F.	.24	1.69		1.93	
	Minimum Charge	Job		510		510	
Barbed Wire							Cost includes material and labor to
	Demolish	L.F.		1.71		1.71	install 3 strand galvanized barbed
	Install	L.F.	.43	.66	.23	1.32	wire fence.
	Demolish and Install	L.F.	.43	2.37	.23	3.03	
	Reinstall	L.F.		.53	.18	.71	
	Minimum Charge	Job		510	400	910	
6' Redwood							Cost includes material and labor to
	Demolish	L.F.		11.30		11.30	install 6' high redwood fencing.
	Install	L.F.	29.50	8.45	2.26	40.21	
	Demolish and Install	L.F.	29.50	19.75	2.26	51.51	
	Reinstall	L.F.		6.74	1.81	8.55	
	Clean	L.F.	.24	1.53		1.77	
	Paint	L.F.	.92	3.97		4.89	
	Minimum Charge	Job		510		510	
6' Red Cedar							Cost includes material and labor to
	Demolish	L.F.		11.30		11.30	install 6' cedar fence.
	Install	L.F.	21	8.45	2.26	31.71	
	Demolish and Install	L.F.	21	19.75	2.26	43.01	
	Reinstall	L.F.		6.74	1.81	8.55	
	Clean	L.F.	.24	1.53		1.77	
	Paint	L.F.	.92	3.97		4.89	
	Minimum Charge	Job		510		510	
6' Shadowbox							Cost includes material and labor to
	Demolish	L.F.		11.30		11.30	install 6' pressure treated pine fence.
	Install	L.F.	19.30	7	1.88	28.18	
	Demolish and Install	L.F.	19.30	18.30	1.88	39.48	
	Reinstall	L.F.		5.62	1.51	7.13	
	Clean	L.F.	.24	1.53		1.77	
	Paint	L.F.	.92	3.97		4.89	
	Minimum Charge	Job		510		510	
3' Picket							Includes material, labor and
	Demolish	L.F.		11.30		11.30	equipment to install pressure treated
	Install	L.F.	8.60	7.50	2.02	18.12	pine picket fence.
	Demolish and Install	L.F.	8.60	18.80	2.02	29.42	
	Reinstall	L.F.		6.02	1.61	7.63	
	Clean	L.F.	.12	1.32		1.44	
	Paint	L.F.	.46	2.16		2.62	
	Minimum Charge	Job		510		510	

Site Work

Fencing	Unit	Material	Labor	Equip.	Total	Specification
Wrought Iron						
Demolish	L.F.		24.50		24.50	Cost includes material and labor to
Install	L.F.	40.50	13.55		54.05	install 5' high wrought iron fence with
Demolish and Install	L.F.	40.50	38.05		78.55	pickets at 4-1/2" O.C.
Reinstall	L.F.		10.85		10.85	
Clean	L.F.		1.32		1.32	
Paint	L.F.	4.15	4.44		8.59	
Minimum Charge	Job		510		510	
Post						
Dirt Based Wood						
Demolish	Ea.		28.50		28.50	Includes material and labor to install
Install	Ea.	34.50	127		161.50	wood post set in earth.
Demolish and Install	Ea.	34.50	155.50		190	
Reinstall	Ea.		101.69		101.69	
Clean	Ea.		4.13		4.13	
Paint	Ea.	1.02	1.51		2.53	
Minimum Charge	Job		510		510	
Concrete Based Wood						
Demolish	Ea.		41	14.90	55.90	Includes material and labor to install
Install	Ea.	20	169		189	wood post set in concrete.
Demolish and Install	Ea.	20	210	14.90	244.90	
Clean	Ea.		4.13		4.13	
Paint	Ea.	1.02	1.51		2.53	
Minimum Charge	Job		510		510	
Block Pilaster						
Demolish	Ea.		157		157	Includes material and labor to install
Install	Ea.	62.50	157		219.50	block pilaster for fence.
Demolish and Install	Ea.	62.50	314		376.50	
Reinstall	Ea.		157.38		157.38	
Paint	Ea.	4.87	16.80		21.67	
Minimum Charge	Job		216		216	
Gate						
Chain Link						
Demolish	Ea.		37.50		37.50	Cost includes material and labor to
Install	Ea.	109	56.50		165.50	install 3' wide chain link pass gate
Demolish and Install	Ea.	109	94		203	and hardware.
Reinstall	Ea.		45.19		45.19	
Clean	Ea.		4.58		4.58	
Minimum Charge	Job		510		510	
Redwood						
Demolish	Ea.		32.50		32.50	Includes material, labor and
Install	Ea.	263	56.50		319.50	equipment to install redwood pass
Demolish and Install	Ea.	263	89		352	gate and hardware.
Reinstall	Ea.		45.19		45.19	
Clean	Ea.		4.58		4.58	
Paint	Ea.	1.53	7.55		9.08	
Minimum Charge	Job		510		510	
Red Cedar						
Demolish	Ea.		32.50		32.50	Cost includes material and labor to
Install	Ea.	99.50	56.50		156	install 3' wide red cedar pass gate
Demolish and Install	Ea.	99.50	89		188.50	and hardware.
Reinstall	Ea.		45.19		45.19	
Clean	Ea.		4.58		4.58	
Paint	Ea.	1.53	7.55		9.08	
Minimum Charge	Job		510		510	

Site Work

Fencing

	Unit	Material	Labor	Equip.	Total	Specification
Wrought Iron						
Demolish	Ea.		24.50		24.50	Includes material, labor and
Install	Ea.	287	127		414	equipment to install wrought iron gate
Demolish and Install	Ea.	287	151.50		438.50	with hardware.
Reinstall	Ea.		101.69		101.69	
Clean	Ea.		4.58		4.58	
Paint	Ea.	8.30	7.55		15.85	
Minimum Charge	Job		510		510	

145

Rough Mechanical

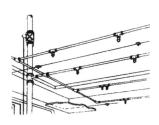

Fire-extinguishing System

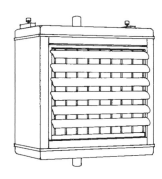

Space Heater

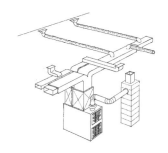

Hot-air Furnace

Plumbing	Unit	Material	Labor	Equip.	Total	Specification
Components						
3/4″ Supply						
Demolish	L.F.		2.86		2.86	Includes material and labor to install
Install	L.F.	5.80	6.80		12.60	type L copper tubing including
Demolish and Install	L.F.	5.80	9.66		15.46	couplings and hangers at 10′ O.C.
Reinstall	L.F.		6.78		6.78	
Minimum Charge	Job		258		258	
1/2″ Fixture Supply						
Install	Ea.	51.50	172		223.50	Includes labor and materials to install supply pipe only to a plumbing fixture.
2″ Single Vent Stack						
Demolish	Ea.		47.50		47.50	Includes material and labor to install
Install	Ea.	27.50	258		285.50	PVC vent stack pipe with couplings at
Demolish and Install	Ea.	27.50	305.50		333	10′ O.C.
Reinstall	Ea.		257.60		257.60	
Minimum Charge	Job		258		258	
4″ Drain						
Demolish	L.F.		3.82		3.82	Includes material and labor to install
Install	L.F.	6.45	8.45		14.90	DVW type PVC pipe with couplings at
Demolish and Install	L.F.	6.45	12.27		18.72	10′ O.C. and 3 strap hangers per 10′
Reinstall	L.F.		8.43		8.43	run.
Minimum Charge	Job		258		258	
1-1/2″ Drain						
Demolish	L.F.		2.86		2.86	Includes material and labor to install
Install	L.F.	1.97	5.50		7.47	DVW type PVC pipe with couplings at
Demolish and Install	L.F.	1.97	8.36		10.33	10′ O.C. and 3 strap hangers per 10′
Reinstall	L.F.		5.48		5.48	run.
Minimum Charge	Job		258		258	
Maintenance						
Flush Out / Unclog						
Clean	Job		258		258	Includes 4 hour minimum labor charge
Minimum Charge	Job		258		258	for a plumber.
Vent Cap						
Large						
Demolish	Ea.		13		13	Includes labor and material to install a
Install	Ea.	43.50	23.50		67	PVC plumbing pipe vent cap.
Demolish and Install	Ea.	43.50	36.50		80	
Clean	Ea.	.56	7.50		8.06	
Paint	Ea.	.21	8.60		8.81	
Minimum Charge	Job		129		129	

147

Rough Mechanical

Plumbing	Unit	Material	Labor	Equip.	Total	Specification
Medium						
Demolish	Ea.		13		13	Includes labor and material to install a
Install	Ea.	37.50	22.50		60	PVC plumbing pipe vent cap.
Demolish and Install	Ea.	37.50	35.50		73	
Clean	Ea.	.56	7.50		8.06	
Paint	Ea.	.21	8.60		8.81	
Minimum Charge	Job		129		129	
Small						
Demolish	Ea.		13		13	Includes material and labor to install a
Install	Ea.	10.10	21.50		31.60	PVC plumbing pipe vent cap.
Demolish and Install	Ea.	10.10	34.50		44.60	
Clean	Ea.	.56	7.50		8.06	
Paint	Ea.	.21	8.60		8.81	
Minimum Charge	Job		129		129	

Plumbing	Unit	Material	Labor	Equip.	Total	Specification
Sump Pump						
1/4 HP						
Demolish	Ea.		51.50		51.50	Includes material and labor to install a
Install	Ea.	165	80.50		245.50	submersible sump pump.
Demolish and Install	Ea.	165	132		297	
Reinstall	Ea.		64.40		64.40	
Clean	Ea.	1.43	20.50		21.93	
Minimum Charge	Job		258		258	
1/3 HP						
Demolish	Ea.		51.50		51.50	Includes material and labor to install a
Install	Ea.	239	86		325	submersible sump pump.
Demolish and Install	Ea.	239	137.50		376.50	
Reinstall	Ea.		68.69		68.69	
Clean	Ea.	1.43	20.50		21.93	
Minimum Charge	Job		258		258	
1/2 HP						
Demolish	Ea.		51.50		51.50	Includes material and labor to install a
Install	Ea.	293	95.50		388.50	submersible sump pump.
Demolish and Install	Ea.	293	147		440	
Reinstall	Ea.		76.33		76.33	
Clean	Ea.	1.43	20.50		21.93	
Minimum Charge	Job		258		258	
Clean and Service						
Clean	Ea.		32		32	Includes labor to clean and service a
Minimum Charge	Job		258		258	sump pump.

Water Heater	Unit	Material	Labor	Equip.	Total	Specification
Electric						
30 Gallon						
Demolish	Ea.		161		161	Includes material and labor to install a
Install	Ea.	995	234		1229	30 gallon electric water heater.
Demolish and Install	Ea.	995	395		1390	
Reinstall	Ea.		187.35		187.35	
Clean	Ea.	1.43	20.50		21.93	
Minimum Charge	Job		258		258	

Rough Mechanical

Water Heater	Unit	Material	Labor	Equip.	Total	Specification
40 Gallon						
Demolish	Ea.		161		161	Includes material and labor to install a
Install	Ea.	1050	258		1308	40 gallon electric water heater.
Demolish and Install	Ea.	1050	419		1469	
Reinstall	Ea.		206.08		206.08	
Clean	Ea.	1.43	20.50		21.93	
Minimum Charge	Job		258		258	
52 Gallon						
Demolish	Ea.		161		161	Includes material and labor to install a
Install	Ea.	1175	258		1433	52 gallon electric water heater.
Demolish and Install	Ea.	1175	419		1594	
Reinstall	Ea.		206.08		206.08	
Clean	Ea.	1.43	20.50		21.93	
Minimum Charge	Job		258		258	
82 Gallon						
Demolish	Ea.		161		161	Includes material and labor to install
Install	Ea.	1825	320		2145	an 82 gallon electric water heater.
Demolish and Install	Ea.	1825	481		2306	
Reinstall	Ea.		257.60		257.60	
Clean	Ea.	1.43	20.50		21.93	
Minimum Charge	Job		258		258	
Gas-fired						
30 Gallon Gas-fired						
Demolish	Ea.		198		198	Includes material and labor to install a
Install	Ea.	1800	258		2058	30 gallon gas water heater.
Demolish and Install	Ea.	1800	456		2256	
Reinstall	Ea.		206.08		206.08	
Clean	Ea.	1.43	20.50		21.93	
Minimum Charge	Job		258		258	
30 Gallon Gas-fired, Flue						
Install	Ea.	65	73.50		138.50	Includes labor and materials to install
Minimum Charge	Job		258		258	a flue for a water heater (up to 50 gallon size).
40 Gallon Gas-fired						
Demolish	Ea.		198		198	Includes material and labor to install a
Install	Ea.	1800	271		2071	40 gallon gas water heater.
Demolish and Install	Ea.	1800	469		2269	
Reinstall	Ea.		216.93		216.93	
Clean	Ea.	1.43	20.50		21.93	
Minimum Charge	Job		258		258	
40 Gallon Gas-fired, Flue						
Install	Ea.	65	73.50		138.50	Includes labor and materials to install
Minimum Charge	Job		258		258	a flue for a water heater (up to 50 gallon size).
50 Gallon Gas-fired						
Demolish	Ea.		198		198	Includes material and labor to install a
Install	Ea.	1875	286		2161	50 gallon gas water heater.
Demolish and Install	Ea.	1875	484		2359	
Reinstall	Ea.		228.98		228.98	
Clean	Ea.	1.43	20.50		21.93	
Minimum Charge	Job		258		258	
50 Gallon Gas-fired, Flue						
Install	Ea.	65	73.50		138.50	Includes labor and materials to install
Minimum Charge	Job		258		258	a flue for a water heater (up to 50 gallon size).

Water Heater

Water Heater	Unit	Material	Labor	Equip.	Total	Specification
Solar Hot Water System						
Water Heater						
Demolish	Ea.		129		129	Includes material and labor to install a
Install	Ea.	1675	345		2020	solar hot water heat exchanger
Demolish and Install	Ea.	1675	474		2149	storage tank, 100 gallon.
Reinstall	Ea.		274.77		274.77	
Minimum Charge	Job		264		264	
Solar Panel						
Demolish	S.F.		1.88		1.88	Includes material and labor to install
Install	S.F.	34.50	3.39		37.89	solar panel.
Demolish and Install	S.F.	34.50	5.27		39.77	
Reinstall	S.F.		2.71		2.71	
Minimum Charge	Job		264		264	
Repair Plumbing						
Install	Job		264		264	Includes 4 hour minimum labor charge
Minimum Charge	Job		264		264	for a steamfitter or pipefitter.
Insulating Wrap						
Demolish	Ea.		5.95		5.95	Includes material and labor to install
Install	Ea.	19.95	16.10		36.05	insulating wrap around a 100 gallon
Demolish and Install	Ea.	19.95	22.05		42	solar hot water heat exchanger
Reinstall	Ea.		12.88		12.88	storage tank.
Minimum Charge	Job		258		258	
Water Softener						
Demolish	Ea.		95.50		95.50	Includes material and labor to install a
Install	Ea.	2075	232		2307	14 GPM water softener.
Demolish and Install	Ea.	2075	327.50		2402.50	
Reinstall	Ea.		185.54		185.54	
Clean	Ea.	1.43	20.50		21.93	
Minimum Charge	Ea.		258		258	

Ductwork

Ductwork	Unit	Material	Labor	Equip.	Total	Specification
Galvanized, Square or Rectangular						
6″						
Demolish	L.F.		2.34		2.34	Includes material and labor to install
Install	L.F.	1.45	11.20		12.65	rectangular ductwork with a greatest
Demolish and Install	L.F.	1.45	13.54		14.99	dimension of 6″.
Reinstall	L.F.		8.97		8.97	
Clean	L.F.		2.69		2.69	
Minimum Charge	Job		252		252	
8″						
Demolish	L.F.		2.34		2.34	Includes material and labor to install
Install	L.F.	1.93	14.40		16.33	rectangular ductwork with a greatest
Demolish and Install	L.F.	1.93	16.74		18.67	dimension of 8″.
Reinstall	L.F.		11.53		11.53	
Clean	L.F.		3.59		3.59	
Minimum Charge	Job		252		252	
10″						
Demolish	L.F.		2.34		2.34	Includes material and labor to install
Install	L.F.	2.41	18.70		21.11	rectangular ductwork with a greatest
Demolish and Install	L.F.	2.41	21.04		23.45	dimension of 10″.
Reinstall	L.F.		14.95		14.95	
Clean	L.F.		4.48		4.48	
Minimum Charge	Job		252		252	

Rough Mechanical

Ductwork	Unit	Material	Labor	Equip.	Total	Specification
Galvanized, Spiral						
4"						
Demolish	L.F.		2.80		2.80	Cost includes material and labor to
Install	L.F.	2.05	2.52		4.57	install 4" diameter spiral pre-formed
Demolish and Install	L.F.	2.05	5.32		7.37	steel galvanized ductwork.
Reinstall	L.F.		2.02		2.02	
Clean	L.F.		1.41		1.41	
Minimum Charge	Job		252		252	
6"						
Demolish	L.F.		2.80		2.80	Cost includes material and labor to
Install	L.F.	2.05	3.24		5.29	install 6" diameter spiral pre-formed
Demolish and Install	L.F.	2.05	6.04		8.09	steel galvanized ductwork.
Reinstall	L.F.		2.59		2.59	
Clean	L.F.		2.11		2.11	
Minimum Charge	Job		252		252	
8"						
Demolish	L.F.		2.80		2.80	Cost includes material and labor to
Install	L.F.	2.73	4.54		7.27	install 8" diameter spiral pre-formed
Demolish and Install	L.F.	2.73	7.34		10.07	steel galvanized ductwork.
Reinstall	L.F.		3.63		3.63	
Clean	L.F.		2.82		2.82	
Minimum Charge	Job		252		252	
10"						
Demolish	L.F.		3.50		3.50	Cost includes material and labor to
Install	L.F.	3.41	5.70		9.11	install 10" diameter spiral pre-formed
Demolish and Install	L.F.	3.41	9.20		12.61	steel galvanized ductwork.
Reinstall	L.F.		4.54		4.54	
Clean	L.F.		3.52		3.52	
Minimum Charge	Job		252		252	
12"						
Demolish	L.F.		4.67		4.67	Cost includes material and labor to
Install	L.F.	4.09	7.55		11.64	install 12" diameter spiral pre-formed
Demolish and Install	L.F.	4.09	12.22		16.31	steel galvanized ductwork.
Reinstall	L.F.		6.05		6.05	
Clean	L.F.		4.23		4.23	
Minimum Charge	Job		252		252	
16"						
Demolish	L.F.		9.35		9.35	Cost includes material and labor to
Install	L.F.	6.75	15.15		21.90	install 16" diameter spiral pre-formed
Demolish and Install	L.F.	6.75	24.50		31.25	steel galvanized ductwork.
Reinstall	L.F.		12.11		12.11	
Clean	L.F.		5.65		5.65	
Minimum Charge	Job		252		252	
Flexible Fiberglass						
6" Non-insulated						
Demolish	L.F.		2.16		2.16	Cost includes material and labor to
Install	L.F.	1.65	3.24		4.89	install 6" diameter non-insulated
Demolish and Install	L.F.	1.65	5.40		7.05	flexible fiberglass fabric on corrosion
Reinstall	L.F.		2.59		2.59	resistant metal ductwork.
Clean	L.F.		2.11		2.11	
Minimum Charge	Job		252		252	
8" Non-insulated						
Demolish	L.F.		3.11		3.11	Cost includes material and labor to
Install	L.F.	2.10	4.54		6.64	install 8" diameter non-insulated
Demolish and Install	L.F.	2.10	7.65		9.75	flexible fiberglass fabric on corrosion
Reinstall	L.F.		3.63		3.63	resistant metal ductwork.
Clean	L.F.		2.82		2.82	
Minimum Charge	Job		252		252	

151

Ductwork	Unit	Material	Labor	Equip.	Total	Specification
10″ Non-insulated						
Demolish	L.F.		4		4	Cost includes material and labor to
Install	L.F.	2.71	5.70		8.41	install 10″ diameter non-insulated
Demolish and Install	L.F.	2.71	9.70		12.41	flexible fiberglass fabric on corrosion
Reinstall	L.F.		4.54		4.54	resistant metal ductwork.
Clean	L.F.		3.52		3.52	
Minimum Charge	Job		252		252	
12″ Non-insulated						
Demolish	L.F.		5.60		5.60	Cost includes material and labor to
Install	L.F.	3.23	7.55		10.78	install 12″ diameter non-insulated
Demolish and Install	L.F.	3.23	13.15		16.38	flexible fiberglass fabric on corrosion
Reinstall	L.F.		6.05		6.05	resistant metal ductwork.
Clean	L.F.		4.23		4.23	
Minimum Charge	Job		252		252	
4″ Insulated						
Demolish	L.F.		1.65		1.65	Cost includes material and labor to
Install	L.F.	2.86	2.67		5.53	install 4″ diameter insulated flexible
Demolish and Install	L.F.	2.86	4.32		7.18	fiberglass fabric on corrosion resistant
Reinstall	L.F.		2.14		2.14	metal ductwork.
Clean	L.F.		1.41		1.41	
Minimum Charge	Job		252		252	
6″ Insulated						
Demolish	L.F.		2.16		2.16	Cost includes material and labor to
Install	L.F.	3.23	3.49		6.72	install 6″ diameter insulated flexible
Demolish and Install	L.F.	3.23	5.65		8.88	fiberglass fabric on corrosion resistant
Reinstall	L.F.		2.79		2.79	metal ductwork.
Clean	L.F.		2.11		2.11	
Minimum Charge	Job		252		252	
8″ Insulated						
Demolish	L.F.		3.11		3.11	Cost includes material and labor to
Install	L.F.	3.84	5.05		8.89	install 8″ diameter insulated flexible
Demolish and Install	L.F.	3.84	8.16		12	fiberglass fabric on corrosion resistant
Reinstall	L.F.		4.04		4.04	metal ductwork.
Clean	L.F.		2.82		2.82	
Minimum Charge	Job		252		252	
10″ Insulated						
Demolish	L.F.		4		4	Cost includes material and labor to
Install	L.F.	4.68	6.50		11.18	install 10″ diameter insulated flexible
Demolish and Install	L.F.	4.68	10.50		15.18	fiberglass fabric on corrosion resistant
Reinstall	L.F.		5.19		5.19	metal ductwork.
Clean	L.F.		3.52		3.52	
Minimum Charge	Job		252		252	
12″ Insulated						
Demolish	L.F.		5.60		5.60	Cost includes material and labor to
Install	L.F.	5.40	9.10		14.50	install 12″ diameter insulated flexible
Demolish and Install	L.F.	5.40	14.70		20.10	fiberglass fabric on corrosion resistant
Reinstall	L.F.		7.26		7.26	metal ductwork.
Clean	L.F.		4.23		4.23	
Minimum Charge	Job		252		252	

Rough Mechanical

Heating	Unit	Material	Labor	Equip.	Total	Specification
Hot Water Baseboard						
Baseboard						
Demolish	L.F.		9.05		9.05	Includes material and labor to install
Install	L.F.	12.30	15.85		28.15	baseboard radiation.
Demolish and Install	L.F.	12.30	24.90		37.20	
Reinstall	L.F.		12.68		12.68	
Clean	L.F.	1.43	5.15		6.58	
Minimum Charge	Job		258		258	
3'						
Demolish	Ea.		28.50		28.50	Includes material and labor to install
Install	Ea.	24	53		77	fin tube baseboard radiation in 3'
Demolish and Install	Ea.	24	81.50		105.50	sections.
Reinstall	Ea.		42.28		42.28	
Clean	Ea.	.41	15.70		16.11	
Minimum Charge	Job		258		258	
4'						
Demolish	Ea.		38		38	Includes material and labor to install
Install	Ea.	32.50	59.50		92	fin tube baseboard radiation in 4'
Demolish and Install	Ea.	32.50	97.50		130	sections.
Reinstall	Ea.		47.56		47.56	
Clean	Ea.	.81	20.50		21.31	
Minimum Charge	Job		258		258	
5'						
Demolish	Ea.		47.50		47.50	Includes material and labor to install
Install	Ea.	40.50	68		108.50	fin tube baseboard radiation in 5'
Demolish and Install	Ea.	40.50	115.50		156	sections.
Reinstall	Ea.		54.35		54.35	
Clean	Ea.	.81	25.50		26.31	
Minimum Charge	Job		258		258	
6'						
Demolish	Ea.		57		57	Includes material and labor to install
Install	Ea.	48.50	79.50		128	fin tube baseboard radiation in 6'
Demolish and Install	Ea.	48.50	136.50		185	sections.
Reinstall	Ea.		63.41		63.41	
Clean	Ea.	.81	30		30.81	
Minimum Charge	Job		258		258	
7'						
Demolish	Ea.		63.50		63.50	Includes material and labor to install
Install	Ea.	56.50	103		159.50	fin tube baseboard radiation in 7'
Demolish and Install	Ea.	56.50	166.50		223	sections.
Reinstall	Ea.		82.43		82.43	
Clean	Ea.	1.02	36.50		37.52	
Minimum Charge	Job		258		258	
9'						
Demolish	Ea.		95.50		95.50	Includes material and labor to install
Install	Ea.	73	147		220	fin tube baseboard radiation in 9'
Demolish and Install	Ea.	73	242.50		315.50	sections.
Reinstall	Ea.		117.76		117.76	
Clean	Ea.	1.63	47		48.63	
Minimum Charge	Job		258		258	

Rough Mechanical

Heating	Unit	Material	Labor	Equip.	Total	Specification
Cabinet Unit Heater						
50 MBH						
Demolish	Ea.		47.50		47.50	Includes material and labor to install a
Install	Ea.	1925	136		2061	cabinet space heater with an output of
Demolish and Install	Ea.	1925	183.50		2108.50	50 MBH.
Reinstall	Ea.		108.71		108.71	
Clean	Ea.	1.02	41.50		42.52	
Minimum Charge	Job		264		264	
75 MBH						
Demolish	Ea.		47.50		47.50	Includes material and labor to install a
Install	Ea.	2100	159		2259	cabinet space heater with an output of
Demolish and Install	Ea.	2100	206.50		2306.50	75 MBH.
Reinstall	Ea.		126.83		126.83	
Clean	Ea.	1.02	41.50		42.52	
Minimum Charge	Job		264		264	
125 MBH						
Demolish	Ea.		57		57	Includes material and labor to install a
Install	Ea.	2475	190		2665	cabinet space heater with an output of
Demolish and Install	Ea.	2475	247		2722	125 MBH.
Reinstall	Ea.		152.19		152.19	
Clean	Ea.	1.02	41.50		42.52	
Minimum Charge	Job		264		264	
175 MBH						
Demolish	Ea.		71.50		71.50	Includes material and labor to install a
Install	Ea.	3225	272		3497	cabinet space heater with an output of
Demolish and Install	Ea.	3225	343.50		3568.50	175 MBH.
Reinstall	Ea.		217.42		217.42	
Clean	Ea.	1.02	41.50		42.52	
Minimum Charge	Job		264		264	

Furnace Unit	Unit	Material	Labor	Equip.	Total	Specification
Wall (gas-fired)						
25 MBH w / Thermostat						
Demolish	Ea.		114		114	Includes material and labor to install a
Install	Ea.	660	182		842	gas fired hot air furnace rated at 25
Demolish and Install	Ea.	660	296		956	MBH.
Reinstall	Ea.		145.28		145.28	
Clean	Ea.	2.04	91.50		93.54	
Minimum Charge	Job		264		264	
35 MBH w / Thermostat						
Demolish	Ea.		114		114	Includes material and labor to install a
Install	Ea.	930	202		1132	gas fired hot air furnace rated at 35
Demolish and Install	Ea.	930	316		1246	MBH.
Reinstall	Ea.		161.42		161.42	
Clean	Ea.	2.04	91.50		93.54	
Minimum Charge	Job		264		264	
50 MBH w / Thermostat						
Demolish	Ea.		114		114	Includes material and labor to install a
Install	Ea.	1300	227		1527	gas fired hot air furnace rated at 50
Demolish and Install	Ea.	1300	341		1641	MBH.
Reinstall	Ea.		181.60		181.60	
Clean	Ea.	2.04	91.50		93.54	
Minimum Charge	Job		264		264	

Rough Mechanical

Furnace Unit	Unit	Material	Labor	Equip.	Total	Specification
60 MBH w / Thermostat						
Demolish	Ea.		114		114	Includes material and labor to install a
Install	Ea.	1425	252		1677	gas fired hot air furnace rated at 60
Demolish and Install	Ea.	1425	366		1791	MBH.
Reinstall	Ea.		201.78		201.78	
Clean	Ea.	2.04	91.50		93.54	
Minimum Charge	Job		264		264	
Floor (gas-fired)						
30 MBH w / Thermostat						
Demolish	Ea.		114		114	Includes material and labor to install a
Install	Ea.	1150	182		1332	gas fired hot air floor furnace rated at
Demolish and Install	Ea.	1150	296		1446	30 MBH.
Reinstall	Ea.		145.28		145.28	
Clean	Ea.	2.04	91.50		93.54	
Minimum Charge	Job		264		264	
50 MBH w / Thermostat						
Demolish	Ea.		114		114	Includes material and labor to install a
Install	Ea.	1275	227		1502	gas fired hot air floor furnace rated at
Demolish and Install	Ea.	1275	341		1616	50 MBH.
Reinstall	Ea.		181.60		181.60	
Clean	Ea.	2.04	91.50		93.54	
Minimum Charge	Job		264		264	
Gas Forced Air						
50 MBH						
Demolish	Ea.		114		114	Includes material and labor to install
Install	Ea.	1250	227		1477	an up flow gas fired hot air furnace
Demolish and Install	Ea.	1250	341		1591	rated at 50 MBH.
Reinstall	Ea.		181.60		181.60	
Clean	Ea.	2.04	91.50		93.54	
Minimum Charge	Job		264		264	
80 MBH						
Demolish	Ea.		114		114	Includes material and labor to install
Install	Ea.	1375	252		1627	an up flow gas fired hot air furnace
Demolish and Install	Ea.	1375	366		1741	rated at 80 MBH.
Reinstall	Ea.		201.78		201.78	
Clean	Ea.	2.04	91.50		93.54	
Minimum Charge	Job		264		264	
100 MBH						
Demolish	Ea.		114		114	Includes material and labor to install
Install	Ea.	1525	284		1809	an up flow gas fired hot air furnace
Demolish and Install	Ea.	1525	398		1923	rated at 100 MBH.
Reinstall	Ea.		227		227	
Clean	Ea.	2.04	91.50		93.54	
Minimum Charge	Job		264		264	
120 MBH						
Demolish	Ea.		114		114	Includes material and labor to install
Install	Ea.	1550	305		1855	an up flow gas fired hot air furnace
Demolish and Install	Ea.	1550	419		1969	rated at 120 MBH.
Reinstall	Ea.		242.13		242.13	
Clean	Ea.	2.04	91.50		93.54	
Minimum Charge	Job		264		264	

Furnace Unit		Unit	Material	Labor	Equip.	Total	Specification
Electric Forced Air							
31.4 MBH							
	Demolish	Ea.		114		114	Includes material and labor to install
	Install	Ea.	2100	255		2355	an electric hot air furnace rated at
	Demolish and Install	Ea.	2100	369		2469	31.4 MBH.
	Reinstall	Ea.		204.34		204.34	
	Clean	Ea.	2.04	91.50		93.54	
	Minimum Charge	Job		264		264	
47.1 MBH							
	Demolish	Ea.		114		114	Includes material and labor to install
	Install	Ea.	2500	267		2767	an electric hot air furnace rated at
	Demolish and Install	Ea.	2500	381		2881	47.1 MBH.
	Reinstall	Ea.		213.84		213.84	
	Clean	Ea.	2.04	91.50		93.54	
	Minimum Charge	Job		264		264	
62.9 MBH							
	Demolish	Ea.		114		114	Includes material and labor to install
	Install	Ea.	2675	287		2962	an electric hot air furnace rated at
	Demolish and Install	Ea.	2675	401		3076	62.9 MBH.
	Reinstall	Ea.		229.88		229.88	
	Clean	Ea.	2.04	91.50		93.54	
	Minimum Charge	Job		264		264	
78.5 MBH							
	Demolish	Ea.		114		114	Includes material and labor to install
	Install	Ea.	3300	295		3595	an electric hot air furnace rated at
	Demolish and Install	Ea.	3300	409		3709	78.5 MBH.
	Reinstall	Ea.		235.77		235.77	
	Clean	Ea.	2.04	91.50		93.54	
	Minimum Charge	Job		264		264	
110 MBH							
	Demolish	Ea.		114		114	Includes material and labor to install
	Install	Ea.	4075	310		4385	an electric hot air furnace rated at
	Demolish and Install	Ea.	4075	424		4499	110 MBH.
	Reinstall	Ea.		248.52		248.52	
	Clean	Ea.	2.04	91.50		93.54	
	Minimum Charge	Job		264		264	
A/C Junction Box							
	Install	Ea.	57	121		178	Includes labor and materials to install
	Minimum Charge	Job		264		264	a heating system emergency shut-off
							switch. Cost includes up to 40 feet of
							non-metallic cable.

Boiler		Unit	Material	Labor	Equip.	Total	Specification
Gas-Fired							
60 MBH							
	Demolish	Ea.		960		960	Includes material and labor to install a
	Install	Ea.	2950	1725		4675	gas fired cast iron boiler rated at 60
	Demolish and Install	Ea.	2950	2685		5635	MBH.
	Reinstall	Ea.		1383.56		1383.56	
	Clean	Ea.	4.07	55		59.07	
	Minimum Charge	Job		264		264	

Boiler

Boiler	Unit	Material	Labor	Equip.	Total	Specification
Gas-Fired						
125 MBH						
Demolish	Ea.		2650		2650	Includes material and labor to install a
Install	Ea.	4750	2125		6875	gas fired cast iron boiler rated at 125
Demolish and Install	Ea.	4750	4775		9525	MBH.
Reinstall	Ea.		1691.02		1691.02	
Clean	Ea.	4.07	55		59.07	
Minimum Charge	Job		264		264	
Gas-Fired						
400 MBH						
Demolish	Ea.		2650		2650	Includes material and labor to install a
Install	Ea.	6125	2975		9100	gas fired cast iron boiler rated at 400
Demolish and Install	Ea.	6125	5625		11750	MBH.
Reinstall	Ea.		2972.50		2972.50	
Clean	Ea.	4.07	55		59.07	
Minimum Charge	Job		264		264	

Heat Pump	Unit	Material	Labor	Equip.	Total	Specification
2 Ton						
Demolish	Ea.		264		264	Includes material and labor to install a
Install	Ea.	1750	475		2225	2 ton heat pump.
Demolish and Install	Ea.	1750	739		2489	
Reinstall	Ea.		380.48		380.48	
Clean	Ea.	4.07	41.50		45.57	
Minimum Charge	Job		264		264	
3 Ton						
Demolish	Ea.		440		440	Includes material and labor to install a
Install	Ea.	2250	795		3045	3 ton heat pump.
Demolish and Install	Ea.	2250	1235		3485	
Reinstall	Ea.		634.13		634.13	
Clean	Ea.	4.07	41.50		45.57	
Minimum Charge	Job		264		264	
4 Ton						
Demolish	Ea.		660		660	Includes material and labor to install a
Install	Ea.	2625	1200		3825	4 ton heat pump.
Demolish and Install	Ea.	2625	1860		4485	
Reinstall	Ea.		951.20		951.20	
Clean	Ea.	4.07	41.50		45.57	
Minimum Charge	Job		264		264	
5 Ton						
Demolish	Ea.		1050		1050	Includes material and labor to install a
Install	Ea.	2825	1900		4725	5 ton heat pump.
Demolish and Install	Ea.	2825	2950		5775	
Reinstall	Ea.		1521.92		1521.92	
Clean	Ea.	4.07	41.50		45.57	
Minimum Charge	Job		264		264	
Thermostat						
Demolish	Ea.		14.70		14.70	Includes material and labor to install
Install	Ea.	209	66		275	an electric thermostat, 2 set back.
Demolish and Install	Ea.	209	80.70		289.70	
Reinstall	Ea.		52.84		52.84	
Clean	Ea.	.08	10.30		10.38	
Minimum Charge	Job		264		264	

Rough Mechanical

Heat Pump	Unit	Material	Labor	Equip.	Total	Specification
A/C Junction Box						
Install	Ea.	505	370		875	Includes labor and materials to install
Minimum Charge	Job		241		241	a heat pump circuit including up to 40 feet of non-metallic cable.

Heating	Unit	Material	Labor	Equip.	Total	Specification
Electric Baseboard						
Demolish	L.F.		6.70		6.70	Includes material and labor to install
Install	L.F.	10.25	16.10		26.35	electric baseboard heating elements.
Demolish and Install	L.F.	10.25	22.80		33.05	
Reinstall	L.F.		12.87		12.87	
Clean	L.F.	1.43	5.15		6.58	
Minimum Charge	Job		241		241	
2' 6"						
Demolish	Ea.		22.50		22.50	Includes material and labor to install a
Install	Ea.	28.50	60.50		89	2' 6" long section of electric
Demolish and Install	Ea.	28.50	83		111.50	baseboard heating elements.
Reinstall	Ea.		48.28		48.28	
Clean	Ea.	.41	12.70		13.11	
Minimum Charge	Job		241		241	
3'						
Demolish	Ea.		24.50		24.50	Includes material and labor to install a
Install	Ea.	34.50	60.50		95	3' long section of electric baseboard
Demolish and Install	Ea.	34.50	85		119.50	heating elements.
Reinstall	Ea.		48.28		48.28	
Clean	Ea.	.41	15.70		16.11	
Minimum Charge	Job		241		241	
4'						
Demolish	Ea.		27		27	Includes material and labor to install a
Install	Ea.	39	72		111	4' long section of electric baseboard
Demolish and Install	Ea.	39	99		138	heating elements.
Reinstall	Ea.		57.65		57.65	
Clean	Ea.	.17	3.44		3.61	
Minimum Charge	Job		241		241	
5'						
Demolish	Ea.		30		30	Includes material and labor to install a
Install	Ea.	51	84.50		135.50	5' long section of electric baseboard
Demolish and Install	Ea.	51	114.50		165.50	heating elements.
Reinstall	Ea.		67.76		67.76	
Clean	Ea.	.81	25.50		26.31	
Minimum Charge	Job		241		241	
6'						
Demolish	Ea.		33.50		33.50	Includes material and labor to install a
Install	Ea.	56	96.50		152.50	6' long section of electric baseboard
Demolish and Install	Ea.	56	130		186	heating elements.
Reinstall	Ea.		77.25		77.25	
Clean	Ea.	.17	4.13		4.30	
Minimum Charge	Job		241		241	
8'						
Demolish	Ea.		44.50		44.50	Includes material and labor to install
Install	Ea.	61.50	121		182.50	an 8' long section of electric
Demolish and Install	Ea.	61.50	165.50		227	baseboard heating elements.
Reinstall	Ea.		96.56		96.56	
Clean	Ea.	.17	5.15		5.32	
Minimum Charge	Job		241		241	

Rough Mechanical

Heating

Heating	Unit	Material	Labor	Equip.	Total	Specification
10'						
Demolish	Ea.		53.50		53.50	Includes material and labor to install a
Install	Ea.	198	146		344	10' long section of electric baseboard
Demolish and Install	Ea.	198	199.50		397.50	heating elements.
Reinstall	Ea.		117.04		117.04	
Clean	Ea.	1.63	51.50		53.13	
Minimum Charge	Job		241		241	
Electric Radiant Wall Panel						
Demolish	Ea.		36		36	Includes material and labor to install
Install	Ea.	470	96.50		566.50	an 18" x 23" electric radiant heat wall
Demolish and Install	Ea.	470	132.50		602.50	panel.
Reinstall	Ea.		77.25		77.25	
Clean	Ea.	.21	10.30		10.51	
Minimum Charge	Job		241		241	

Heating Control

Heating Control	Unit	Material	Labor	Equip.	Total	Specification
Thermostat						
Demolish	Ea.		14.70		14.70	Includes material and labor to install
Install	Ea.	209	66		275	an electric thermostat, 2 set back.
Demolish and Install	Ea.	209	80.70		289.70	
Reinstall	Ea.		52.84		52.84	
Clean	Ea.	.08	10.30		10.38	
Minimum Charge	Job		264		264	

Air Conditioning

Air Conditioning	Unit	Material	Labor	Equip.	Total	Specification
Self-contained						
2 Ton						
Demolish	Ea.		475		475	Includes material and labor to install a
Install	Ea.	4300	820		5120	rooftop air conditioning unit, 2 ton.
Demolish and Install	Ea.	4300	1295		5595	No hoisting is included.
Reinstall	Ea.		820.71		820.71	
Clean	Ea.	4.07	41.50		45.57	
Minimum Charge	Job		264		264	
3 Ton						
Demolish	Ea.		475		475	Includes material and labor to install a
Install	Ea.	4450	1075		5525	rooftop air conditioning unit, 3 ton.
Demolish and Install	Ea.	4450	1550		6000	No hoisting is included.
Reinstall	Ea.		1082.14		1082.14	
Clean	Ea.	4.07	41.50		45.57	
Minimum Charge	Job		264		264	
4 Ton						
Demolish	Ea.		475		475	Includes material and labor to install a
Install	Ea.	5425	1250		6675	rooftop air conditioning unit, 4 ton.
Demolish and Install	Ea.	5425	1725		7150	No hoisting is included.
Reinstall	Ea.		1254.88		1254.88	
Clean	Ea.	4.07	41.50		45.57	
Minimum Charge	Job		264		264	
5 Ton						
Demolish	Ea.		475		475	Includes material and labor to install a
Install	Ea.	5475	1350		6825	rooftop air conditioning unit, 5 ton.
Demolish and Install	Ea.	5475	1825		7300	No hoisting is included.
Reinstall	Ea.		1354.99		1354.99	
Clean	Ea.	4.07	41.50		45.57	
Minimum Charge	Job		264		264	

Rough Mechanical

Air Conditioning	Unit	Material	Labor	Equip.	Total	Specification
Compressor						
2 Ton						
Demolish	Ea.		440		440	Includes material and labor to install a
Install	Ea.	310	264		574	refrigeration compressor.
Demolish and Install	Ea.	310	704		1014	
Reinstall	Ea.		211.36		211.36	
Clean	Ea.	4.07	41.50		45.57	
Minimum Charge	Job		264		264	
2.5 Ton						
Demolish	Ea.		440		440	Includes material and labor to install a
Install	Ea.	365	286		651	refrigeration compressor.
Demolish and Install	Ea.	365	726		1091	
Reinstall	Ea.		228.99		228.99	
Clean	Ea.	4.07	41.50		45.57	
Minimum Charge	Job		264		264	
3 Ton						
Demolish	Ea.		440		440	Includes material and labor to install a
Install	Ea.	400	299		699	refrigeration compressor.
Demolish and Install	Ea.	400	739		1139	
Reinstall	Ea.		238.96		238.96	
Clean	Ea.	4.07	41.50		45.57	
Minimum Charge	Job		264		264	
4 Ton						
Demolish	Ea.		440		440	Includes material and labor to install a
Install	Ea.	510	240		750	refrigeration compressor.
Demolish and Install	Ea.	510	680		1190	
Reinstall	Ea.		192.15		192.15	
Clean	Ea.	4.07	41.50		45.57	
Minimum Charge	Job		264		264	
5 Ton						
Demolish	Ea.		440		440	Includes material and labor to install a
Install	Ea.	560	325		885	refrigeration compressor.
Demolish and Install	Ea.	560	765		1325	
Reinstall	Ea.		261.75		261.75	
Clean	Ea.	4.07	41.50		45.57	
Minimum Charge	Job		264		264	
Condensing Unit						
2 Ton Air-cooled						
Demolish	Ea.		126		126	Includes material and labor to install a
Install	Ea.	1425	455		1880	2 ton condensing unit.
Demolish and Install	Ea.	1425	581		2006	
Reinstall	Ea.		362.36		362.36	
Clean	Ea.	4.07	41.50		45.57	
Minimum Charge	Job		264		264	
2.5 Ton Air-cooled						
Demolish	Ea.		147		147	Includes material and labor to install a
Install	Ea.	1575	560		2135	2.5 ton condensing unit.
Demolish and Install	Ea.	1575	707		2282	
Reinstall	Ea.		447.62		447.62	
Clean	Ea.	4.07	41.50		45.57	
Minimum Charge	Job		264		264	
3 Ton Air-cooled						
Demolish	Ea.		176		176	Includes material and labor to install a
Install	Ea.	1625	730		2355	3 ton condensing unit.
Demolish and Install	Ea.	1625	906		2531	
Reinstall	Ea.		585.35		585.35	
Clean	Ea.	4.07	41.50		45.57	
Minimum Charge	Job		264		264	

Rough Mechanical

Air Conditioning

	Unit	Material	Labor	Equip.	Total	Specification
4 Ton Air-cooled						
Demolish	Ea.		264		264	Includes material and labor to install a
Install	Ea.	2150	1050		3200	4 ton condensing unit.
Demolish and Install	Ea.	2150	1314		3464	
Reinstall	Ea.		845.51		845.51	
Clean	Ea.	4.07	41.50		45.57	
Minimum Charge	Job		264		264	
5 Ton Air-cooled						
Demolish	Ea.		395		395	Includes material and labor to install a
Install	Ea.	2550	1575		4125	5 ton condensing unit.
Demolish and Install	Ea.	2550	1970		4520	
Reinstall	Ea.		1268.27		1268.27	
Clean	Ea.	4.07	41.50		45.57	
Minimum Charge	Job		264		264	
Air Handler						
1,200 CFM						
Demolish	Ea.		315		315	Includes material and labor to install a
Install	Ea.	4475	730		5205	1200 CFM capacity air handling unit.
Demolish and Install	Ea.	4475	1045		5520	
Reinstall	Ea.		585.35		585.35	
Clean	Ea.	4.07	41.50		45.57	
Minimum Charge	Job		264		264	
2,000 CFM						
Demolish	Ea.		315		315	Includes material and labor to install a
Install	Ea.	6750	865		7615	2000 CFM capacity air handling unit.
Demolish and Install	Ea.	6750	1180		7930	
Reinstall	Ea.		691.78		691.78	
Clean	Ea.	4.07	41.50		45.57	
Minimum Charge	Job		264		264	
3,000 CFM						
Demolish	Ea.		350		350	Includes material and labor to install a
Install	Ea.	9875	950		10825	3000 CFM capacity air handling unit.
Demolish and Install	Ea.	9875	1300		11175	
Reinstall	Ea.		760.96		760.96	
Clean	Ea.	4.07	41.50		45.57	
Minimum Charge	Job		264		264	
Evaporative Cooler						
4,300 CFM, 1/3 HP						
Demolish	Ea.		127		127	Includes material and labor to install
Install	Ea.	790	284		1074	an evaporative cooler.
Demolish and Install	Ea.	790	411		1201	
Reinstall	Ea.		227		227	
Clean	Ea.	1.63	41.50		43.13	
Minimum Charge	Job		264		264	
4,700 CFM, 1/2 HP						
Demolish	Ea.		127		127	Includes material and labor to install
Install	Ea.	790	293		1083	an evaporative cooler.
Demolish and Install	Ea.	790	420		1210	
Reinstall	Ea.		234.32		234.32	
Clean	Ea.	1.63	41.50		43.13	
Minimum Charge	Job		264		264	
5,600 CFM, 3/4 HP						
Demolish	Ea.		127		127	Includes material and labor to install
Install	Ea.	1400	305		1705	an evaporative cooler.
Demolish and Install	Ea.	1400	432		1832	
Reinstall	Ea.		242.13		242.13	
Clean	Ea.	1.63	41.50		43.13	
Minimum Charge	Job		264		264	

For customer support on your Contractor's Pricing Guide: Residential Repair & Remodeling, call 888.606.7279.

Air Conditioning

	Unit	Material	Labor	Equip.	Total	Specification
Master Cool Brand						
Demolish	Ea.		127		127	Includes material and labor to install
Install	Ea.	790	315		1105	an evaporative cooler.
Demolish and Install	Ea.	790	442		1232	
Reinstall	Ea.		250.48		250.48	
Clean	Ea.	1.63	41.50		43.13	
Minimum Charge	Job		264		264	
Motor						
Demolish	Ea.		71.50		71.50	Includes material and labor to install
Install	Ea.	218	221		439	an evaporative cooler motor.
Demolish and Install	Ea.	218	292.50		510.50	
Reinstall	Ea.		176.52		176.52	
Clean	Ea.	1.63	41.50		43.13	
Minimum Charge	Job		264		264	
Water Pump						
Demolish	Ea.		71.50		71.50	Includes material and labor to install
Install	Ea.	61	126		187	an evaporative cooler water pump.
Demolish and Install	Ea.	61	197.50		258.50	
Reinstall	Ea.		100.88		100.88	
Clean	Ea.	1.63	41.50		43.13	
Minimum Charge	Job		264		264	
Fan Belt						
Demolish	Ea.		14.30		14.30	Includes material and labor to install
Install	Ea.	44	25		69	an evaporative cooler fan belt.
Demolish and Install	Ea.	44	39.30		83.30	
Reinstall	Ea.		20.18		20.18	
Minimum Charge	Job		264		264	
Pads (4 ea.)						
Demolish	Ea.		14.30		14.30	Includes material and labor to install
Install	Ea.	25	63		88	pre-fabricated roof pads for an
Demolish and Install	Ea.	25	77.30		102.30	evaporative cooler.
Reinstall	Ea.		50.44		50.44	
Clean	Ea.	1.63	41.50		43.13	
Minimum Charge	Job		264		264	
Service call						
Install	Job		252		252	Includes 4 hour minimum labor charge for a sheetmetal worker.
Window Mounted						
Low Output BTU						
Install	Ea.	420	57		477	Includes material and labor to install a
Reinstall	Ea.		45.40		45.40	window air conditioning unit.
Clean	Ea.	.62	10.30		10.92	
Minimum Charge	Job		264		264	
Mid-range Output BTU						
Demolish	Ea.		40		40	Includes material and labor to install a
Install	Ea.	525	75.50		600.50	window air conditioning unit.
Demolish and Install	Ea.	525	115.50		640.50	
Reinstall	Ea.		60.53		60.53	
Clean	Ea.	.62	10.30		10.92	
Minimum Charge	Job		264		264	
High Output BTU						
Demolish	Ea.		50.50		50.50	Includes material and labor to install a
Install	Ea.	650	75.50		725.50	window air conditioning unit.
Demolish and Install	Ea.	650	126		776	
Reinstall	Ea.		60.53		60.53	
Clean	Ea.	.62	10.30		10.92	
Minimum Charge	Job		264		264	

Rough Mechanical

Air Conditioning

Air Conditioning	Unit	Material	Labor	Equip.	Total	Specification
Hotel-type Hot / Cold						
Demolish	Ea.		84		84	Includes material and labor to install
Install	Ea.	1625	315		1940	packaged air conditioning unit,
Demolish and Install	Ea.	1625	399		2024	15000 BTU cooling, 13900 BTU
Reinstall	Ea.		253.65		253.65	heating.
Clean	Ea.	.81	20.50		21.31	
Minimum Charge	Job		264		264	
Thermostat						
Demolish	Ea.		14.70		14.70	Includes material and labor to install
Install	Ea.	209	66		275	an electric thermostat, 2 set back.
Demolish and Install	Ea.	209	80.70		289.70	
Reinstall	Ea.		52.84		52.84	
Clean	Ea.	.08	10.30		10.38	
Minimum Charge	Job		264		264	
Concrete Pad						
Demolish	S.F.		2.29		2.29	Includes material and labor to install
Install	S.F.	2.70	5.20		7.90	air conditioner pad.
Demolish and Install	S.F.	2.70	7.49		10.19	
Reinstall	S.F.		4.15		4.15	
Clean	S.F.	.08	.66		.74	
Paint	S.F.	.15	.39		.54	
Minimum Charge	Job		264		264	
A/C Junction Box						
Install	Ea.	158	138		296	Includes labor and materials to install
Minimum Charge	Job		241		241	an air conditioning circuit including up to 40 feet of non-metallic cable.

Sprinkler System

Sprinkler System	Unit	Material	Labor	Equip.	Total	Specification
Exposed						
Wet Type						
Demolish	SF Flr.		.23		.23	Includes material and labor to install
Install	SF Flr.	2.42	2.21		4.63	an exposed wet pipe fire sprinkler
Demolish and Install	SF Flr.	2.42	2.44		4.86	system.
Reinstall	SF Flr.		2.21		2.21	
Minimum Charge	Job		258		258	
Dry Type						
Demolish	SF Flr.		.23		.23	Includes material and labor to install
Install	SF Flr.	2.78	2.21		4.99	an exposed dry pipe fire sprinkler
Demolish and Install	SF Flr.	2.78	2.44		5.22	system.
Reinstall	SF Flr.		2.21		2.21	
Minimum Charge	Job		258		258	
Concealed						
Wet Type						
Demolish	SF Flr.		.46		.46	Includes material and labor to install a
Install	SF Flr.	2.42	3.02		5.44	concealed wet pipe fire sprinkler
Demolish and Install	SF Flr.	2.42	3.48		5.90	system.
Reinstall	SF Flr.		3.02		3.02	
Minimum Charge	Job		258		258	
Dry Type						
Demolish	SF Flr.		.46		.46	Includes material and labor to install a
Install	SF Flr.	2.78	3.02		5.80	concealed dry pipe fire sprinkler
Demolish and Install	SF Flr.	2.78	3.48		6.26	system.
Reinstall	SF Flr.		3.02		3.02	
Minimum Charge	Job		258		258	

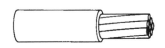
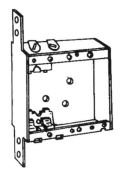

| Junction Boxes | Cable | Switch Box |

Electrical Per S.F.		Unit	Material	Labor	Equip.	Total	Specification
Residential							
Wiring, Lighting and Fixtures							
	Install	S.F.	1.58	2.01		3.59	Includes labor and material to replace house wiring, lighting fixtures, and built-in appliances. Apply S.F. cost to floor area including garages but not basements.
	Minimum Charge	Job		241		241	
Wiring Only							
	Install	S.F.	.18	.79		.97	Includes labor and material to replace house wiring only. Apply S.F. cost to floor area including garages but not basements.

Main Service		Unit	Material	Labor	Equip.	Total	Specification
Complete Panel							
100 Amp.							
	Demolish	Ea.		199		199	Cost includes material and labor to install 100 Amp. residential service panel including 12 breaker interior panel, 10 single pole breakers and hook-up. Wiring is not included.
	Install	Ea.	535	405		940	
	Demolish and Install	Ea.	535	604		1139	
	Reinstall	Ea.		324.57		324.57	
	Clean	Ea.	.41	82.50		82.91	
	Minimum Charge	Job		241		241	
150 Amp.							
	Demolish	Ea.		199		199	Cost includes material and labor to install 150 Amp. residential service, incl. 20 breaker exterior panel, main switch, 1 GFI and 14 single pole breakers, meter socket including hookup. Wiring is not included.
	Install	Ea.	775	470		1245	
	Demolish and Install	Ea.	775	669		1444	
	Reinstall	Ea.		374.99		374.99	
	Clean	Ea.	.41	82.50		82.91	
	Minimum Charge	Job		241		241	
200 Amp.							
	Demolish	Ea.		199		199	Cost includes material and labor to install 200 Amp. residential service, incl. 20 breaker exterior panel, main switch, 1 GFI and 18 single pole breakers, meter socket including hookup. Wiring is not included.
	Install	Ea.	1150	535		1685	
	Demolish and Install	Ea.	1150	734		1884	
	Reinstall	Ea.		429.16		429.16	
	Clean	Ea.	.41	82.50		82.91	
	Minimum Charge	Job		241		241	

Rough Electrical

Main Service		Unit	Material	Labor	Equip.	Total	Specification
Masthead							
	Demolish	Ea.		33.50		33.50	Includes material and labor to install
	Install	Ea.	62	201		263	galvanized steel conduit including
	Demolish and Install	Ea.	62	234.50		296.50	masthead.
	Reinstall	Ea.		160.93		160.93	
	Minimum Charge	Job		241		241	
Allowance to Repair							
	Install	Job		241		241	Minimum labor charge for residential
	Minimum Charge	Job		241		241	wiring.

For customer support on your Contractor's Pricing Guide: Residential Repair & Remodeling, call 888.606.7279.

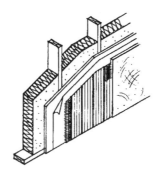

Wall Insulation

Floor Insulation		Unit	Material	Labor	Equip.	Total	Specification
Kraft Faced Batt							
3-1/2" (R-13)							
	Demolish	S.F.		.15		.15	Cost includes material and labor to
	Install	S.F.	.44	.65		1.09	install 3-1/2" (R-13) kraft paper or foil
	Demolish and Install	S.F.	.44	.80		1.24	faced roll and batt fiberglass
	Minimum Charge	Job		227		227	insulation, joists 16" O.C.
6" (R-19)							
	Demolish	S.F.		.15		.15	Cost includes material and labor to
	Install	S.F.	.56	.76		1.32	install 6" (R-19) kraft paper or foil
	Demolish and Install	S.F.	.56	.91		1.47	faced roll and batt fiberglass
	Minimum Charge	Job		227		227	insulation, joists 16" O.C.
10" (R-30)							
	Demolish	S.F.		.15		.15	Cost includes material and labor to
	Install	S.F.	.80	.91		1.71	install 10" (R-30) kraft paper-faced roll
	Demolish and Install	S.F.	.80	1.06		1.86	and batt fiberglass insulation, joists
	Minimum Charge	Job		227		227	16" O.C.
12" (R-38)							
	Demolish	S.F.		.15		.15	Cost includes material and labor to
	Install	S.F.	1.16	.96		2.12	install 12" (R-38) kraft paper-faced roll
	Demolish and Install	S.F.	1.16	1.11		2.27	and batt fiberglass insulation, joists
	Minimum Charge	Job		227		227	16" O.C.
Foil Faced Batt							
3-1/2" (R-13)							
	Demolish	S.F.		.15		.15	Cost includes material and labor to
	Install	S.F.	.44	.65		1.09	install 3-1/2" (R-13) kraft paper or foil
	Demolish and Install	S.F.	.44	.80		1.24	faced roll and batt fiberglass
	Minimum Charge	Job		227		227	insulation, joists 16" O.C.
6" (R-19)							
	Demolish	S.F.		.15		.15	Cost includes material and labor to
	Install	S.F.	.56	.76		1.32	install 6" (R-19) kraft paper or foil
	Demolish and Install	S.F.	.56	.91		1.47	faced roll and batt fiberglass
	Minimum Charge	Job		227		227	insulation, joists 16" O.C.
Unfaced Batt							
3-1/2" (R-13)							
	Demolish	S.F.		.15		.15	Cost includes material and labor to
	Install	S.F.	.36	.76		1.12	install 3-1/2" (R-13) unfaced roll and
	Demolish and Install	S.F.	.36	.91		1.27	batt fiberglass insulation, joists 16"
	Minimum Charge	Job		227		227	O.C.

Insulation

Floor Insulation

		Unit	Material	Labor	Equip.	Total	Specification
6" (R-19)							
	Demolish	S.F.		.15		.15	Cost includes material and labor to
	Install	S.F.	.43	.91		1.34	install 6" (R-19) unfaced roll and batt
	Demolish and Install	S.F.	.43	1.06		1.49	fiberglass insulation, joists 16" O.C.
	Minimum Charge	Job		227		227	
10" (R-30)							
	Demolish	S.F.		.15		.15	Cost includes material and labor to
	Install	S.F.	.66	1.01		1.67	install 10" (R-30) unfaced roll and batt
	Demolish and Install	S.F.	.66	1.16		1.82	fiberglass insulation, joists 16" O.C.
	Minimum Charge	Job		227		227	
12" (R-38)							
	Demolish	S.F.		.15		.15	Cost includes material and labor to
	Install	S.F.	.83	1.07		1.90	install 12" (R-38) unfaced roll and batt
	Demolish and Install	S.F.	.83	1.22		2.05	fiberglass insulation, joists 16" O.C.
	Minimum Charge	Job		227		227	
Rigid Board							
1/2"							
	Demolish	S.F.		.15		.15	Includes material and labor to install
	Install	S.F.	.34	.57		.91	rigid foam insulation board foil faced
	Demolish and Install	S.F.	.34	.72		1.06	two sides.
	Minimum Charge	Job		227		227	
3/4"							
	Demolish	S.F.		.15		.15	Includes material and labor to install
	Install	S.F.	.42	.57		.99	rigid foam insulation board foil faced
	Demolish and Install	S.F.	.42	.72		1.14	two sides.
	Minimum Charge	Job		227		227	
1-1/2"							
	Demolish	S.F.		.15		.15	Includes material and labor to install
	Install	S.F.	.65	.61		1.26	rigid foam insulation board foil faced
	Demolish and Install	S.F.	.65	.76		1.41	two sides.
	Minimum Charge	Job		227		227	
Support Wire							
	Demolish	S.F.		.06		.06	Includes material and labor to install
	Install	S.F.	.02	.13		.15	friction fit support wire.
	Demolish and Install	S.F.	.02	.19		.21	
	Minimum Charge	Job		227		227	
Vapor Barrier							
	Demolish	S.F.		.08		.08	Cost includes material and labor to
	Install	S.F.	.03	.12		.15	install 4 mil polyethylene vapor barrier
	Demolish and Install	S.F.	.03	.20		.23	10' wide sheets with 6" overlaps.
	Minimum Charge	Job		227		227	

Wall Insulation

		Unit	Material	Labor	Equip.	Total	Specification
Kraft Faced Batt							
3-1/2" (R-13)							
	Demolish	S.F.		.15		.15	Cost includes material and labor to
	Install	S.F.	.36	.34		.70	install 3-1/2" (R-13) kraft paper-faced
	Demolish and Install	S.F.	.36	.49		.85	roll and batt fiberglass insulation, joist
	Minimum Charge	Job		227		227	or studs 16" O.C.
6" (R-19)							
	Demolish	S.F.		.15		.15	Cost includes material and labor to
	Install	S.F.	.48	.34		.82	install 6" (R-19) kraft paper-faced roll
	Demolish and Install	S.F.	.48	.49		.97	and batt fiberglass insulation, joist or
	Minimum Charge	Job		227		227	studs 16" O.C.

Insulation

Wall Insulation		Unit	Material	Labor	Equip.	Total	Specification
10″ (R-30)							
	Demolish	S.F.		.15		.15	Cost includes material and labor to
	Install	S.F.	.80	.40		1.20	install 10″ (R-30) kraft paper-faced roll
	Demolish and Install	S.F.	.80	.55		1.35	and batt fiberglass insulation, joist or
	Minimum Charge	Job		227		227	studs 16″ O.C.
12″ (R-38)							
	Demolish	S.F.		.15		.15	Cost includes material and labor to
	Install	S.F.	1.16	.40		1.56	install 12″ (R-38) kraft paper-faced roll
	Demolish and Install	S.F.	1.16	.55		1.71	and batt fiberglass insulation, joist or
	Minimum Charge	Job		227		227	studs 16″ O.C.
Foil Faced Batt							
3-1/2″ (R-13)							
	Demolish	S.F.		.15		.15	Cost includes material and labor to
	Install	S.F.	.52	.34		.86	install 3-1/2″ (R-13) foil faced roll and
	Demolish and Install	S.F.	.52	.49		1.01	batt fiberglass insulation, joist or studs
	Minimum Charge	Job		227		227	16″ O.C.
6″ (R-19)							
	Demolish	S.F.		.15		.15	Cost includes material and labor to
	Install	S.F.	.68	.34		1.02	install 6″ (R-19) foil faced roll and batt
	Demolish and Install	S.F.	.68	.49		1.17	fiberglass insulation, joist or studs 16″
	Minimum Charge	Job		227		227	O.C.
Unfaced Batt							
3-1/2″ (R-13)							
	Demolish	S.F.		.15		.15	Cost includes material and labor to
	Install	S.F.	.36	.34		.70	install 3-1/2″ (R-13) unfaced roll and
	Demolish and Install	S.F.	.36	.49		.85	batt fiberglass insulation, joist or studs
	Minimum Charge	Job		227		227	16″ O.C.
6″ (R-19)							
	Demolish	S.F.		.15		.15	Cost includes material and labor to
	Install	S.F.	.43	.40		.83	install 6″ (R-19) unfaced roll and batt
	Demolish and Install	S.F.	.43	.55		.98	fiberglass insulation, joist or studs 16″
	Minimum Charge	Job		227		227	O.C.
10″ (R-30)							
	Demolish	S.F.		.15		.15	Cost includes material and labor to
	Install	S.F.	.66	.40		1.06	install 10″ (R-30) unfaced roll and batt
	Demolish and Install	S.F.	.66	.55		1.21	fiberglass insulation, joist or studs 16″
	Minimum Charge	Job		227		227	O.C.
12″ (R-38)							
	Demolish	S.F.		.15		.15	Cost includes material and labor to
	Install	S.F.	.83	.45		1.28	install 12″ (R-38) unfaced roll and batt
	Demolish and Install	S.F.	.83	.60		1.43	fiberglass insulation, joist or studs 16″
	Minimum Charge	Job		227		227	O.C.
Blown							
R-13							
	Demolish	S.F.		.15		.15	Includes material and labor to install
	Install	S.F.	.29	1.47	.65	2.41	fiberglass blown-in insulation in wall
	Demolish and Install	S.F.	.29	1.62	.65	2.56	with wood siding.
	Minimum Charge	Job		510	226	736	
R-19							
	Demolish	S.F.		.15		.15	Includes material and labor to install
	Install	S.F.	.44	1.69	.75	2.88	fiberglass blown-in insulation in wall
	Demolish and Install	S.F.	.44	1.84	.75	3.03	with wood siding.
	Minimum Charge	Job		510	226	736	

For customer support on your Contractor's Pricing Guide: Residential Repair & Remodeling, call 888.606.7279.

Insulation

Wall Insulation

Wall Insulation	Unit	Material	Labor	Equip.	Total	Specification
Rigid Board						
1/2"						
Demolish	S.F.		.18		.18	Includes material and labor to install
Install	S.F.	.34	.57		.91	rigid foam insulation board, foil faced,
Demolish and Install	S.F.	.34	.75		1.09	two sides.
Minimum Charge	Job		227		227	
3/4"						
Demolish	S.F.		.18		.18	Includes material and labor to install
Install	S.F.	.42	.57		.99	rigid foam insulation board, foil faced,
Demolish and Install	S.F.	.42	.75		1.17	two sides.
Minimum Charge	Job		227		227	
1-1/2"						
Demolish	S.F.		.18		.18	Includes material and labor to install
Install	S.F.	.65	.62		1.27	rigid foam insulation board, foil faced,
Demolish and Install	S.F.	.65	.80		1.45	two sides.
Minimum Charge	Job		227		227	
Vapor Barrier						
Demolish	S.F.		.05		.05	Cost includes material and labor to
Install	S.F.	.03	.12		.15	install 4 mil polyethylene vapor barrier
Demolish and Install	S.F.	.03	.17		.20	10' wide sheets with 6" overlaps.
Minimum Charge	Job		227		227	

Ceiling Insulation

Ceiling Insulation	Unit	Material	Labor	Equip.	Total	Specification
Kraft Faced Batt						
3-1/2" (R-13)						
Demolish	S.F.		.15		.15	Cost includes material and labor to
Install	S.F.	.36	.34		.70	install 3-1/2" (R-13) kraft paper-faced
Demolish and Install	S.F.	.36	.49		.85	roll and batt fiberglass insulation, joist
Minimum Charge	Job		227		227	or studs 16" O.C.
6" (R-19)						
Demolish	S.F.		.15		.15	Cost includes material and labor to
Install	S.F.	.48	.34		.82	install 6" (R-19) kraft paper-faced roll
Demolish and Install	S.F.	.48	.49		.97	and batt fiberglass insulation, joist or
Minimum Charge	Job		227		227	studs 16" O.C.
10" (R-30)						
Demolish	S.F.		.15		.15	Cost includes material and labor to
Install	S.F.	.80	.40		1.20	install 10" (R-30) kraft paper-faced roll
Demolish and Install	S.F.	.80	.55		1.35	and batt fiberglass insulation, joist or
Minimum Charge	Job		227		227	studs 16" O.C.
12" (R-38)						
Demolish	S.F.		.15		.15	Cost includes material and labor to
Install	S.F.	1.16	.40		1.56	install 12" (R-38) kraft paper-faced roll
Demolish and Install	S.F.	1.16	.55		1.71	and batt fiberglass insulation, joist or
Minimum Charge	Job		227		227	studs 16" O.C.
Foil Faced Batt						
3-1/2" (R-13)						
Demolish	S.F.		.15		.15	Cost includes material and labor to
Install	S.F.	.52	.34		.86	install 3-1/2" (R-13) foil faced roll and
Demolish and Install	S.F.	.52	.49		1.01	batt fiberglass insulation, joist or studs
Minimum Charge	Job		227		227	16" O.C.

For customer support on your Contractor's Pricing Guide: Residential Repair & Remodeling, call 888.606.7279.

Insulation

Ceiling Insulation		Unit	Material	Labor	Equip.	Total	Specification
6" (R-19)							
	Demolish	S.F.		.15		.15	Cost includes material and labor to
	Install	S.F.	.68	.34		1.02	install 6" (R-19) foil faced roll and batt
	Demolish and Install	S.F.	.68	.49		1.17	fiberglass insulation, joist or studs 16"
	Minimum Charge	Job		227		227	O.C.
Unfaced Batt							
3-1/2" (R-13)							
	Demolish	S.F.		.15		.15	Cost includes material and labor to
	Install	S.F.	.36	.34		.70	install 3-1/2" (R-13) unfaced roll and
	Demolish and Install	S.F.	.36	.49		.85	batt fiberglass insulation, joist or studs
	Minimum Charge	Job		227		227	16" O.C.
6" (R-19)							
	Demolish	S.F.		.15		.15	Cost includes material and labor to
	Install	S.F.	.43	.40		.83	install 6" (R-19) unfaced roll and batt
	Demolish and Install	S.F.	.43	.55		.98	fiberglass insulation, joist or studs 16"
	Minimum Charge	Job		227		227	O.C.
10" (R-30)							
	Demolish	S.F.		.15		.15	Cost includes material and labor to
	Install	S.F.	.66	.40		1.06	install 10" (R-30) unfaced roll and batt
	Demolish and Install	S.F.	.66	.55		1.21	fiberglass insulation, joist or studs 16"
	Minimum Charge	Job		227		227	O.C.
12" (R-38)							
	Demolish	S.F.		.15		.15	Cost includes material and labor to
	Install	S.F.	.83	.45		1.28	install 12" (R-38) unfaced roll and batt
	Demolish and Install	S.F.	.83	.60		1.43	fiberglass insulation, joist or studs 16"
	Minimum Charge	Job		227		227	O.C.
Blown							
5"							
	Demolish	S.F.		.46		.46	Includes material and labor to install
	Install	S.F.	.21	.27	.12	.60	fiberglass blown-in insulation.
	Demolish and Install	S.F.	.21	.73	.12	1.06	
	Minimum Charge	Job		510	226	736	
6"							
	Demolish	S.F.		.46		.46	Includes material and labor to install
	Install	S.F.	.29	.34	.15	.78	fiberglass blown-in insulation.
	Demolish and Install	S.F.	.29	.80	.15	1.24	
	Minimum Charge	Job		510	226	736	
9"							
	Demolish	S.F.		.46		.46	Includes material and labor to install
	Install	S.F.	.36	.46	.21	1.03	fiberglass blown-in insulation.
	Demolish and Install	S.F.	.36	.92	.21	1.49	
	Minimum Charge	Job		510	226	736	
12"							
	Demolish	S.F.		.46		.46	Includes material and labor to install
	Install	S.F.	.51	.68	.30	1.49	fiberglass blown-in insulation.
	Demolish and Install	S.F.	.51	1.14	.30	1.95	
	Minimum Charge	Job		510	226	736	
Rigid Board							
1/2"							
	Demolish	S.F.		.25		.25	Includes material and labor to install
	Install	S.F.	.34	.57		.91	rigid foam insulation board, foil faced,
	Demolish and Install	S.F.	.34	.82		1.16	two sides.
	Minimum Charge	Job		227		227	

Insulation

Ceiling Insulation

Ceiling Insulation	Unit	Material	Labor	Equip.	Total	Specification
3/4"						
Demolish	S.F.		.25		.25	Includes material and labor to install
Install	S.F.	.42	.57		.99	rigid foam insulation board, foil faced,
Demolish and Install	S.F.	.42	.82		1.24	two sides.
Minimum Charge	Job		227		227	
1-1/2"						
Demolish	S.F.		.25		.25	Includes material and labor to install
Install	S.F.	.65	.62		1.27	rigid foam insulation board, foil faced,
Demolish and Install	S.F.	.65	.87		1.52	two sides.
Minimum Charge	Job		227		227	

Exterior Insulation

Exterior Insulation	Unit	Material	Labor	Equip.	Total	Specification
Rigid Board						
1/2"						
Demolish	S.F.		.14		.14	Includes material and labor to install
Install	S.F.	.34	.57		.91	rigid foam insulation board foil faced
Demolish and Install	S.F.	.34	.71		1.05	two sides.
Minimum Charge	Job		227		227	
3/4"						
Demolish	S.F.		.14		.14	Includes material and labor to install
Install	S.F.	.42	.57		.99	rigid foam insulation board foil faced
Demolish and Install	S.F.	.42	.71		1.13	two sides.
Minimum Charge	Job		227		227	
1-1/2"						
Demolish	S.F.		.14		.14	Includes material and labor to install
Install	S.F.	.65	.61		1.26	rigid foam insulation board foil faced
Demolish and Install	S.F.	.65	.75		1.40	two sides.
Minimum Charge	Job		227		227	
Vapor Barrier						
Demolish	S.F.		.08		.08	Cost includes material and labor to
Install	S.F.	.03	.12		.15	install 4 mil polyethylene vapor barrier
Demolish and Install	S.F.	.03	.20		.23	10' wide sheets with 6" overlaps.
Minimum Charge	Job		227		227	

Roof Insulation Board

Roof Insulation Board	Unit	Material	Labor	Equip.	Total	Specification
Fiberglass						
3/4" Fiberglass						
Demolish	Sq.		48		48	Includes material and labor to install
Install	Sq.	67	26		93	fiberglass board insulation 3/4" thick,
Demolish and Install	Sq.	67	74		141	R-2.80 and C-0.36 values.
Minimum Charge	Job		227		227	
1" Fiberglass						
Demolish	Sq.		48		48	Includes material and labor to install
Install	Sq.	95	29		124	fiberglass board insulation 1" thick,
Demolish and Install	Sq.	95	77		172	R-4.20 and C-0.24 values.
Minimum Charge	Job		227		227	
1-3/8" Fiberglass						
Demolish	Sq.		48		48	Includes material and labor to install
Install	Sq.	151	32		183	fiberglass board insulation 1-3/8"
Demolish and Install	Sq.	151	80		231	thick, R-5.30 and C-0.19 values.
Minimum Charge	Job		227		227	

Insulation

Roof Insulation Board		Unit	Material	Labor	Equip.	Total	Specification
1-5/8" Fiberglass							
	Demolish	Sq.		48		48	Includes material and labor to install
	Install	Sq.	156	33.50		189.50	fiberglass board insulation 1-5/8"
	Demolish and Install	Sq.	156	81.50		237.50	thick, R-6.70 and C-0.15 values.
	Minimum Charge	Job		227		227	
2-1/4" Fiberglass							
	Demolish	Sq.		48		48	Includes material and labor to install
	Install	Sq.	185	37		222	fiberglass board insulation 2-1/4"
	Demolish and Install	Sq.	185	85		270	thick, R-8.30 and C-0.12 values.
	Minimum Charge	Job		227		227	
Perlite							
1" Perlite							
	Demolish	Sq.		48		48	Includes material and labor to install
	Install	Sq.	58.50	29		87.50	perlite board insulation 1" thick,
	Demolish and Install	Sq.	58.50	77		135.50	R-2.80 and C-0.36 values.
	Minimum Charge	Job		227		227	
4" Perlite							
	Demolish	Sq.		48		48	Includes material and labor to install
	Install	Sq.	233	56		289	perlite board insulation 4" thick,
	Demolish and Install	Sq.	233	104		337	R-10.0 and C-0.10 values.
	Minimum Charge	Job		227		227	
1-1/2" Perlite							
	Demolish	Sq.		48		48	Includes material and labor to install
	Install	Sq.	86	32		118	perlite board insulation 1-1/2" thick,
	Demolish and Install	Sq.	86	80		166	R-4.20 and C-0.24 values.
	Minimum Charge	Job		227		227	
2" Perlite							
	Demolish	Sq.		48		48	Includes material and labor to install
	Install	Sq.	117	39		156	perlite board insulation 2" thick,
	Demolish and Install	Sq.	117	87		204	R-5.30 and C-0.19 values.
	Minimum Charge	Job		227		227	
2-1/2" Perlite							
	Demolish	Sq.		48		48	Includes material and labor to install
	Install	Sq.	154	43.50		197.50	perlite board insulation 2-1/2" thick,
	Demolish and Install	Sq.	154	91.50		245.50	R-6.70 and C-0.15 values.
	Minimum Charge	Job		227		227	
3" Perlite							
	Demolish	Sq.		48		48	Includes material and labor to install
	Install	Sq.	175	46.50		221.50	perlite board insulation 3" thick,
	Demolish and Install	Sq.	175	94.50		269.50	R-8.30 and C-0.12 values.
	Minimum Charge	Job		227		227	
Polystyrene							
2" Polystyrene							
	Demolish	Sq.		48		48	Includes material and labor to install
	Install	Sq.	57	37		94	extruded polystyrene board, R5.0/ 2"
	Demolish and Install	Sq.	57	85		142	in thickness @25PSI compressive
	Minimum Charge	Job		227		227	strength.
1" Polystyrene							
	Demolish	Sq.		48		48	Includes material and labor to install
	Install	Sq.	28.50	25		53.50	extruded polystyrene board, R5.0/ 1"
	Demolish and Install	Sq.	28.50	73		101.50	in thickness @25PSI compressive
	Minimum Charge	Job		227		227	strength.
1-1/2" Polystyrene							
	Demolish	Sq.		48		48	Includes material and labor to install
	Install	Sq.	43	29		72	extruded polystyrene board, R5.0/
	Demolish and Install	Sq.	43	77		120	1-1/2" in thickness @25PSI
	Minimum Charge	Job		227		227	compressive strength.

Insulation

Roof Insulation Board		Unit	Material	Labor	Equip.	Total	Specification
Urethane							
3/4" Urethane							
	Demolish	Sq.		48		48	Includes material and labor to install
	Install	Sq.	50.50	22		72.50	urethane board insulation 3/4" thick,
	Demolish and Install	Sq.	50.50	70		120.50	R-5.30 and C-0.19 values.
	Minimum Charge	Job		227		227	
1" Urethane							
	Demolish	Sq.		48		48	Includes material and labor to install
	Install	Sq.	57	23.50		80.50	urethane board insulation 1" thick,
	Demolish and Install	Sq.	57	71.50		128.50	R-6.70 and C-0.15 values.
	Minimum Charge	Job		227		227	
1-1/2" Urethane							
	Demolish	Sq.		48		48	Includes material and labor to install
	Install	Sq.	75	27		102	urethane board insulation 1-1/2"
	Demolish and Install	Sq.	75	75		150	thick, R-10.1 and C-0.09 values.
	Minimum Charge	Job		227		227	
2" Urethane							
	Demolish	Sq.		48		48	Includes material and labor to install
	Install	Sq.	91.50	32		123.50	urethane board insulation 2" thick,
	Demolish and Install	Sq.	91.50	80		171.50	R-13.4 and C-0.06 values.
	Minimum Charge	Job		227		227	

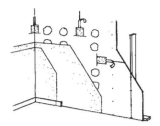

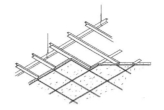

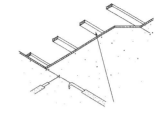

Gypsum Wallboard   **Tile Ceiling**   **Gypsum Wallboard Ceiling**

Wall Plaster	Unit	Material	Labor	Equip.	Total	Specification
Interior						
Cement						
Demolish	S.Y.		6.10		6.10	Includes material and labor to install
Install	S.Y.	14.95	26	2.09	43.04	cement plaster on interior wood
Demolish and Install	S.Y.	14.95	32.10	2.09	49.14	framing with metal lath.
Clean	S.Y.		3		3	
Paint	S.Y.	2.08	5.05		7.13	
Minimum Charge	Job		206		206	
Acoustic						
Demolish	S.Y.		6.10		6.10	Includes material and labor to install
Install	S.Y.	11.55	32	2.08	45.63	acoustical cement plaster on interior
Demolish and Install	S.Y.	11.55	38.10	2.08	51.73	wood framing with metal lath.
Clean	S.Y.		3		3	
Paint	S.Y.	2.08	5.05		7.13	
Minimum Charge	Job		206		206	
Gypsum						
Demolish	S.Y.		6.10		6.10	Includes material and labor to install
Install	S.Y.	10.40	27	1.77	39.17	gypsum wall or ceiling plaster on
Demolish and Install	S.Y.	10.40	33.10	1.77	45.27	interior wood framing with metal lath.
Clean	S.Y.		3		3	
Paint	S.Y.	2.08	5.05		7.13	
Minimum Charge	Job		206		206	
Stucco						
Demolish	S.Y.		14.65		14.65	Includes material and labor to install
Install	S.Y.	7.05	37.50	2.45	47	stucco on wire mesh over wood
Demolish and Install	S.Y.	7.05	52.15	2.45	61.65	framing.
Clean	S.Y.		3		3	
Paint	S.Y.	2.08	5.05		7.13	
Minimum Charge	Job		206		206	
Thin Coat						
Veneer						
Demolish	S.Y.		6.10		6.10	Includes materials and labor to install
Install	S.Y.	.99	4.83	.39	6.21	thin coat veneer plaster.
Demolish and Install	S.Y.	.99	10.93	.39	12.31	
Clean	S.Y.		3		3	
Paint	S.Y.	2.08	5.05		7.13	
Minimum Charge	Job		206		206	

Walls / Ceilings

Wall Plaster

Wall Plaster	Unit	Material	Labor	Equip.	Total	Specification
Lath						
Metal						
Demolish	S.Y.		1.44		1.44	Includes material and labor to install
Install	S.Y.	4.75	5.70		10.45	wire lath on steel framed walls.
Demolish and Install	S.Y.	4.75	7.14		11.89	
Clean	S.Y.		3		3	
Minimum Charge	Job		101		101	
Gypsum						
Demolish	S.Y.		1.44		1.44	Includes material and labor to install
Install	S.Y.	2.67	5.35		8.02	gypsum lath on walls.
Demolish and Install	S.Y.	2.67	6.79		9.46	
Clean	S.Y.		3		3	
Minimum Charge	Job		206		206	
Repairs						
Fill in Small Cracks						
Install	L.F.	.07	3.29		3.36	Includes labor and material to fill in
Minimum Charge	Job		206		206	hairline cracks up to 1/8" with filler, sand and prep for paint.
Repair up to 1" Stock						
Install	L.F.	.30	4.11		4.41	Includes labor and material to fill in
Minimum Charge	Job		206		206	hairline cracks up to 1" with filler, sand and prep for paint.
Patch Small Hole (2")						
Install	Ea.	.07	4.33		4.40	Includes labor and material to patch
Minimum Charge	Job		206		206	up to a 2" diameter hole in plaster.
Patch Large Hole						
Install	Ea.	8.85	137		145.85	Includes labor and material to patch
Minimum Charge	Job		206		206	up to a 20" diameter hole in plaster.
Patch Plaster Section						
Install	Ea.	2.30	4.11		6.41	Includes labor and material to repair
Minimum Charge	Job		206		206	section of plaster wall, area less than 20 S.F.
Repair Inside Corner						
Install	L.F.	3.54	10.30		13.84	Includes material and labor to patch
Minimum Charge	Job		206		206	plaster corners.

Ceiling Plaster

Ceiling Plaster	Unit	Material	Labor	Equip.	Total	Specification
Interior						
Cement						
Demolish	S.Y.		6.10		6.10	Includes material and labor to install
Install	S.Y.	14.95	26	2.09	43.04	cement plaster on interior wood
Demolish and Install	S.Y.	14.95	32.10	2.09	49.14	framing with metal lath.
Clean	S.Y.		3.14		3.14	
Paint	S.Y.	2.08	5.05		7.13	
Minimum Charge	Job		206		206	
Acoustic						
Demolish	S.Y.		6.10		6.10	Includes material and labor to install
Install	S.Y.	11.55	32	2.08	45.63	acoustical cement plaster on interior
Demolish and Install	S.Y.	11.55	38.10	2.08	51.73	wood framing with metal lath.
Clean	S.Y.		3.14		3.14	
Paint	S.Y.	2.08	5.05		7.13	
Minimum Charge	Job		206		206	

For customer support on your Contractor's Pricing Guide: Residential Repair & Remodeling, call 888.606.7279.

Walls / Ceilings

Ceiling Plaster

	Unit	Material	Labor	Equip.	Total	Specification
Gypsum						
Demolish	S.Y.		6.10		6.10	Includes material and labor to install
Install	S.Y.	10.40	27	1.77	39.17	gypsum wall or ceiling plaster on
Demolish and Install	S.Y.	10.40	33.10	1.77	45.27	interior wood framing with metal lath.
Clean	S.Y.		3.14		3.14	
Paint	S.Y.	2.08	5.05		7.13	
Minimum Charge	Job		206		206	
Stucco						
Demolish	S.Y.		14.65		14.65	Includes material and labor to install
Install	S.Y.	7.05	37.50	2.45	47	stucco on wire mesh over wood
Demolish and Install	S.Y.	7.05	52.15	2.45	61.65	framing.
Clean	S.Y.		3.14		3.14	
Paint	S.Y.	2.08	5.05		7.13	
Minimum Charge	Job		206		206	

Thin Coat
Veneer

	Unit	Material	Labor	Equip.	Total	Specification
Demolish	S.Y.		6.10		6.10	Includes materials and labor to install
Install	S.Y.	.99	4.83	.39	6.21	thin coat veneer plaster.
Demolish and Install	S.Y.	.99	10.93	.39	12.31	
Clean	S.Y.		3.14		3.14	
Paint	S.Y.	2.08	5.05		7.13	
Minimum Charge	Job		206		206	

Lath
Metal

	Unit	Material	Labor	Equip.	Total	Specification
Demolish	S.Y.		1.44		1.44	Includes material and labor to install
Install	S.Y.	4.75	7.15		11.90	wire lath on steel framed ceilings.
Demolish and Install	S.Y.	4.75	8.59		13.34	
Minimum Charge	Job		206		206	

Gypsum

	Unit	Material	Labor	Equip.	Total	Specification
Demolish	S.Y.		1.44		1.44	Includes material and labor to install
Install	S.Y.	2.67	4.46		7.13	gypsum lath on ceilings.
Demolish and Install	S.Y.	2.67	5.90		8.57	
Minimum Charge	Job		206		206	

Repairs
Fill in Small Cracks

	Unit	Material	Labor	Equip.	Total	Specification
Install	S.F.	.07	.75		.82	Includes labor and material to fill in
Minimum Charge	Job		410		410	hairline cracks up to 1/8" with filler, sand and prep for paint.

Gypsum Wallboard

	Unit	Material	Labor	Equip.	Total	Specification
Finished						
1/4"						
Demolish	S.F.		.37		.37	Cost includes material and labor to
Install	S.F.	.34	1.37		1.71	install 1/4" drywall, including joint
Demolish and Install	S.F.	.34	1.74		2.08	taping, premixed compound and
Clean	S.F.		.30		.30	screws.
Paint	S.F.	.25	.33		.58	
Minimum Charge	Job		227		227	
3/8"						
Demolish	S.F.		.37		.37	Cost includes material and labor to
Install	S.F.	.33	1.39		1.72	install 3/8" drywall including joint
Demolish and Install	S.F.	.33	1.76		2.09	tape, premixed compound and
Clean	S.F.		.30		.30	screws.
Paint	S.F.	.25	.33		.58	
Minimum Charge	Job		227		227	

For customer support on your Contractor's Pricing Guide: Residential Repair & Remodeling, call 888.606.7279.

Walls / Ceilings

Gypsum Wallboard

Gypsum Wallboard	Unit	Material	Labor	Equip.	Total	Specification
1/2″						
Demolish	S.F.		.37		.37	Cost includes material and labor to
Install	S.F.	.31	1.40		1.71	install 1/2″ drywall including joint
Demolish and Install	S.F.	.31	1.77		2.08	tape, premixed compound and
Clean	S.F.		.30		.30	screws.
Paint	S.F.	.25	.33		.58	
Minimum Charge	Job		227		227	
1/2″ Water Resistant						
Demolish	S.F.		.37		.37	Cost includes material and labor to
Install	S.F.	.35	1.40		1.75	install 1/2″ WR drywall including joint
Demolish and Install	S.F.	.35	1.77		2.12	tape, premixed compound and
Clean	S.F.		.30		.30	screws.
Paint	S.F.	.25	.33		.58	
Minimum Charge	Job		227		227	
1/2″ Fire-rated						
Demolish	S.F.		.37		.37	Cost includes material and labor to
Install	S.F.	.34	1.40		1.74	install 1/2″ FR drywall including
Demolish and Install	S.F.	.34	1.77		2.11	screws. Taping and finishing not
Clean	S.F.		.30		.30	included.
Paint	S.F.	.25	.33		.58	
Minimum Charge	Job		227		227	
5/8″						
Demolish	S.F.		.37		.37	Cost includes material and labor to
Install	S.F.	.36	1.35		1.71	install 5/8″ drywall including joint
Demolish and Install	S.F.	.36	1.72		2.08	tape, premixed compound and
Clean	S.F.		.30		.30	screws.
Paint	S.F.	.25	.33		.58	
Minimum Charge	Job		227		227	
5/8″ Fire-rated						
Demolish	S.F.		.37		.37	Cost includes material and labor to
Install	S.F.	.37	1.59		1.96	install 5/8″ FR drywall, including joint
Demolish and Install	S.F.	.37	1.96		2.33	taping, premixed compound and
Clean	S.F.		.30		.30	screws.
Paint	S.F.	.25	.33		.58	
Minimum Charge	Job		227		227	
2-hour Fire-rated						
Demolish	S.F.		.92		.92	Cost includes material and labor to
Install	S.F.	.80	3.15		3.95	install 2 layers of 5/8″ type X drywall
Demolish and Install	S.F.	.80	4.07		4.87	including joint taping, premixed
Clean	S.F.		.30		.30	compound and screws.
Paint	S.F.	.25	.33		.58	
Minimum Charge	Job		227		227	
Unfinished						
1/4″						
Demolish	S.F.		.37		.37	Cost includes material and labor to
Install	S.F.	.30	.48		.78	install 1/4″ drywall including screws.
Demolish and Install	S.F.	.30	.85		1.15	Taping and finishing not included.
Clean	S.F.		.30		.30	
Minimum Charge	Job		227		227	
3/8″						
Demolish	S.F.		.37		.37	Cost includes material and labor to
Install	S.F.	.29	.48		.77	install 3/8″ drywall including screws.
Demolish and Install	S.F.	.29	.85		1.14	Tape, finish or texture is not included.
Clean	S.F.		.30		.30	
Minimum Charge	Job		227		227	

For customer support on your Contractor's Pricing Guide: Residential Repair & Remodeling, call 888.606.7279.

Walls / Ceilings

Gypsum Wallboard		Unit	Material	Labor	Equip.	Total	Specification
1/2″							
	Demolish	S.F.		.37		.37	Cost includes material and labor to
	Install	S.F.	.27	.48		.75	install 1/2″ drywall including screws.
	Demolish and Install	S.F.	.27	.85		1.12	Taping and finishing not included.
	Clean	S.F.		.30		.30	
	Minimum Charge	Job		227		227	
1/2″ Water Resistant							
	Demolish	S.F.		.37		.37	Cost includes material and labor to
	Install	S.F.	.31	.48		.79	install 1/2″ WR drywall including
	Demolish and Install	S.F.	.31	.85		1.16	screws. Taping and finishing are not
	Clean	S.F.		.30		.30	included.
	Minimum Charge	Job		227		227	
1/2″ Fire-rated							
	Demolish	S.F.		.37		.37	Cost includes labor and material to
	Install	S.F.	.27	.33		.60	install 1/2″ FR drywall including
	Demolish and Install	S.F.	.27	.70		.97	screws. Tape, finish or texture is not
	Clean	S.F.		.30		.30	included.
	Minimum Charge	Job		227		227	
5/8″							
	Demolish	S.F.		.37		.37	Cost includes material and labor to
	Install	S.F.	.31	.54		.85	install 5/8″ drywall including screws.
	Demolish and Install	S.F.	.31	.91		1.22	Taping and finishing not included.
	Clean	S.F.		.30		.30	
	Minimum Charge	Job		227		227	
5/8″ Fire-rated							
	Demolish	S.F.		.37		.37	Cost includes material and labor to
	Install	S.F.	.32	.54		.86	install 5/8″ FR drywall including
	Demolish and Install	S.F.	.32	.91		1.23	screws. Taping and finishing not
	Clean	S.F.		.30		.30	included.
	Minimum Charge	Job		227		227	
2-hour Fire-rated							
	Demolish	S.F.		.92		.92	Cost includes material and labor to
	Install	S.F.	.80	3.15		3.95	install 2 layers of 5/8″ type X drywall
	Demolish and Install	S.F.	.80	4.07		4.87	including joint taping, premixed
	Clean	S.F.		.30		.30	compound and screws.
	Minimum Charge	Job		227		227	
Accessories							
Tape & Finish							
	Install	S.F.	.06	.45		.51	Includes labor and material for taping,
	Minimum Charge	Job		227		227	sanding and finish.
Texture							
Sprayed							
	Install	S.F.	.04	.54		.58	Includes labor and material to install
	Minimum Charge	Job		227		227	by spray, texture finish.
Trowel							
	Install	S.F.		.27		.27	Includes labor and material to apply
	Minimum Charge	Job		227		227	hand troweled texture.
Patch Hole in Gypsum Wallboard							
	Install	S.F.	.52	60.50		61.02	Includes labor and material to patch a
	Minimum Charge	Job		227		227	one square foot area of damaged
							drywall.
Furring Strips							
	Demolish	S.F.		.09		.09	Cost includes material and labor to
	Install	S.F.	.20	1.01		1.21	install 1″ x 2″ furring strips, 16″ O.C.
	Demolish and Install	S.F.	.20	1.10		1.30	per S.F. of surface area to be covered.
	Minimum Charge	Job		114		114	

179

Walls / Ceilings

Ceramic Wall Tile	Unit	Material	Labor	Equip.	Total	Specification
Thin Set						
2" x 2"						
Demolish	S.F.		1.22		1.22	Cost includes material and labor to
Install	S.F.	6.80	6.70		13.50	install 2" x 2" ceramic tile in organic
Demolish and Install	S.F.	6.80	7.92		14.72	adhesive including cutting and grout.
Clean	S.F.		.63		.63	
Minimum Charge	Job		202		202	
4-1/4" x 4-1/4"						
Demolish	S.F.		1.22		1.22	Cost includes material and labor to
Install	S.F.	2.74	3.78		6.52	install 4-1/4" x 4-1/4" ceramic tile in
Demolish and Install	S.F.	2.74	5		7.74	organic adhesive including grout.
Clean	S.F.		.63		.63	
Minimum Charge	Job		202		202	
6" x 6"						
Demolish	S.F.		1.22		1.22	Cost includes material and labor to
Install	S.F.	3.81	4.11		7.92	install 6" x 6" ceramic tile in organic
Demolish and Install	S.F.	3.81	5.33		9.14	adhesive including cutting and grout.
Clean	S.F.		.63		.63	
Minimum Charge	Job		202		202	
8" x 8"						
Demolish	S.F.		1.22		1.22	Cost includes material and labor to
Install	S.F.	5.25	4.23		9.48	install 8" x 8" ceramic tile in organic
Demolish and Install	S.F.	5.25	5.45		10.70	adhesive including cutting and grout.
Clean	S.F.		.63		.63	
Minimum Charge	Job		202		202	
Thick Set						
2" x 2"						
Demolish	S.F.		1.47		1.47	Cost includes material and labor to
Install	S.F.	7	8.05		15.05	install 2" x 2" ceramic tile in mortar
Demolish and Install	S.F.	7	9.52		16.52	bed including cutting and grout.
Clean	S.F.		.63		.63	
Minimum Charge	Job		202		202	
4-1/4" x 4-1/4"						
Demolish	S.F.		1.47		1.47	Cost includes material and labor to
Install	S.F.	10.35	8.05		18.40	install 4-1/4" x 4-1/4" ceramic tile in
Demolish and Install	S.F.	10.35	9.52		19.87	mortar bed including cutting and
Clean	S.F.		.63		.63	grout.
Minimum Charge	Job		202		202	
6" x 6"						
Demolish	S.F.		1.47		1.47	Cost includes material and labor to
Install	S.F.	11	7.35		18.35	install 6" x 6" ceramic tile in mortar
Demolish and Install	S.F.	11	8.82		19.82	bed including cutting and grout.
Clean	S.F.		.63		.63	
Minimum Charge	Job		202		202	
8" x 8"						
Demolish	S.F.		1.47		1.47	Cost includes material and labor to
Install	S.F.	10.25	6.70		16.95	install 8" x 8" ceramic tile in mortar
Demolish and Install	S.F.	10.25	8.17		18.42	bed including cutting and grout.
Clean	S.F.		.63		.63	
Minimum Charge	Job		202		202	
Re-grout						
Install	S.F.	.15	4.03		4.18	Includes material and labor to regrout
Minimum Charge	Job		202		202	wall tiles.

Walls / Ceilings

Wall Tile

	Unit	Material	Labor	Equip.	Total	Specification
Mirror						
Custom Cut Wall						
Demolish	S.F.		.70		.70	Includes material and labor to install
Install	S.F.	2.67	4.52		7.19	distortion-free float glass mirror, 1/8"
Demolish and Install	S.F.	2.67	5.22		7.89	to 3/16" thick with cut and polished
Clean	S.F.	.03	.19		.22	edges.
Minimum Charge	Job		220		220	
Beveled Wall						
Demolish	S.F.		.70		.70	Includes material and labor to install
Install	S.F.	7.05	8		15.05	distortion-free float glass mirror, 1/8"
Demolish and Install	S.F.	7.05	8.70		15.75	to 3/16" thick with beveled edges.
Clean	S.F.	.03	.19		.22	
Minimum Charge	Job		220		220	
Smoked Wall						
Demolish	S.F.		.70		.70	Includes material and labor to install
Install	S.F.	9.55	8		17.55	distortion-free smoked float glass
Demolish and Install	S.F.	9.55	8.70		18.25	mirror, 1/8" to 3/16" thick with cut
Clean	S.F.	.03	.19		.22	and polished edges.
Minimum Charge	Job		220		220	
Marble						
Good Grade						
Demolish	S.F.		2.29		2.29	Cost includes material and labor to
Install	S.F.	9.90	17.55		27.45	install 3/8" x 12" x 12" marble tile in
Demolish and Install	S.F.	9.90	19.84		29.74	mortar bed with grout.
Clean	S.F.		.63		.63	
Minimum Charge	Job		202		202	
Better Grade						
Demolish	S.F.		2.29		2.29	Cost includes material and labor to
Install	S.F.	11	17.55		28.55	install 3/8" x 12" x 12" marble, in
Demolish and Install	S.F.	11	19.84		30.84	mortar bed with grout.
Clean	S.F.		.63		.63	
Minimum Charge	Job		202		202	
Premium Grade						
Demolish	S.F.		2.29		2.29	Cost includes material and labor to
Install	S.F.	18.80	17.55		36.35	install 3/8" x 12" x 12" marble, in
Demolish and Install	S.F.	18.80	19.84		38.64	mortar bed with grout.
Clean	S.F.		.63		.63	
Minimum Charge	Job		202		202	
Glass Block						
Demolish	S.F.		1.83		1.83	Cost includes material and labor to
Install	S.F.	30.50	17.45		47.95	install 3-7/8" thick, 6" x 6" smooth
Demolish and Install	S.F.	30.50	19.28		49.78	face glass block including mortar and
Clean	S.F.		.63		.63	cleaning.
Minimum Charge	Job		395		395	
Cork						
Demolish	S.F.		.28		.28	Cost includes material and labor to
Install	S.F.	5.15	1.58		6.73	install 3/16" thick cork tile and
Demolish and Install	S.F.	5.15	1.86		7.01	installation.
Clean	S.F.		.63		.63	
Minimum Charge	Job		190		190	
Re-grout						
Install	S.F.	.15	4.03		4.18	Includes material and labor to regrout
Minimum Charge	Job		202		202	wall tiles.

Walls / Ceilings

Wall Paneling

	Unit	Material	Labor	Equip.	Total	Specification
Plywood						
Good Grade						
Demolish	S.F.		.37		.37	Cost includes material and labor to
Install	S.F.	1.60	1.82		3.42	install 4' x 8' x 1/4" paneling,
Demolish and Install	S.F.	1.60	2.19		3.79	including nails and adhesive.
Clean	S.F.		.30		.30	
Paint	S.F.	.33	.72		1.05	
Minimum Charge	Job		227		227	
Better Grade						
Demolish	S.F.		.37		.37	Cost includes material and labor to
Install	S.F.	1.32	2.16		3.48	install 4' x 8' x 1/4" plywood wall
Demolish and Install	S.F.	1.32	2.53		3.85	panel, including nails and adhesive.
Clean	S.F.		.30		.30	
Paint	S.F.	.33	.72		1.05	
Minimum Charge	Job		227		227	
Premium Grade						
Demolish	S.F.		.37		.37	Cost includes material and labor to
Install	S.F.	1.38	2.59		3.97	install 4' x 8' x 1/4" plywood wall
Demolish and Install	S.F.	1.38	2.96		4.34	panel, including nails and adhesive.
Clean	S.F.		.30		.30	
Paint	S.F.	.33	.72		1.05	
Minimum Charge	Job		227		227	
Hardboard						
Embossed Paneling						
Demolish	S.F.		.37		.37	Includes material and labor to install
Install	S.F.	.73	1.82		2.55	hardboard embossed paneling
Demolish and Install	S.F.	.73	2.19		2.92	including adhesives.
Clean	S.F.		.30		.30	
Minimum Charge	Job		227		227	
Plastic Molding						
Demolish	L.F.		.37		.37	Includes material and labor to install
Install	L.F.	.85	1.14		1.99	plastic trim for paneling.
Demolish and Install	L.F.	.85	1.51		2.36	
Clean	L.F.	.03	.19		.22	
Minimum Charge	Job		227		227	
Hardboard						
Embossed Brick						
Demolish	S.F.		.37		.37	Cost includes material and labor to
Install	S.F.	.95	2.02		2.97	install 4' x 8' non-ceramic mineral
Demolish and Install	S.F.	.95	2.39		3.34	brick-like veneer wall panels including
Clean	S.F.		.30		.30	adhesive.
Paint	S.F.	.23	.57		.80	
Minimum Charge	Job		227		227	
Embossed Ceramic Tile						
Demolish	S.F.		.37		.37	Includes material and labor to install
Install	S.F.	.73	1.82		2.55	hardboard embossed paneling
Demolish and Install	S.F.	.73	2.19		2.92	including adhesives.
Clean	S.F.		.30		.30	
Paint	S.F.	.23	.57		.80	
Minimum Charge	Job		227		227	
Pegboard						
Demolish	S.F.		.37		.37	Cost includes material and labor to
Install	S.F.	.53	1.82		2.35	install 4' x 8' x 1/4" pegboard panel,
Demolish and Install	S.F.	.53	2.19		2.72	including adhesive.
Clean	S.F.		.30		.30	
Paint	S.F.	.23	.57		.80	
Minimum Charge	Job		227		227	

Walls / Ceilings

Wall Paneling

	Unit	Material	Labor	Equip.	Total	Specification
Furring Strips						
Demolish	S.F.		.09		.09	Cost includes material and labor to
Install	S.F.	.20	1.01		1.21	install 1" x 2" furring strips, 16" O.C.
Demolish and Install	S.F.	.20	1.10		1.30	per S.F. of surface area to be covered.
Minimum Charge	Job		227		227	
Molding						
Unfinished Base						
Demolish	L.F.		.61		.61	Cost includes material and labor to
Install	L.F.	.57	1.37		1.94	install 1/2" x 2-1/2" pine or birch
Demolish and Install	L.F.	.57	1.98		2.55	base molding.
Reinstall	L.F.		1.10		1.10	
Clean	L.F.	.03	.19		.22	
Paint	L.F.	.04	.41		.45	
Minimum Charge	Job		227		227	
Unfinished Corner						
Demolish	L.F.		.61		.61	Includes material and labor to install
Install	L.F.	1.71	1.89		3.60	unfinished pine corner molding.
Demolish and Install	L.F.	1.71	2.50		4.21	
Reinstall	L.F.		1.51		1.51	
Clean	L.F.	.03	.19		.22	
Paint	L.F.	.14	.60		.74	
Minimum Charge	Job		227		227	
Unfinished Cove						
Demolish	L.F.		.61		.61	Cost includes material and labor to
Install	L.F.	1.98	1.71		3.69	install 11/16" x 2-3/4" pine cove
Demolish and Install	L.F.	1.98	2.32		4.30	molding.
Reinstall	L.F.		1.37		1.37	
Clean	L.F.	.03	.19		.22	
Paint	L.F.	.14	.60		.74	
Minimum Charge	Job		227		227	
Unfinished Crown						
Demolish	L.F.		.73		.73	Includes material and labor to install
Install	L.F.	2.22	1.82		4.04	unfinished pine crown molding.
Demolish and Install	L.F.	2.22	2.55		4.77	
Reinstall	L.F.		1.45		1.45	
Clean	L.F.	.04	.25		.29	
Paint	L.F.	.10	.41		.51	
Minimum Charge	Job		227		227	
Prefinished Base						
Demolish	L.F.		.61		.61	Cost includes material and labor to
Install	L.F.	1.28	1.46		2.74	install 9/16" x 3-5/16" prefinished
Demolish and Install	L.F.	1.28	2.07		3.35	base molding.
Reinstall	L.F.		1.16		1.16	
Clean	L.F.	.03	.19		.22	
Paint	L.F.	.04	.41		.45	
Minimum Charge	Job		227		227	
Prefinished Corner						
Demolish	L.F.		.61		.61	Includes material and labor to install
Install	L.F.	2.64	1.89		4.53	inside or outside prefinished corner
Demolish and Install	L.F.	2.64	2.50		5.14	molding.
Reinstall	L.F.		1.51		1.51	
Clean	L.F.	.03	.19		.22	
Paint	L.F.	.14	.60		.74	
Minimum Charge	Job		227		227	

Walls / Ceilings

Wall Paneling	Unit	Material	Labor	Equip.	Total	Specification
Prefinished Crown						
Demolish	L.F.		.73		.73	Cost includes material and labor to
Install	L.F.	1.84	1.82		3.66	install 3/4" x 3-13/16" prefinished
Demolish and Install	L.F.	1.84	2.55		4.39	crown molding.
Reinstall	L.F.		1.45		1.45	
Clean	L.F.	.04	.25		.29	
Paint	L.F.	.10	.41		.51	
Minimum Charge	Job		227		227	

Wallcovering	Unit	Material	Labor	Equip.	Total	Specification
Surface Prep						
Patch Walls						
Install	S.F.	.11	4.11		4.22	Includes labor and material to repair
Minimum Charge	Job		190		190	section of plaster wall, area less than 20 S.F.
Sizing						
Install	S.F.	.09	.19		.28	Includes labor and materials to apply
Minimum Charge	Job		190		190	wall sizing.
Good Grade						
Walls						
Demolish	S.F.		.70		.70	Includes material and labor to install
Install	S.F.	1.03	.79		1.82	pre-pasted, vinyl-coated wallcovering.
Demolish and Install	S.F.	1.03	1.49		2.52	
Clean	S.F.	.01	.07		.08	
Minimum Charge	Job		190		190	
Ceiling						
Demolish	S.F.		.70		.70	Includes material and labor to install
Install	S.F.	1.03	.79		1.82	pre-pasted, vinyl-coated wallcovering.
Demolish and Install	S.F.	1.03	1.49		2.52	
Clean	S.F.	.03	.19		.22	
Minimum Charge	Job		190		190	
Better Grade						
Walls						
Demolish	S.F.		.70		.70	Includes material and labor to install
Install	S.F.	1.18	.79		1.97	pre-pasted, vinyl-coated wallcovering.
Demolish and Install	S.F.	1.18	1.49		2.67	
Clean	S.F.	.01	.07		.08	
Minimum Charge	Job		190		190	
Ceiling						
Demolish	S.F.		.70		.70	Includes material and labor to install
Install	S.F.	1.18	.79		1.97	pre-pasted, vinyl-coated wallcovering.
Demolish and Install	S.F.	1.18	1.49		2.67	
Clean	S.F.	.03	.19		.22	
Minimum Charge	Job		190		190	
Premium Grade						
Walls						
Demolish	S.F.		.88		.88	Includes material and labor to install
Install	S.F.	1.27	.79		2.06	pre-pasted, vinyl-coated wallcovering.
Demolish and Install	S.F.	1.27	1.67		2.94	
Clean	S.F.	.01	.07		.08	
Minimum Charge	Job		190		190	
Ceiling						
Demolish	S.F.		.88		.88	Includes material and labor to install
Install	S.F.	1.27	.79		2.06	pre-pasted, vinyl-coated wallcovering.
Demolish and Install	S.F.	1.27	1.67		2.94	
Clean	S.F.	.03	.19		.22	
Minimum Charge	Job		190		190	

Walls / Ceilings

Wallcovering

	Unit	Material	Labor	Equip.	Total	Specification
Mural						
Pre-printed						
Demolish	Ea.		115		115	Includes material and labor to install
Install	Ea.	282	286		568	wallcovering with a pre-printed mural.
Demolish and Install	Ea.	282	401		683	
Clean	Ea.		8.25		8.25	
Minimum Charge	Job		190		190	
Hand-painted						
Demolish	Ea.		115		115	Includes material and labor to install
Install	Ea.	540	575		1115	wallcovering with a hand-painted
Demolish and Install	Ea.	540	690		1230	mural.
Clean	Ea.		8.25		8.25	
Minimum Charge	Job		190		190	
Vinyl-coated						
Good Grade						
Demolish	Ea.		20.50		20.50	Includes material and labor to install
Install	Ea.	17.50	26.50		44	pre-pasted, vinyl-coated wallcovering.
Demolish and Install	Ea.	17.50	47		64.50	Based upon 30 S.F. rolls.
Clean	Ea.		8.25		8.25	
Minimum Charge	Job		190		190	
Better Grade						
Demolish	Ea.		20.50		20.50	Includes material and labor to install
Install	Ea.	27	26.50		53.50	pre-pasted, vinyl-coated wallcovering.
Demolish and Install	Ea.	27	47		74	Based upon 30 S.F. rolls.
Clean	Ea.		8.25		8.25	
Minimum Charge	Job		190		190	
Premium Grade						
Demolish	Ea.		20.50		20.50	Includes material and labor to install
Install	Ea.	32.50	26.50		59	pre-pasted, vinyl-coated wallcovering.
Demolish and Install	Ea.	32.50	47		79.50	Based upon 30 S.F. rolls.
Clean	Ea.		8.25		8.25	
Minimum Charge	Job		190		190	
Blank Underliner						
Demolish	Ea.		20.50		20.50	Includes material and labor to install
Install	Ea.	9.90	27		36.90	blank stock liner paper or bridging
Demolish and Install	Ea.	9.90	47.50		57.40	paper. Based upon 30 S.F. rolls.
Clean	Ea.		8.25		8.25	
Minimum Charge	Job		190		190	
Solid Vinyl						
Good Grade						
Demolish	Roll		23.50		23.50	Includes material and labor to install
Install	Ea.	25	26.50		51.50	pre-pasted, vinyl wallcovering. Based
Demolish and Install	Ea.	25	50		75	upon 30 S.F. rolls.
Clean	Ea.		8.25		8.25	
Minimum Charge	Job		190		190	
Better Grade						
Demolish	Roll		23.50		23.50	Includes material and labor to install
Install	Ea.	32.50	26.50		59	pre-pasted, vinyl wallcovering. Based
Demolish and Install	Ea.	32.50	50		82.50	upon 30 S.F. rolls.
Clean	Ea.		8.25		8.25	
Minimum Charge	Job		190		190	
Premium Grade						
Demolish	Roll		23.50		23.50	Includes material and labor to install
Install	Ea.	52.50	26.50		79	pre-pasted, vinyl wallcovering. Based
Demolish and Install	Ea.	52.50	50		102.50	upon 30 S.F. rolls.
Clean	Ea.		8.25		8.25	
Minimum Charge	Job		190		190	

Walls / Ceilings

Wallcovering

Wallcovering		Unit	Material	Labor	Equip.	Total	Specification
Blank Underliner							
	Demolish	Ea.		20.50		20.50	Includes material and labor to install
	Install	Ea.	9.90	27		36.90	blank stock liner paper or bridging
	Demolish and Install	Ea.	9.90	47.50		57.40	paper. Based upon 30 S.F. rolls.
	Clean	Ea.		8.25		8.25	
	Minimum Charge	Job		190		190	
Grasscloth / String							
Good Grade							
	Demolish	Ea.		20.50		20.50	Includes material and labor to install
	Install	Ea.	55.50	34.50		90	grasscloth or string wallcovering.
	Demolish and Install	Ea.	55.50	55		110.50	Based upon 30 S.F. rolls.
	Clean	Ea.		11		11	
	Minimum Charge	Job		190		190	
Better Grade							
	Demolish	Ea.		20.50		20.50	Includes material and labor to install
	Install	Ea.	62	34.50		96.50	grasscloth or string wallcovering.
	Demolish and Install	Ea.	62	55		117	Based upon 30 S.F. rolls.
	Clean	Ea.		11		11	
	Minimum Charge	Job		190		190	
Premium Grade							
	Demolish	Ea.		20.50		20.50	Includes material and labor to install
	Install	Ea.	70	34.50		104.50	grasscloth or string wallcovering.
	Demolish and Install	Ea.	70	55		125	Based upon 30 S.F. rolls.
	Clean	Ea.		11		11	
	Minimum Charge	Job		190		190	
Blank Underliner							
	Demolish	Ea.		20.50		20.50	Includes material and labor to install
	Install	Ea.	9.90	27		36.90	blank stock liner paper or bridging
	Demolish and Install	Ea.	9.90	47.50		57.40	paper. Based upon 30 S.F. rolls.
	Clean	Ea.		8.25		8.25	
	Minimum Charge	Job		190		190	
Designer							
Premium Grade							
	Demolish	S.F.		.88		.88	Includes material and labor to install
	Install	Ea.	117	40		157	paperback vinyl, untrimmed and
	Demolish and Install	Ea.	117	40.88		157.88	hand-painted. Based upon 30 S.F.
	Clean	Ea.		8.25		8.25	rolls.
	Minimum Charge	Job		190		190	
Blank Underliner							
	Demolish	Ea.		20.50		20.50	Includes material and labor to install
	Install	Ea.	9.90	27		36.90	blank stock liner paper or bridging
	Demolish and Install	Ea.	9.90	47.50		57.40	paper. Based upon 30 S.F. rolls.
	Clean	Ea.		8.25		8.25	
	Minimum Charge	Job		190		190	
Foil							
	Demolish	Ea.		20.50		20.50	Includes material and labor to install
	Install	Ea.	28.50	34.50		63	foil coated wallcovering. Based upon
	Demolish and Install	Ea.	28.50	55		83.50	30 S.F. rolls.
	Clean	Ea.		8.25		8.25	
	Minimum Charge	Job		190		190	
Wallcovering Border							
Good Grade							
	Demolish	L.F.		1.33		1.33	Includes material and labor install
	Install	L.F.	1.71	1.46		3.17	vinyl-coated border 4" to 6" wide.
	Demolish and Install	L.F.	1.71	2.79		4.50	
	Clean	L.F.		.12		.12	
	Minimum Charge	Job		190		190	

Walls / Ceilings

Wallcovering

	Unit	Material	Labor	Equip.	Total	Specification
Better Grade						
Demolish	L.F.		1.33		1.33	Includes material and labor install
Install	L.F.	3.06	1.46		4.52	vinyl-coated border 4" to 6" wide.
Demolish and Install	L.F.	3.06	2.79		5.85	
Clean	L.F.		.12		.12	
Minimum Charge	Job		190		190	
Premium Grade						
Demolish	L.F.		1.33		1.33	Includes material and labor install
Install	L.F.	3.81	1.46		5.27	vinyl-coated border 4" to 6" wide.
Demolish and Install	L.F.	3.81	2.79		6.60	
Clean	L.F.		.12		.12	
Minimum Charge	Job		190		190	

Acoustical Ceiling

	Unit	Material	Labor	Equip.	Total	Specification
Complete System						
2' x 2' Square Edge						
Demolish	S.F.		.61		.61	Includes material and labor to install
Install	S.F.	2.26	1.40		3.66	T-bar grid acoustical ceiling system,
Demolish and Install	S.F.	2.26	2.01		4.27	random pinhole 5/8" thick square
Reinstall	S.F.		1.12		1.12	edge tile.
Clean	S.F.	.04	.25		.29	
Minimum Charge	Job		227		227	
2' x 2' Reveal Edge						
Demolish	S.F.		.61		.61	Includes material and labor to install
Install	S.F.	2.70	1.82		4.52	T-bar grid acoustical ceiling system,
Demolish and Install	S.F.	2.70	2.43		5.13	random pinhole 5/8" thick reveal
Reinstall	S.F.		1.45		1.45	edge tile.
Clean	S.F.	.04	.25		.29	
Minimum Charge	Job		227		227	
2' x 2' Fire-rated						
Demolish	S.F.		.61		.61	Includes material and labor to install
Install	S.F.	3.77	1.40		5.17	T-bar grid FR acoustical ceiling system,
Demolish and Install	S.F.	3.77	2.01		5.78	random pinhole 5/8" thick square
Reinstall	S.F.		1.12		1.12	edge tile.
Clean	S.F.	.04	.25		.29	
Minimum Charge	Job		227		227	
2' x 4' Square Edge						
Demolish	S.F.		.61		.61	Cost includes material and labor to
Install	S.F.	2.67	1.19		3.86	install 2' x 4' T-bar grid acoustical
Demolish and Install	S.F.	2.67	1.80		4.47	ceiling system, random pinhole 5/8"
Reinstall	S.F.		.96		.96	thick square edge tile.
Clean	S.F.	.04	.25		.29	
Minimum Charge	Job		227		227	
2' x 4' Reveal Edge						
Demolish	S.F.		.61		.61	Includes material and labor to install
Install	S.F.	2.41	1.65		4.06	T-bar grid acoustical ceiling system,
Demolish and Install	S.F.	2.41	2.26		4.67	random pinhole 5/8" thick reveal
Reinstall	S.F.		1.32		1.32	edge tile.
Clean	S.F.	.04	.25		.29	
Minimum Charge	Job		227		227	
2' x 4' Fire-rated						
Demolish	S.F.		.61		.61	Includes material and labor to install
Install	S.F.	3.77	1.40		5.17	T-bar grid FR acoustical ceiling system,
Demolish and Install	S.F.	3.77	2.01		5.78	random pinhole 5/8" thick square
Reinstall	S.F.		1.12		1.12	edge tile.
Clean	S.F.	.04	.25		.29	
Minimum Charge	Job		227		227	

187

Walls / Ceilings

Acoustical Ceiling		Unit	Material	Labor	Equip.	Total	Specification
T-bar Grid							
2' x 2'							
	Demolish	S.F.		.72		.72	Includes material and labor to install
	Install	S.F.	1.09	.70		1.79	T-bar grid acoustical ceiling
	Demolish and Install	S.F.	1.09	1.42		2.51	suspension system.
	Reinstall	S.F.		.56		.56	
	Clean	S.F.	.04	.25		.29	
	Minimum Charge	Job		206		206	
2' x 4'							
	Demolish	S.F.		.72		.72	Includes material and labor to install
	Install	S.F.	.85	.57		1.42	T-bar grid acoustical ceiling
	Demolish and Install	S.F.	.85	1.29		2.14	suspension system.
	Reinstall	S.F.		.45		.45	
	Clean	S.F.	.04	.25		.29	
	Minimum Charge	Job		206		206	
Tile Panels							
2' x 2' Square Edge							
	Demolish	S.F.		.20		.20	Includes material and labor to install
	Install	S.F.	1.06	.73		1.79	square edge panels 5/8" thick with
	Demolish and Install	S.F.	1.06	.93		1.99	pinhole pattern. T-bar grid not
	Reinstall	S.F.		.58		.58	included.
	Clean	S.F.	.04	.25		.29	
	Minimum Charge	Job		227		227	
2' x 2' Reveal Edge							
	Demolish	S.F.		.20		.20	Includes material and labor to install
	Install	S.F.	1.14	.97		2.11	reveal edge panels 5/8" thick with
	Demolish and Install	S.F.	1.14	1.17		2.31	pinhole pattern. T-bar grid not
	Reinstall	S.F.		.77		.77	included.
	Clean	S.F.	.04	.25		.29	
	Minimum Charge	Job		227		227	
2' x 2' Fire-rated							
	Demolish	S.F.		.20		.20	Includes material and labor to install
	Install	S.F.	1.25	.67		1.92	FR square edge panels 5/8" thick with
	Demolish and Install	S.F.	1.25	.87		2.12	pinhole pattern. T-bar grid not
	Reinstall	S.F.		.54		.54	included.
	Clean	S.F.	.04	.25		.29	
	Minimum Charge	Job		227		227	
2' x 4' Square Edge							
	Demolish	S.F.		.20		.20	Includes material and labor to install
	Install	S.F.	1.06	.73		1.79	square edge panels 5/8" thick with
	Demolish and Install	S.F.	1.06	.93		1.99	pinhole pattern. T-bar grid not
	Reinstall	S.F.		.58		.58	included.
	Clean	S.F.	.04	.25		.29	
	Minimum Charge	Job		227		227	
2' x 4' Reveal Edge							
	Demolish	S.F.		.20		.20	Includes material and labor to install
	Install	S.F.	1.14	.97		2.11	reveal edge panels 5/8" thick with
	Demolish and Install	S.F.	1.14	1.17		2.31	pinhole pattern. T-bar grid not
	Reinstall	S.F.		.77		.77	included.
	Clean	S.F.	.04	.25		.29	
	Minimum Charge	Job		227		227	
2' x 4' Fire-rated							
	Demolish	S.F.		.20		.20	Includes material and labor to install
	Install	S.F.	1.25	.67		1.92	FR square edge panels 5/8" thick with
	Demolish and Install	S.F.	1.25	.87		2.12	pinhole pattern. T-bar grid not
	Reinstall	S.F.		.54		.54	included.
	Clean	S.F.	.04	.25		.29	
	Minimum Charge	Job		227		227	

Acoustical Ceiling	Unit	Material	Labor	Equip.	Total	Specification
Bleach Out Stains						
Install	S.F.	.13	.33		.46	Includes labor and material to apply
Minimum Charge	Job		227		227	laundry type chlorine bleach to remove light staining of acoustic tile material.
Adhesive Tile						
Good Grade						
Demolish	S.F.		.81		.81	Cost includes material and labor to
Install	S.F.	1.23	1.51		2.74	install 12" x 12" tile with embossed
Demolish and Install	S.F.	1.23	2.32		3.55	texture pattern. Furring strips not
Clean	S.F.	.03	.19		.22	included.
Minimum Charge	Job		227		227	
Better Grade						
Demolish	S.F.		.81		.81	Cost includes material and labor to
Install	S.F.	1.44	1.51		2.95	install 12" x 12" tile with embossed
Demolish and Install	S.F.	1.44	2.32		3.76	texture pattern. Furring strips not
Clean	S.F.	.03	.19		.22	included.
Minimum Charge	Job		227		227	
Premium Grade						
Demolish	S.F.		.81		.81	Cost includes material and labor to
Install	S.F.	5.45	1.51		6.96	install 12" x 12" tile with embossed
Demolish and Install	S.F.	5.45	2.32		7.77	texture pattern. Furring strips not
Clean	S.F.	.03	.19		.22	included.
Minimum Charge	Job		227		227	
Bleach Out Stains						
Install	S.F.	.13	.33		.46	Includes labor and material to apply
Minimum Charge	Job		227		227	laundry type chlorine bleach to remove light staining of acoustic tile material.
Stapled Tile						
Good Grade						
Demolish	S.F.		.49		.49	Cost includes material and labor to
Install	S.F.	1.23	1.51		2.74	install 12" x 12" tile with embossed
Demolish and Install	S.F.	1.23	2		3.23	texture pattern. Furring strips not
Clean	S.F.	.03	.19		.22	included.
Minimum Charge	Job		227		227	
Better Grade						
Demolish	S.F.		.49		.49	Cost includes material and labor to
Install	S.F.	1.44	1.51		2.95	install 12" x 12" tile with embossed
Demolish and Install	S.F.	1.44	2		3.44	texture pattern. Furring strips not
Clean	S.F.	.03	.19		.22	included.
Minimum Charge	Job		227		227	
Premium Grade						
Demolish	S.F.		.49		.49	Cost includes material and labor to
Install	S.F.	5.45	1.51		6.96	install 12" x 12" tile with embossed
Demolish and Install	S.F.	5.45	2		7.45	texture pattern. Furring strips not
Clean	S.F.	.03	.19		.22	included.
Minimum Charge	Job		227		227	
Bleach Out Stains						
Install	S.F.	.13	.33		.46	Includes labor and material to apply
Minimum Charge	Job		227		227	laundry type chlorine bleach to remove light staining of acoustic tile material.

189

Acoustical Ceiling	Unit	Material	Labor	Equip.	Total	Specification
Blown						
Scrape Off Existing						
Install	S.F.		.68		.68	Includes labor and material to remove
Minimum Charge	Job		227		227	existing blown acoustical ceiling material.
Seal Ceiling Gypsum Wallboard						
Install	S.F.	.07	.33		.40	Includes labor and material to seal
Minimum Charge	Job		189		189	existing drywall with shellac-based material.
Reblow Existing						
Install	S.F.	.04	.54		.58	Includes labor and material to install
Minimum Charge	Job		545	74.50	619.50	by spray, texture finish.
Bleach Out Stains						
Install	S.F.	.13	.33		.46	Includes labor and material to apply
Minimum Charge	Job		227		227	laundry type chlorine bleach to remove light staining of acoustic tile material.
New Blown						
Install	S.F.	.04	.54		.58	Includes labor and material to install
Minimum Charge	Job		189		189	by spray, texture finish.
Apply Glitter To						
Install	S.F.	.22	.17		.39	Includes labor and material to apply
Minimum Charge	Job		189		189	glitter material to existing or new drywall.
Furring Strips						
Demolish	S.F.		.09		.09	Cost includes material and labor to
Install	S.F.	.20	1.01		1.21	install 1" x 2" furring strips, 16" O.C.
Demolish and Install	S.F.	.20	1.10		1.30	per S.F. of surface area to be covered.
Minimum Charge	Job		114		114	

| Finished Stair | Bi-Fold Door | Raised Panel Door |

Hollow Core Door	Unit	Material	Labor	Equip.	Total	Specification
Pre-hung Wood						
1' 6" x 6' 8"						
Demolish	Ea.		42		42	Cost includes material and labor to
Install	Ea.	151	25		176	install 1' 6" x 6' 8" pre-hung 1-3/8"
Demolish and Install	Ea.	151	67		218	lauan door with split pine jamb
Reinstall	Ea.		25.22		25.22	including casing, stops, and hinges.
Clean	Ea.	4.07	13.75		17.82	
Paint	Ea.	10.30	54		64.30	
Minimum Charge	Job		227		227	
2' x 6' 8"						
Demolish	Ea.		42		42	Cost includes material and labor to
Install	Ea.	165	26.50		191.50	install 2' x 6' 8" pre-hung 1-3/8"
Demolish and Install	Ea.	165	68.50		233.50	lauan door with split pine jamb
Reinstall	Ea.		26.71		26.71	including casing, stops, and hinges.
Clean	Ea.	4.07	13.75		17.82	
Paint	Ea.	10.30	54		64.30	
Minimum Charge	Job		227		227	
2' 4" x 6' 8"						
Demolish	Ea.		42		42	Cost includes material and labor to
Install	Ea.	169	26.50		195.50	install 2' 4" x 6' 8" pre-hung 1-3/8"
Demolish and Install	Ea.	169	68.50		237.50	lauan door with split pine jamb
Reinstall	Ea.		26.71		26.71	including casing, stops, and hinges.
Clean	Ea.	4.07	13.75		17.82	
Paint	Ea.	10.30	54		64.30	
Minimum Charge	Job		227		227	
2' 6" x 6' 8"						
Demolish	Ea.		42		42	Cost includes material and labor to
Install	Ea.	174	26.50		200.50	install 2' 6" x 6' 8" pre-hung 1-3/8"
Demolish and Install	Ea.	174	68.50		242.50	lauan door with split pine jamb
Reinstall	Ea.		26.71		26.71	including casing, stops, and hinges.
Clean	Ea.	4.07	13.75		17.82	
Paint	Ea.	10.30	54		64.30	
Minimum Charge	Job		227		227	
2' 8" x 6' 8"						
Demolish	Ea.		42		42	Cost includes material and labor to
Install	Ea.	169	26.50		195.50	install 2' 8" x 6' 8" pre-hung 1-3/8"
Demolish and Install	Ea.	169	68.50		237.50	lauan door with split pine jamb,
Reinstall	Ea.		26.71		26.71	including casing, stop and hinges.
Clean	Ea.	4.07	13.75		17.82	
Paint	Ea.	10.30	54		64.30	
Minimum Charge	Job		227		227	

Finish Carpentry

Hollow Core Door

	Unit	Material	Labor	Equip.	Total	Specification
3' x 6' 8"						
Demolish	Ea.		42		42	Cost includes material and labor to
Install	Ea.	175	28.50		203.50	install 3' x 6' 8" pre-hung 1-3/8"
Demolish and Install	Ea.	175	70.50		245.50	lauan door with split pine jamb
Reinstall	Ea.		28.38		28.38	including casing, stops, and hinges.
Clean	Ea.	4.07	13.75		17.82	
Paint	Ea.	10.30	54		64.30	
Minimum Charge	Job		227		227	
Pocket Type						
Demolish	Ea.		63		63	Cost includes material and labor to
Install	Ea.	44	50.50		94.50	install 2' 8" x 6' 8" lauan door.
Demolish and Install	Ea.	44	113.50		157.50	
Reinstall	Ea.		50.44		50.44	
Clean	Ea.	4.07	13.75		17.82	
Paint	Ea.	10.30	54		64.30	
Minimum Charge	Job		227		227	
Stain						
Install	Ea.	7.25	31.50		38.75	Includes labor and material to stain
Minimum Charge	Job		189		189	single door and trim on both sides.
Pre-hung Masonite						
1' 6" x 6' 8"						
Demolish	Ea.		42		42	Cost includes material and labor to
Install	Ea.	226	25		251	install 1' 6" x 6' 8" pre-hung 1-3/8"
Demolish and Install	Ea.	226	67		293	masonite door with split pine jamb
Reinstall	Ea.		25.22		25.22	including casing, stops, and hinges.
Clean	Ea.	4.07	13.75		17.82	
Paint	Ea.	10.30	54		64.30	
Minimum Charge	Job		227		227	
2' x 6' 8"						
Demolish	Ea.		42		42	Cost includes material and labor to
Install	Ea.	223	26.50		249.50	install 2' x 6' 8" pre-hung 1-3/8"
Demolish and Install	Ea.	223	68.50		291.50	masonite door with split pine jamb
Reinstall	Ea.		26.71		26.71	including casing, stops, and hinges.
Clean	Ea.	4.07	13.75		17.82	
Paint	Ea.	10.30	54		64.30	
Minimum Charge	Job		227		227	
2' 4" x 6' 8"						
Demolish	Ea.		42		42	Cost includes material and labor to
Install	Ea.	226	26.50		252.50	install 2' 4" x 6' 8" pre-hung 1-3/8"
Demolish and Install	Ea.	226	68.50		294.50	masonite door with split pine jamb
Reinstall	Ea.		26.71		26.71	including casing, stops, and hinges.
Clean	Ea.	4.07	13.75		17.82	
Paint	Ea.	10.30	54		64.30	
Minimum Charge	Job		227		227	
2' 8" x 6' 8"						
Demolish	Ea.		42		42	Cost includes material and labor to
Install	Ea.	228	26.50		254.50	install 2' 8" x 6' 8" pre-hung 1-3/8"
Demolish and Install	Ea.	228	68.50		296.50	masonite door with split pine jamb
Reinstall	Ea.		26.71		26.71	including casing, stops, and hinges.
Clean	Ea.	4.07	13.75		17.82	
Paint	Ea.	10.30	54		64.30	
Minimum Charge	Job		227		227	
3' x 6' 8"						
Demolish	Ea.		42		42	Cost includes material and labor to
Install	Ea.	235	28.50		263.50	install 3' x 6' 8" pre-hung 1-3/8"
Demolish and Install	Ea.	235	70.50		305.50	masonite door with split pine jamb
Reinstall	Ea.		28.38		28.38	including casing, stops, and hinges.
Clean	Ea.	4.07	13.75		17.82	
Paint	Ea.	10.30	54		64.30	
Minimum Charge	Job		227		227	

For customer support on your Contractor's Pricing Guide: Residential Repair & Remodeling, call 888.606.7279.

Finish Carpentry

Hollow Core Door

	Unit	Material	Labor	Equip.	Total	Specification
Pocket Type						
Demolish	Ea.		63		63	Cost includes material and labor to
Install	Ea.	44	50.50		94.50	install 2' 8" x 6' 8" lauan door.
Demolish and Install	Ea.	44	113.50		157.50	
Reinstall	Ea.		50.44		50.44	
Clean	Ea.	4.07	13.75		17.82	
Paint	Ea.	10.30	54		64.30	
Minimum Charge	Job		227		227	
Stain						
Install	Ea.	7.25	31.50		38.75	Includes labor and material to stain
Minimum Charge	Job		189		189	single door and trim on both sides.
Hollow Core Door Only						
1' 6" x 6' 8"						
Demolish	Ea.		9		9	Cost includes material and labor to
Install	Ea.	56	24		80	install 1' 6" x 6' 8" lauan hollow core
Demolish and Install	Ea.	56	33		89	door slab.
Reinstall	Ea.		23.89		23.89	
Clean	Ea.	1.69	10.30		11.99	
Paint	Ea.	4.92	47		51.92	
Minimum Charge	Job		227		227	
2' x 6' 8"						
Demolish	Ea.		9		9	Cost includes material and labor to
Install	Ea.	55	25		80	install 2' x 6' 8" lauan hollow core
Demolish and Install	Ea.	55	34		89	door slab.
Reinstall	Ea.		25.22		25.22	
Clean	Ea.	1.69	10.30		11.99	
Paint	Ea.	4.92	47		51.92	
Minimum Charge	Job		227		227	
2' 4" x 6' 8"						
Demolish	Ea.		9		9	Cost includes material and labor to
Install	Ea.	59.50	25		84.50	install 2' 4" x 6' 8" lauan hollow core
Demolish and Install	Ea.	59.50	34		93.50	door slab.
Reinstall	Ea.		25.22		25.22	
Clean	Ea.	1.69	10.30		11.99	
Paint	Ea.	4.92	47		51.92	
Minimum Charge	Job		227		227	
2' 8" x 6' 8"						
Demolish	Ea.		9		9	Cost includes material and labor to
Install	Ea.	61	25		86	install 2' 8" x 6' 8" lauan hollow core
Demolish and Install	Ea.	61	34		95	door slab.
Reinstall	Ea.		25.22		25.22	
Clean	Ea.	1.69	10.30		11.99	
Paint	Ea.	4.92	47		51.92	
Minimum Charge	Job		227		227	
3' x 6' 8"						
Demolish	Ea.		9		9	Cost includes material and labor to
Install	Ea.	63.50	26.50		90	install 3' x 6' 8" lauan hollow core
Demolish and Install	Ea.	63.50	35.50		99	door slab.
Reinstall	Ea.		26.71		26.71	
Clean	Ea.	1.69	10.30		11.99	
Paint	Ea.	4.92	47		51.92	
Minimum Charge	Job		227		227	
Pocket Type						
Demolish	Ea.		9		9	Cost includes material and labor to
Install	Ea.	44	50.50		94.50	install 2' 8" x 6' 8" lauan door.
Demolish and Install	Ea.	44	59.50		103.50	
Reinstall	Ea.		50.44		50.44	
Clean	Ea.	1.69	10.30		11.99	
Paint	Ea.	4.92	47		51.92	
Minimum Charge	Job		227		227	

193

Finish Carpentry

Hollow Core Door	Unit	Material	Labor	Equip.	Total	Specification
Stain						
Install	Ea.	4.03	42		46.03	Includes labor and materials to stain a
Minimum Charge	Job		189		189	flush style door on both sides by brush.
Casing Trim						
Single Width						
Demolish	Ea.		8		8	Cost includes material and labor to
Install	Opng.	26.50	38		64.50	install 11/16" x 2-1/2" pine ranch
Demolish and Install	Opng.	26.50	46		72.50	style casing for one side of a standard
Reinstall	Opng.		30.27		30.27	door opening.
Clean	Opng.	.57	3.30		3.87	
Paint	Ea.	4.73	7.55		12.28	
Minimum Charge	Job		227		227	
Double Width						
Demolish	Ea.		9.50		9.50	Cost includes material and labor to
Install	Opng.	31	45.50		76.50	install 11/16" x 2-1/2" pine ranch
Demolish and Install	Opng.	31	55		86	style door casing for one side of a
Reinstall	Opng.		36.32		36.32	double door opening.
Clean	Opng.	.68	6.60		7.28	
Paint	Ea.	4.73	8.60		13.33	
Minimum Charge	Job		227		227	
Hardware						
Doorknob w / Lock						
Demolish	Ea.		18		18	Includes material and labor to install
Install	Ea.	40	28.50		68.50	privacy lockset.
Demolish and Install	Ea.	40	46.50		86.50	
Reinstall	Ea.		22.70		22.70	
Clean	Ea.	.41	10.30		10.71	
Minimum Charge	Job		227		227	
Doorknob						
Demolish	Ea.		18		18	Includes material and labor to install
Install	Ea.	39	28.50		67.50	residential passage lockset, keyless
Demolish and Install	Ea.	39	46.50		85.50	bored type.
Reinstall	Ea.		22.70		22.70	
Clean	Ea.	.41	10.30		10.71	
Minimum Charge	Job		227		227	
Deadbolt						
Demolish	Ea.		15.75		15.75	Includes material and labor to install a
Install	Ea.	70	32.50		102.50	deadbolt.
Demolish and Install	Ea.	70	48.25		118.25	
Reinstall	Ea.		25.94		25.94	
Clean	Ea.	.41	10.30		10.71	
Minimum Charge	Job		227		227	
Lever Handle						
Demolish	Ea.		18		18	Includes material and labor to install
Install	Ea.	51.50	45.50		97	residential passage lockset, keyless
Demolish and Install	Ea.	51.50	63.50		115	bored type with lever handle.
Reinstall	Ea.		36.32		36.32	
Clean	Ea.	.41	10.30		10.71	
Minimum Charge	Job		227		227	
Closer						
Demolish	Ea.		9.45		9.45	Includes material and labor to install
Install	Ea.	174	70		244	pneumatic light duty closer for interior
Demolish and Install	Ea.	174	79.45		253.45	type doors.
Reinstall	Ea.		55.88		55.88	
Clean	Ea.	.21	13.75		13.96	
Minimum Charge	Job		227		227	

194

Finish Carpentry

Hollow Core Door

	Unit	Material	Labor	Equip.	Total	Specification
Push Plate						
Demolish	Ea.		14		14	Includes material and labor to install
Install	Ea.	28	38		66	bronze push-pull plate.
Demolish and Install	Ea.	28	52		80	
Reinstall	Ea.		30.27		30.27	
Clean	Ea.	.41	10.30		10.71	
Minimum Charge	Job		227		227	
Kickplate						
Demolish	Ea.		14		14	Includes material and labor to install
Install	Ea.	25	30.50		55.50	aluminum kickplate, 10" x 28".
Demolish and Install	Ea.	25	44.50		69.50	
Reinstall	Ea.		24.21		24.21	
Clean	Ea.	.41	10.30		10.71	
Minimum Charge	Job		227		227	
Exit Sign						
Demolish	Ea.		24.50		24.50	Includes material and labor to install a
Install	Ea.	129	60.50		189.50	wall mounted interior electric exit sign.
Demolish and Install	Ea.	129	85		214	
Reinstall	Ea.		48.28		48.28	
Clean	Ea.	.03	7.60		7.63	
Minimum Charge	Job		227		227	
Jamb						
Demolish	Ea.		31.50		31.50	Includes material and labor to install
Install	Ea.	119	12.25		131.25	flat pine jamb with square cut heads
Demolish and Install	Ea.	119	43.75		162.75	and rabbeted sides for 6' 8" high and
Clean	Ea.	.13	8.25		8.38	3-9/16" door including trim sets for
Paint	Ea.	7.55	25		32.55	both sides.
Minimum Charge	Job		227		227	
Shave & Refit						
Install	Ea.		38		38	Includes labor to shave and rework
Minimum Charge	Job		227		227	door to fit opening at the job site.

Solid Core Door

	Unit	Material	Labor	Equip.	Total	Specification
Pre-hung Wood						
2' 4" x 6' 8"						
Demolish	Ea.		42		42	Cost includes material and labor to
Install	Ea.	252	26.50		278.50	install 2' 4" x 6' 8" pre-hung 1-3/4"
Demolish and Install	Ea.	252	68.50		320.50	door, 4-1/2" split jamb, including
Reinstall	Ea.		26.71		26.71	casing and stop, hinges, aluminum sill,
Clean	Ea.	4.07	13.75		17.82	weatherstripping.
Paint	Ea.	11.55	54		65.55	
Minimum Charge	Job		227		227	
2' 6" x 6' 8"						
Demolish	Ea.		42		42	Cost includes material and labor to
Install	Ea.	250	26.50		276.50	install 2' 6" x 6' 8" pre-hung 1-3/4"
Demolish and Install	Ea.	250	68.50		318.50	door, 4-1/2" split jamb, including
Reinstall	Ea.		26.71		26.71	casing and stop, hinges, aluminum sill,
Clean	Ea.	4.07	13.75		17.82	weatherstripping.
Paint	Ea.	11.55	54		65.55	
Minimum Charge	Job		227		227	

Finish Carpentry

Solid Core Door

Solid Core Door	Unit	Material	Labor	Equip.	Total	Specification
2' 8" x 6' 8"						
Demolish	Ea.		42		42	Cost includes material and labor to
Install	Ea.	243	26.50		269.50	install 2' 8" x 6' 8" pre-hung 1-3/4"
Demolish and Install	Ea.	243	68.50		311.50	door, 4-1/2" split jamb, including
Reinstall	Ea.		26.71		26.71	casing and stop, hinges, aluminum sill,
Clean	Ea.	4.07	13.75		17.82	weatherstripping.
Paint	Ea.	11.55	54		65.55	
Minimum Charge	Job		227		227	
3' x 6' 8"						
Demolish	Ea.		42		42	Cost includes material and labor to
Install	Ea.	252	28.50		280.50	install 3' x 6' 8" pre-hung 1-3/4"
Demolish and Install	Ea.	252	70.50		322.50	door, including casing and stop,
Reinstall	Ea.		28.38		28.38	hinges, jamb, aluminum sill,
Clean	Ea.	4.07	13.75		17.82	weatherstripped.
Paint	Ea.	11.55	54		65.55	
Minimum Charge	Job		227		227	
3' 6" x 6' 8"						
Demolish	Ea.		42		42	Cost includes material and labor to
Install	Ea.	395	30.50		425.50	install 3' 6" x 6' 8" pre-hung 1-3/4"
Demolish and Install	Ea.	395	72.50		467.50	door, 4-1/2" split jamb, including
Reinstall	Ea.		30.27		30.27	casing and stop, hinges, aluminum sill,
Clean	Ea.	4.07	13.75		17.82	weatherstripping.
Paint	Ea.	11.55	54		65.55	
Minimum Charge	Job		227		227	
Raised Panel						
Demolish	Ea.		42		42	Cost includes material and labor to
Install	Ea.	485	25		510	install 3' x 6' 8" pre-hung 1-3/4"
Demolish and Install	Ea.	485	67		552	door, jamb, including casing and
Reinstall	Ea.		25.22		25.22	stop, hinges, aluminum sill,
Clean	Ea.	4.07	13.75		17.82	weatherstripping.
Paint	Ea.	11.55	126		137.55	
Minimum Charge	Job		227		227	
Stain						
Install	Ea.	7.25	31.50		38.75	Includes labor and material to stain
Minimum Charge	Job		189		189	single door and trim on both sides.

Bifold Door

Bifold Door	Unit	Material	Labor	Equip.	Total	Specification
Wood, Single						
3'						
Demolish	Ea.		31.50		31.50	Cost includes material and labor to
Install	Ea.	78.50	70		148.50	install 3' wide solid wood bi-fold door
Demolish and Install	Ea.	78.50	101.50		180	not including jamb, trim, track, and
Reinstall	Ea.		69.85		69.85	hardware.
Clean	Ea.	4.07	13.75		17.82	
Paint	Ea.	10.30	54		64.30	
Minimum Charge	Job		227		227	
5'						
Demolish	Ea.		31.50		31.50	Includes material and labor to install
Install	Ea.	116	82.50		198.50	pair of 2' 6" wide solid wood bi-fold
Demolish and Install	Ea.	116	114		230	doors not including jamb, trim, track,
Reinstall	Ea.		82.55		82.55	and hardware.
Clean	Ea.	1.78	23.50		25.28	
Paint	Ea.	34	94.50		128.50	
Minimum Charge	Job		227		227	

For customer support on your Contractor's Pricing Guide: Residential Repair & Remodeling, call 888.606.7279.

Bifold Door	Unit	Material	Labor	Equip.	Total	Specification
6'						
Demolish	Ea.		31.50		31.50	Includes material and labor to install
Install	Ea.	141	91		232	pair of 3' wide solid wood bi-fold
Demolish and Install	Ea.	141	122.50		263.50	doors not including jamb, trim, track,
Reinstall	Ea.		90.80		90.80	and hardware.
Clean	Ea.	2.04	27.50		29.54	
Paint	Ea.	34	94.50		128.50	
Minimum Charge	Job		227		227	
8'						
Demolish	Ea.		31.50		31.50	Includes material and labor to install
Install	Ea.	215	101		316	pair of 4' wide solid wood bi-fold
Demolish and Install	Ea.	215	132.50		347.50	doors not including jamb, trim, track,
Reinstall	Ea.		100.89		100.89	and hardware.
Clean	Ea.	2.81	33		35.81	
Paint	Ea.	17.75	94.50		112.25	
Minimum Charge	Job		227		227	
Mirrored						
6'						
Demolish	Ea.		31.50		31.50	Includes material and labor to install
Install	Ea.	420	50.50		470.50	pair of 3' wide mirrored bi-fold doors
Demolish and Install	Ea.	420	82		502	including jamb, trim, track, and
Reinstall	Ea.		50.44		50.44	hardware.
Clean	Ea.	2.04	13.75		15.79	
Minimum Charge	Job		227		227	
8'						
Demolish	Ea.		31.50		31.50	Includes material and labor to install
Install	Ea.	700	75.50		775.50	pair of 4' wide mirrored bi-fold doors
Demolish and Install	Ea.	700	107		807	including jamb, trim, track, and
Reinstall	Ea.		75.67		75.67	hardware.
Clean	Ea.	2.81	16.50		19.31	
Minimum Charge	Job		227		227	
Louvered						
3'						
Demolish	Ea.		31.50		31.50	Cost includes material and labor to
Install	Ea.	231	70		301	install 3' wide louvered bi-fold door
Demolish and Install	Ea.	231	101.50		332.50	not including jamb, trim, track, and
Reinstall	Ea.		69.85		69.85	hardware.
Clean	Ea.	4.07	13.75		17.82	
Paint	Ea.	10.30	54		64.30	
Minimum Charge	Job		227		227	
5'						
Demolish	Ea.		31.50		31.50	Includes material and labor to install
Install	Ea.	288	82.50		370.50	pair of 2' 6" wide louvered wood
Demolish and Install	Ea.	288	114		402	bi-fold doors not including jamb, trim,
Reinstall	Ea.		82.55		82.55	track, and hardware.
Clean	Ea.	1.78	23.50		25.28	
Paint	Ea.	34	94.50		128.50	
Minimum Charge	Job		227		227	
6'						
Demolish	Ea.		31.50		31.50	Includes material and labor to install
Install	Ea.	320	91		411	pair of 3' wide louvered wood bi-fold
Demolish and Install	Ea.	320	122.50		442.50	doors not including jamb, trim, track,
Reinstall	Ea.		90.80		90.80	and hardware.
Clean	Ea.	2.04	27.50		29.54	
Paint	Ea.	34	94.50		128.50	
Minimum Charge	Job		227		227	

Bifold Door	Unit	Material	Labor	Equip.	Total	Specification
8'						
Demolish	Ea.		31.50		31.50	Includes material and labor to install
Install	Ea.	490	91		581	pair of 4' wide louvered wood bi-fold
Demolish and Install	Ea.	490	122.50		612.50	doors not including jamb, trim, track,
Reinstall	Ea.		90.80		90.80	and hardware.
Clean	Ea.	9.70	23.50		33.20	
Paint	Ea.	17.75	94.50		112.25	
Minimum Charge	Job		227		227	
Accordion Type						
Demolish	S.F.		1.08		1.08	Includes material and labor to install
Install	Ea.	45.50	91		136.50	vinyl folding accordion closet doors,
Demolish and Install	Ea.	45.50	92.08		137.58	including track and frame.
Reinstall	Ea.		90.80		90.80	
Clean	S.F.		.40		.40	
Minimum Charge	Job		227		227	
Casing Trim						
Single Width						
Demolish	Ea.		8		8	Cost includes material and labor to
Install	Opng.	26.50	38		64.50	install 11/16" x 2-1/2" pine ranch
Demolish and Install	Opng.	26.50	46		72.50	style casing for one side of a standard
Reinstall	Opng.		30.27		30.27	door opening.
Clean	Opng.	.57	3.30		3.87	
Paint	Ea.	4.73	7.55		12.28	
Minimum Charge	Job		227		227	
Double Width						
Demolish	Ea.		9.50		9.50	Cost includes material and labor to
Install	Opng.	31	45.50		76.50	install 11/16" x 2-1/2" pine ranch
Demolish and Install	Opng.	31	55		86	style door casing for one side of a
Reinstall	Opng.		36.32		36.32	double door opening.
Clean	Opng.	.68	6.60		7.28	
Paint	Ea.	4.73	8.60		13.33	
Minimum Charge	Job		227		227	
Jamb						
Demolish	Ea.		31.50		31.50	Includes material and labor to install
Install	Ea.	119	12.25		131.25	flat pine jamb with square cut heads
Demolish and Install	Ea.	119	43.75		162.75	and rabbeted sides for 6' 8" high and
Clean	Ea.	.13	8.25		8.38	3-9/16" door including trim sets for
Paint	Ea.	7.55	25		32.55	both sides.
Minimum Charge	Job		114		114	
Stain						
Install	Ea.	7.25	31.50		38.75	Includes labor and material to stain
Minimum Charge	Job		189		189	single door and trim on both sides.
Shave & Refit						
Install	Ea.		38		38	Includes labor to shave and rework
Minimum Charge	Job		114		114	door to fit opening at the job site.

For customer support on your Contractor's Pricing Guide: Residential Repair & Remodeling, call 888.606.7279.

Finish Carpentry

Louver Door		Unit	Material	Labor	Equip.	Total	Specification
Full							
2' x 6' 8"							
	Demolish	Ea.		9		9	Cost includes material and labor to
	Install	Ea.	279	45.50		324.50	install 2' x 6' 8" pine full louver door,
	Demolish and Install	Ea.	279	54.50		333.50	jamb, hinges.
	Reinstall	Ea.		45.40		45.40	
	Clean	Ea.	7.55	41.50		49.05	
	Paint	Ea.	10.30	54		64.30	
	Minimum Charge	Job		227		227	
2' 8" x 6' 8"							
	Demolish	Ea.		9		9	Cost includes material and labor to
	Install	Ea.	310	45.50		355.50	install 2' 8" x 6' 8" pine full louver
	Demolish and Install	Ea.	310	54.50		364.50	door, jamb, hinges.
	Reinstall	Ea.		45.40		45.40	
	Clean	Ea.	7.55	41.50		49.05	
	Paint	Ea.	10.30	54		64.30	
	Minimum Charge	Job		227		227	
3' x 6' 8"							
	Demolish	Ea.		9		9	Cost includes material and labor to
	Install	Ea.	325	48		373	install 3' x 6' 8" pine full louver door,
	Demolish and Install	Ea.	325	57		382	jamb, hinges.
	Reinstall	Ea.		47.79		47.79	
	Clean	Ea.	7.55	41.50		49.05	
	Paint	Ea.	10.30	54		64.30	
	Minimum Charge	Job		227		227	
Half							
2' x 6' 8"							
	Demolish	Ea.		9		9	Cost includes material and labor to
	Install	Ea.	390	45.50		435.50	install 2' x 6' 8" pine half louver door,
	Demolish and Install	Ea.	390	54.50		444.50	jamb, hinges.
	Reinstall	Ea.		45.40		45.40	
	Clean	Ea.	7.55	41.50		49.05	
	Paint	Ea.	10.30	54		64.30	
	Minimum Charge	Job		227		227	
2' 8" x 6' 8"							
	Demolish	Ea.		9		9	Cost includes material and labor to
	Install	Ea.	415	45.50		460.50	install 2' 8" x 6' 8" pine half louver
	Demolish and Install	Ea.	415	54.50		469.50	door, jamb, hinges.
	Reinstall	Ea.		45.40		45.40	
	Clean	Ea.	7.55	41.50		49.05	
	Paint	Ea.	10.30	54		64.30	
	Minimum Charge	Job		227		227	
3' x 6' 8"							
	Demolish	Ea.		9		9	Cost includes material and labor to
	Install	Ea.	430	48		478	install 3' x 6' 8" pine half louver door,
	Demolish and Install	Ea.	430	57		487	jamb, hinges.
	Reinstall	Ea.		47.79		47.79	
	Clean	Ea.	7.55	41.50		49.05	
	Paint	Ea.	10.30	54		64.30	
	Minimum Charge	Job		227		227	
Jamb							
	Demolish	Ea.		31.50		31.50	Includes material and labor to install
	Install	Ea.	119	12.25		131.25	flat pine jamb with square cut heads
	Demolish and Install	Ea.	119	43.75		162.75	and rabbeted sides for 6' 8" high and
	Clean	Ea.	.13	8.25		8.38	3-9/16" door including trim sets for
	Paint	Ea.	7.55	25		32.55	both sides.
	Minimum Charge	Job		114		114	

For customer support on your Contractor's Pricing Guide: Residential Repair & Remodeling, call 888.606.7279.

Finish Carpentry

Louver Door	Unit	Material	Labor	Equip.	Total	Specification
Stain						
Install	Ea.	7.25	31.50		38.75	Includes labor and material to stain single door and trim on both sides.
Minimum Charge	Job		189		189	
Shave & Refit						
Install	Ea.		38		38	Includes labor to shave and rework door to fit opening at the job site.
Minimum Charge	Job		189		189	

Bypass Sliding Door	Unit	Material	Labor	Equip.	Total	Specification
Wood						
5'						
Demolish	Ea.		42		42	Includes material and labor to install pair of 2' 6" wide lauan hollow core wood bi-pass doors with jamb, track hardware.
Install	Opng.	206	82.50		288.50	
Demolish and Install	Opng.	206	124.50		330.50	
Reinstall	Opng.		82.55		82.55	
Clean	Ea.	1.78	23.50		25.28	
Paint	Ea.	34	94.50		128.50	
Minimum Charge	Job		227		227	
6'						
Demolish	Ea.		42		42	Includes material and labor to install pair of 3' wide lauan hollow core wood bi-pass doors with jamb, track and hardware.
Install	Opng.	230	91		321	
Demolish and Install	Opng.	230	133		363	
Reinstall	Opng.		90.80		90.80	
Clean	Ea.	2.04	27.50		29.54	
Paint	Ea.	34	94.50		128.50	
Minimum Charge	Job		227		227	
8'						
Demolish	Ea.		42		42	Includes material and labor to install pair of 4' wide lauan hollow core wood bi-pass doors with jamb, track and hardware.
Install	Opng.	345	114		459	
Demolish and Install	Opng.	345	156		501	
Reinstall	Opng.		113.50		113.50	
Clean	Ea.	2.81	33		35.81	
Paint	Ea.	17.75	94.50		112.25	
Minimum Charge	Job		227		227	
Mirrored						
6'						
Demolish	Ea.		42		42	Includes material and labor to install pair of 3' wide mirrored bi-pass doors with jamb, trim, track hardware.
Install	Opng.	310	91		401	
Demolish and Install	Opng.	310	133		443	
Reinstall	Opng.		90.80		90.80	
Clean	Ea.	2.04	13.75		15.79	
Minimum Charge	Job		227		227	
8'						
Demolish	Ea.		42		42	Includes material and labor to install pair of 4' wide mirrored bi-pass doors with jamb, trim, track hardware.
Install	Opng.	485	101		586	
Demolish and Install	Opng.	485	143		628	
Reinstall	Opng.		100.89		100.89	
Clean	Ea.	2.81	16.50		19.31	
Minimum Charge	Job		227		227	

Finish Carpentry

Bypass Sliding Door	Unit	Material	Labor	Equip.	Total	Specification
Casing Trim						
Single Width						
Demolish	Ea.		8		8	Cost includes material and labor to
Install	Opng.	26.50	38		64.50	install 11/16" x 2-1/2" pine ranch
Demolish and Install	Opng.	26.50	46		72.50	style casing for one side of a standard
Reinstall	Opng.		30.27		30.27	door opening.
Clean	Opng.	.57	3.30		3.87	
Paint	Ea.	4.73	7.55		12.28	
Minimum Charge	Job		227		227	
Double Width						
Demolish	Ea.		9.50		9.50	Cost includes material and labor to
Install	Opng.	31	45.50		76.50	install 11/16" x 2-1/2" pine ranch
Demolish and Install	Opng.	31	55		86	style door casing for one side of a
Reinstall	Opng.		36.32		36.32	double door opening.
Clean	Opng.	.68	6.60		7.28	
Paint	Ea.	4.73	8.60		13.33	
Minimum Charge	Job		227		227	
Jamb						
Demolish	Ea.		31.50		31.50	Includes material and labor to install
Install	Ea.	119	12.25		131.25	flat pine jamb with square cut heads
Demolish and Install	Ea.	119	43.75		162.75	and rabbeted sides for 6' 8" high and
Clean	Ea.	.13	8.25		8.38	3-9/16" door including trim sets for
Paint	Ea.	7.55	25		32.55	both sides.
Minimum Charge	Job		114		114	
Stain						
Install	Ea.	7.25	31.50		38.75	Includes labor and material to stain
Minimum Charge	Job		189		189	single door and trim on both sides.
Shave & Refit						
Install	Ea.		38		38	Includes labor to shave and rework
Minimum Charge	Job		114		114	door to fit opening at the job site.

Base Molding	Unit	Material	Labor	Equip.	Total	Specification
One Piece						
1-5/8"						
Demolish	L.F.		.61		.61	Includes material and labor to install
Install	L.F.	.35	1.34		1.69	pine or birch base molding.
Demolish and Install	L.F.	.35	1.95		2.30	
Reinstall	L.F.		1.07		1.07	
Clean	L.F.	.03	.19		.22	
Paint	L.F.	.04	.41		.45	
Minimum Charge	Job		227		227	
2-1/2"						
Demolish	L.F.		.61		.61	Cost includes material and labor to
Install	L.F.	.57	1.37		1.94	install 1/2" x 2-1/2" pine or birch
Demolish and Install	L.F.	.57	1.98		2.55	base molding.
Reinstall	L.F.		1.10		1.10	
Clean	L.F.	.03	.19		.22	
Paint	L.F.	.04	.41		.45	
Minimum Charge	Job		227		227	

Finish Carpentry

Base Molding		Unit	Material	Labor	Equip.	Total	Specification
3-1/2"							
	Demolish	L.F.		.61		.61	Cost includes material and labor to
	Install	L.F.	.87	1.46		2.33	install 9/16" x 3-1/2" pine or birch
	Demolish and Install	L.F.	.87	2.07		2.94	base molding.
	Reinstall	L.F.		1.16		1.16	
	Clean	L.F.	.03	.19		.22	
	Paint	L.F.	.08	.45		.53	
	Minimum Charge	Job		227		227	
6"							
	Demolish	L.F.		.61		.61	Cost includes material and labor to
	Install	L.F.	.64	1.46		2.10	install 1" x 6" pine or birch base
	Demolish and Install	L.F.	.64	2.07		2.71	molding.
	Reinstall	L.F.		1.16		1.16	
	Clean	L.F.	.03	.28		.31	
	Paint	L.F.	.10	.57		.67	
	Minimum Charge	Job		227		227	
8"							
	Demolish	L.F.		.73		.73	Cost includes material and labor to
	Install	L.F.	.82	1.46		2.28	install 1" x 8" pine or birch base
	Demolish and Install	L.F.	.82	2.19		3.01	molding.
	Reinstall	L.F.		1.16		1.16	
	Clean	L.F.	.04	.33		.37	
	Paint	L.F.	.11	.63		.74	
	Minimum Charge	Job		227		227	
Premium / Custom Grade							
	Demolish	L.F.		.73		.73	Includes material and labor to install
	Install	L.F.	4.58	4.37		8.95	custom 2 or 3 piece oak or other
	Demolish and Install	L.F.	4.58	5.10		9.68	hardwood, stain grade trim.
	Reinstall	L.F.		3.49		3.49	
	Clean	L.F.	.04	.33		.37	
	Paint	L.F.	.12	.63		.75	
	Minimum Charge	Job		227		227	
One Piece Oak							
1-5/8"							
	Demolish	L.F.		.61		.61	Cost includes material and labor to
	Install	L.F.	.92	1.34		2.26	install 1/2" x 1-5/8" oak base
	Demolish and Install	L.F.	.92	1.95		2.87	molding.
	Reinstall	L.F.		1.07		1.07	
	Clean	L.F.	.03	.19		.22	
	Paint	L.F.	.14	.56		.70	
	Minimum Charge	Job		227		227	
2-1/2"							
	Demolish	L.F.		.61		.61	Cost includes material and labor to
	Install	L.F.	1.49	1.37		2.86	install 1/2" x 2-1/2" oak base
	Demolish and Install	L.F.	1.49	1.98		3.47	molding.
	Reinstall	L.F.		1.10		1.10	
	Clean	L.F.	.03	.19		.22	
	Paint	L.F.	.14	.57		.71	
	Minimum Charge	Job		227		227	
3-1/2"							
	Demolish	L.F.		.61		.61	Cost includes material and labor to
	Install	L.F.	1.82	1.40		3.22	install 1/2" x 3-1/2" oak base
	Demolish and Install	L.F.	1.82	2.01		3.83	molding.
	Reinstall	L.F.		1.12		1.12	
	Clean	L.F.	.03	.19		.22	
	Paint	L.F.	.14	.58		.72	
	Minimum Charge	Job		227		227	

Finish Carpentry

Base Molding

	Unit	Material	Labor	Equip.	Total	Specification
6"						
Demolish	L.F.		.61		.61	Cost includes material and labor to
Install	L.F.	3.24	1.46		4.70	install 1" x 6" oak base molding.
Demolish and Install	L.F.	3.24	2.07		5.31	
Reinstall	L.F.		1.16		1.16	
Clean	L.F.	.04	.33		.37	
Paint	L.F.	.15	.60		.75	
Minimum Charge	Job		227		227	
8"						
Demolish	L.F.		.73		.73	Cost includes material and labor to
Install	L.F.	4.47	1.46		5.93	install 1" x 8" oak base molding.
Demolish and Install	L.F.	4.47	2.19		6.66	
Reinstall	L.F.		1.16		1.16	
Clean	L.F.	.04	.33		.37	
Paint	L.F.	.15	.63		.78	
Minimum Charge	Job		227		227	

Vinyl Base Molding

	Unit	Material	Labor	Equip.	Total	Specification
2-1/2"						
Demolish	L.F.		.61		.61	Cost includes material and labor to
Install	L.F.	1.25	1.28		2.53	install 2-1/2" vinyl or rubber cove
Demolish and Install	L.F.	1.25	1.89		3.14	base including adhesive.
Reinstall	L.F.		1.28		1.28	
Clean	L.F.		.30		.30	
Minimum Charge	Job		215		215	
4"						
Demolish	L.F.		.61		.61	Cost includes material and labor to
Install	Ea.	2.48	1.28		3.76	install 4" vinyl or rubber cove base
Demolish and Install	Ea.	2.48	1.89		4.37	including adhesive.
Reinstall	Ea.		1.28		1.28	
Clean	L.F.		.30		.30	
Minimum Charge	Job		215		215	
6"						
Demolish	L.F.		.61		.61	Cost includes material and labor to
Install	L.F.	1.33	1.28		2.61	install 6" vinyl or rubber cove base
Demolish and Install	L.F.	1.33	1.89		3.22	including adhesive.
Reinstall	L.F.		1.28		1.28	
Clean	L.F.		.30		.30	
Minimum Charge	Job		215		215	

Prefinished

	Unit	Material	Labor	Equip.	Total	Specification
Demolish	L.F.		.61		.61	Cost includes material and labor to
Install	L.F.	1.28	1.46		2.74	install 9/16" x 3-5/16" prefinished
Demolish and Install	L.F.	1.28	2.07		3.35	base molding.
Reinstall	L.F.		1.16		1.16	
Clean	L.F.	.03	.19		.22	
Paint	L.F.	.08	.45		.53	
Minimum Charge	Job		227		227	

Shoe

	Unit	Material	Labor	Equip.	Total	Specification
Demolish	L.F.		.61		.61	Includes material and labor to install
Install	L.F.	.35	1.34		1.69	pine or birch base molding.
Demolish and Install	L.F.	.35	1.95		2.30	
Reinstall	L.F.		1.07		1.07	
Clean	L.F.	.03	.19		.22	
Paint	L.F.	.14	.60		.74	
Minimum Charge	Job		227		227	

For customer support on your Contractor's Pricing Guide: Residential Repair & Remodeling, call 888.606.7279.

Finish Carpentry

Base Molding

Base Molding		Unit	Material	Labor	Equip.	Total	Specification
Carpeted							
	Demolish	L.F.		.24		.24	Includes material and labor to install
	Install	L.F.	.99	2.44		3.43	carpet baseboard with adhesive.
	Demolish and Install	L.F.	.99	2.68		3.67	
	Reinstall	L.F.		1.95		1.95	
	Clean	L.F.		.12		.12	
	Minimum Charge	Job		202		202	
Ceramic Tile							
	Demolish	L.F.		1.33		1.33	Includes material and labor to install
	Install	L.F.	4.48	5.60		10.08	thin set ceramic tile cove base
	Demolish and Install	L.F.	4.48	6.93		11.41	including grout.
	Clean	L.F.		.63		.63	
	Minimum Charge	Job		202		202	
Quarry Tile							
	Demolish	L.F.		1.33		1.33	Includes material and labor to install
	Install	L.F.	6.45	6.55		13	quarry tile base including grout.
	Demolish and Install	L.F.	6.45	7.88		14.33	
	Clean	L.F.		.63		.63	
	Minimum Charge	Job		202		202	
Caulk and Renail							
	Install	L.F.	.01	.71		.72	Includes labor and material to renail
	Minimum Charge	Job		227		227	base and caulk where necessary.

Molding

Molding		Unit	Material	Labor	Equip.	Total	Specification
Oak Base							
	Demolish	L.F.		.61		.61	Cost includes material and labor to
	Install	L.F.	1.49	1.37		2.86	install 1/2" x 2-1/2" oak base
	Demolish and Install	L.F.	1.49	1.98		3.47	molding.
	Reinstall	L.F.		1.10		1.10	
	Clean	L.F.	.03	.19		.22	
	Paint	L.F.	.14	.57		.71	
	Minimum Charge	Job		227		227	
Shoe							
	Demolish	L.F.		.61		.61	Includes material and labor to install
	Install	L.F.	.35	1.34		1.69	pine or birch base molding.
	Demolish and Install	L.F.	.35	1.95		2.30	
	Reinstall	L.F.		1.07		1.07	
	Clean	L.F.	.03	.19		.22	
	Paint	L.F.	.14	.60		.74	
	Minimum Charge	Job		227		227	
Quarter Round							
	Demolish	L.F.		.61		.61	Includes material and labor to install
	Install	L.F.	.55	1.78		2.33	pine base quarter round molding.
	Demolish and Install	L.F.	.55	2.39		2.94	
	Reinstall	L.F.		1.42		1.42	
	Clean	L.F.	.03	.19		.22	
	Paint	L.F.	.12	.94		1.06	
	Minimum Charge	Job		227		227	
Chair Rail							
	Demolish	L.F.		.61		.61	Cost includes material and labor to
	Install	L.F.	1.74	1.68		3.42	install 5/8" x 2-1/2" oak chair rail
	Demolish and Install	L.F.	1.74	2.29		4.03	molding.
	Reinstall	L.F.		1.35		1.35	
	Clean	L.F.	.03	.19		.22	
	Paint	L.F.	.12	.94		1.06	
	Minimum Charge	Job		227		227	

Finish Carpentry

Molding

Molding		Unit	Material	Labor	Equip.	Total	Specification
Crown							
	Demolish	L.F.		.61		.61	Includes material and labor to install
	Install	L.F.	2.22	1.82		4.04	unfinished pine crown molding.
	Demolish and Install	L.F.	2.22	2.43		4.65	
	Reinstall	L.F.		1.45		1.45	
	Clean	L.F.	.04	.25		.29	
	Paint	L.F.	.12	.94		1.06	
	Minimum Charge	Job		227		227	
Cove							
	Demolish	L.F.		.61		.61	Cost includes material and labor to
	Install	L.F.	1.98	1.71		3.69	install 11/16" x 2-3/4" pine cove
	Demolish and Install	L.F.	1.98	2.32		4.30	molding.
	Reinstall	L.F.		1.37		1.37	
	Clean	L.F.	.03	.19		.22	
	Paint	L.F.	.12	.94		1.06	
	Minimum Charge	Job		227		227	
Corner							
	Demolish	L.F.		.61		.61	Includes material and labor to install
	Install	L.F.	1.89	.71		2.60	inside or outside pine corner molding.
	Demolish and Install	L.F.	1.89	1.32		3.21	
	Reinstall	L.F.		.57		.57	
	Clean	L.F.	.03	.19		.22	
	Paint	L.F.	.12	.94		1.06	
	Minimum Charge	Job		227		227	
Picture							
	Demolish	L.F.		.61		.61	Cost includes material and labor to
	Install	L.F.	1.54	1.89		3.43	install 9/16" x 2-1/2" pine casing.
	Demolish and Install	L.F.	1.54	2.50		4.04	
	Reinstall	L.F.		1.51		1.51	
	Clean	L.F.	.03	.19		.22	
	Paint	L.F.	.12	.94		1.06	
	Minimum Charge	Job		227		227	
Single Cased Opening							
	Demolish	Ea.		31.50		31.50	Includes material and labor to install
	Install	Ea.	26.50	28.50		55	flat pine jamb with square cut heads
	Demolish and Install	Ea.	26.50	60		86.50	and rabbeted sides for 6' 8" high
	Clean	Ea.	.29	6.90		7.19	opening including trim sets for both
	Paint	Ea.	1.46	9.45		10.91	sides.
	Minimum Charge	Job		227		227	
Double Cased Opening							
	Demolish	Ea.		31.50		31.50	Includes material and labor to install
	Install	Ea.	53	32.50		85.50	flat pine jamb with square cut heads
	Demolish and Install	Ea.	53	64		117	and rabbeted sides for 6' 8" high
	Clean	Ea.	.33	8.05		8.38	opening including trim sets for both
	Paint	Ea.	1.68	11.10		12.78	sides.
	Minimum Charge	Job		227		227	

Window Trim Set

Window Trim Set		Unit	Material	Labor	Equip.	Total	Specification
Single							
	Demolish	Ea.		8.15		8.15	Cost includes material and labor to
	Install	Opng.	40.50	45.50		86	install 11/16" x 2-1/2" pine ranch
	Demolish and Install	Opng.	40.50	53.65		94.15	style window casing.
	Reinstall	Opng.		36.32		36.32	
	Clean	Ea.	.41	4.71		5.12	
	Paint	Ea.	1.36	9		10.36	
	Minimum Charge	Job		227		227	

Finish Carpentry

Window Trim Set

	Unit	Material	Labor	Equip.	Total	Specification
Double						
Demolish	Ea.		9.15		9.15	Cost includes material and labor to
Install	Ea.	55	57		112	install 11/16" x 2-1/2" pine ranch
Demolish and Install	Ea.	55	66.15		121.15	style trim for casing one side of a
Reinstall	Ea.		45.40		45.40	double window opening.
Clean	Ea.	.73	5.50		6.23	
Paint	Ea.	2.29	15.10		17.39	
Minimum Charge	Job		227		227	
Triple						
Demolish	Ea.		10.50		10.50	Cost includes material and labor to
Install	Opng.	71	75.50		146.50	install 11/16" x 2-1/2" pine ranch
Demolish and Install	Opng.	71	86		157	style trim for casing one side of a
Reinstall	Opng.		60.53		60.53	triple window opening.
Clean	Ea.	1.02	6.60		7.62	
Paint	Ea.	3.21	21		24.21	
Minimum Charge	Job		227		227	
Window Casing Per L.F.						
Demolish	L.F.		.61		.61	Cost includes material and labor to
Install	L.F.	1.55	1.82		3.37	install 11/16" x 2-1/2" pine ranch
Demolish and Install	L.F.	1.55	2.43		3.98	style trim for casing.
Reinstall	L.F.		1.45		1.45	
Clean	L.F.	.03	.19		.22	
Paint	L.F.	.12	.94		1.06	
Minimum Charge	Job		227		227	

Window Sill

	Unit	Material	Labor	Equip.	Total	Specification
Wood						
Demolish	L.F.		.61		.61	Includes material and labor to install
Install	L.F.	2.28	2.27		4.55	flat, wood window stool.
Demolish and Install	L.F.	2.28	2.88		5.16	
Reinstall	L.F.		1.82		1.82	
Clean	L.F.	.03	.19		.22	
Paint	L.F.	.12	.94		1.06	
Minimum Charge	Job		227		227	
Marble						
Demolish	L.F.		.92		.92	Includes material and labor to install
Install	L.F.	12.20	9.25		21.45	marble window sills, 6" x 3/4" thick.
Demolish and Install	L.F.	12.20	10.17		22.37	
Reinstall	L.F.		7.41		7.41	
Clean	L.F.	.03	.19		.22	
Minimum Charge	Job		202		202	

Shelving

	Unit	Material	Labor	Equip.	Total	Specification
Pine						
18"						
Demolish	L.F.		1.05		1.05	Cost includes material and labor to
Install	L.F.	8.60	4.78		13.38	install 1" thick custom shelving board
Demolish and Install	L.F.	8.60	5.83		14.43	18" wide.
Reinstall	L.F.		3.82		3.82	
Clean	L.F.	.07	.55		.62	
Paint	L.F.	1.51	.59		2.10	
Minimum Charge	Job		227		227	

Finish Carpentry

Shelving		Unit	Material	Labor	Equip.	Total	Specification
24″							
	Demolish	L.F.		1.05		1.05	Cost includes material and labor to
	Install	L.F.	11.45	5.35		16.80	install 1″ thick custom shelving board.
	Demolish and Install	L.F.	11.45	6.40		17.85	
	Reinstall	L.F.		4.27		4.27	
	Clean	L.F.	.08	.83		.91	
	Paint	L.F.	2.05	.79		2.84	
	Minimum Charge	Job		227		227	
Particle Board							
	Demolish	L.F.		.96		.96	Cost includes material and labor to
	Install	L.F.	1.35	2.39		3.74	install 3/4″ thick particle shelving
	Demolish and Install	L.F.	1.35	3.35		4.70	board 18″ wide.
	Reinstall	L.F.		1.91		1.91	
	Clean	L.F.	.07	.55		.62	
	Paint	S.F.	.14	.88		1.02	
	Minimum Charge	Job		227		227	
Plywood							
	Demolish	L.F.		.96		.96	Cost includes material and labor to
	Install	L.F.	.99	7		7.99	install 3/4″ thick plywood shelving
	Demolish and Install	L.F.	.99	7.96		8.95	board.
	Reinstall	L.F.		5.59		5.59	
	Clean	L.F.	.07	.55		.62	
	Paint	S.F.	.14	.88		1.02	
	Minimum Charge	Job		227		227	
Hardwood (oak)							
	Demolish	L.F.		.96		.96	Includes material and labor to install
	Install	L.F.	5.90	3.13		9.03	oak shelving, 1″ x 12″, incl. cleats and
	Demolish and Install	L.F.	5.90	4.09		9.99	bracing.
	Reinstall	L.F.		2.50		2.50	
	Clean	L.F.	.07	.55		.62	
	Paint	S.F.	.14	.88		1.02	
	Minimum Charge	Job		227		227	
Custom Bookcase							
	Demolish	L.F.		12.20		12.20	Includes material and labor to install
	Install	L.F.	44	45.50		89.50	custom modular bookcase unit with
	Demolish and Install	L.F.	44	57.70		101.70	clear pine faced frames, shelves 12″
	Reinstall	L.F.		36.32		36.32	O.C., 7′ high, 8″ deep.
	Clean	L.F.	1.63	11.80		13.43	
	Paint	L.F.	1.55	29		30.55	
	Minimum Charge	Job		227		227	
Glass							
	Demolish	L.F.		.96		.96	Includes material and labor to install
	Install	L.F.	36	3.13		39.13	glass shelving, 1″ x 12″, incl. cleats
	Demolish and Install	L.F.	36	4.09		40.09	and bracing.
	Reinstall	L.F.		2.50		2.50	
	Clean	L.F.	.08	.51		.59	
	Minimum Charge	Job		220		220	

Closet Shelving		Unit	Material	Labor	Equip.	Total	Specification
Closet Shelf and Rod							
	Demolish	L.F.		.96		.96	Cost includes material and labor to
	Install	L.F.	3.04	7.55		10.59	install 1″ thick custom shelving board
	Demolish and Install	L.F.	3.04	8.51		11.55	18″ wide and 1″ diameter clothes pole
	Reinstall	L.F.		6.05		6.05	with brackets 3′ O.C.
	Clean	L.F.	.04	.41		.45	
	Paint	L.F.	.23	.94		1.17	
	Minimum Charge	Job		227		227	

Finish Carpentry

Closet Shelving

	Unit	Material	Labor	Equip.	Total	Specification
Clothing Rod						
Demolish	L.F.		.64		.64	Includes material and labor to install
Install	L.F.	2.42	2.27		4.69	fir closet pole, 1-5/8" diameter.
Demolish and Install	L.F.	2.42	2.91		5.33	
Reinstall	L.F.		1.82		1.82	
Clean	L.F.	.04	.57		.61	
Paint	L.F.	.21	1.03		1.24	
Minimum Charge	Job		227		227	

Stair Assemblies

	Unit	Material	Labor	Equip.	Total	Specification
Straight Hardwood						
Demolish	Ea.		440		440	Includes material and labor to install
Install	Flight	1250	305		1555	factory cut and assembled straight
Demolish and Install	Flight	1250	745		1995	closed box stairs with oak treads and
Clean	Flight	4.07	28.50		32.57	prefinished stair rail with balusters.
Paint	Flight	21	50.50		71.50	
Minimum Charge	Job		227		227	
Clear Oak Tread						
Demolish	Ea.		11.45		11.45	Includes material and labor to install
Install	Ea.	106	25		131	1-1/4" thick clear oak treads per riser.
Demolish and Install	Ea.	106	36.45		142.45	
Clean	Ea.	.13	2.87		3	
Paint	Ea.	.70	2.80		3.50	
Minimum Charge	Job		227		227	
Landing						
Demolish	S.F.		3.82		3.82	Includes material and labor to install
Install	S.F.	12.80	25		37.80	clear oak landing.
Demolish and Install	S.F.	12.80	28.82		41.62	
Clean	S.F.		.41		.41	
Paint	S.F.	.44	.93		1.37	
Minimum Charge	Job		227		227	
Refinish / Stain						
Install	S.F.	1.47	2.54		4.01	Includes labor and material to refinish
Minimum Charge	Job		189		189	wood floor.
Spiral Hardwood						
Demolish	Ea.		440		440	Cost includes material and labor to
Install	Flight	3775	605		4380	install 4' - 6' diameter spiral stairs with
Demolish and Install	Flight	3775	1045		4820	oak treads, factory cut and assembled
Clean	Flight	4.07	28.50		32.57	with double handrails.
Paint	Flight	21	50.50		71.50	
Minimum Charge	Job		227		227	

Stairs

	Unit	Material	Labor	Equip.	Total	Specification
Disappearing						
Demolish	Ea.		23		23	Includes material and labor to install a
Install	Ea.	268	130		398	folding staircase.
Demolish and Install	Ea.	268	153		421	
Reinstall	Ea.		103.77		103.77	
Clean	Ea.	.62	20.50		21.12	
Minimum Charge	Job		227		227	

Finish Carpentry

Stair Components

	Unit	Material	Labor	Equip.	Total	Specification
Handrail w / Balusters						
Demolish	L.F.		3.06		3.06	Includes material and labor to install
Install	L.F.	42.50	9.45		51.95	prefinished assembled stair rail with
Demolish and Install	L.F.	42.50	12.51		55.01	brackets, turned balusters and newel.
Clean	L.F.	.81	2.28		3.09	
Paint	L.F.	2.22	4.44		6.66	
Minimum Charge	Job		227		227	
Bannister						
Demolish	L.F.		3.06		3.06	Includes material and labor to install
Install	L.F.	39.50	7.55		47.05	built-up oak railings.
Demolish and Install	L.F.	39.50	10.61		50.11	
Clean	L.F.	.81	2.06		2.87	
Paint	L.F.	2.22	3.43		5.65	
Minimum Charge	Job		227		227	

For customer support on your Contractor's Pricing Guide: Residential Repair & Remodeling, call 888.606.7279.

Wall and Base Cabinets

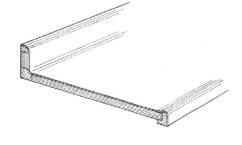

Countertop

Wall Cabinets	Unit	Material	Labor	Equip.	Total	Specification
Good Grade Laminated						
12″ Wide, 1 Door						
Demolish	Ea.		9.15		9.15	Includes material and labor to install
Install	Ea.	194	41.50		235.50	good modular unit with melamine
Demolish and Install	Ea.	194	50.65		244.65	laminated to particle board, in
Reinstall	Ea.		33.02		33.02	textured colors or wood grain print
Clean	Ea.	.62	6.35		6.97	finish, hinges and pulls.
Paint	Ea.	.67	13.50		14.17	
Minimum Charge	Job		227		227	
15″ Wide, 1 Door						
Demolish	Ea.		11.45		11.45	Includes material and labor to install
Install	Ea.	202	42.50		244.50	good modular unit with melamine
Demolish and Install	Ea.	202	53.95		255.95	laminated to particle board, in
Reinstall	Ea.		33.94		33.94	textured colors or wood grain print
Clean	Ea.	.76	7.85		8.61	finish, hinges and pulls.
Paint	Ea.	.84	17.15		17.99	
Minimum Charge	Job		227		227	
18″ Wide, 1 Door						
Demolish	Ea.		13.85		13.85	Includes material and labor to install
Install	Ea.	223	43.50		266.50	good modular unit with melamine
Demolish and Install	Ea.	223	57.35		280.35	laminated to particle board, in
Reinstall	Ea.		34.76		34.76	textured colors or wood grain print
Clean	Ea.	.91	9.45		10.36	finish, hinges and pulls.
Paint	Ea.	.99	19.85		20.84	
Minimum Charge	Job		227		227	
21″ Wide, 1 Door						
Demolish	Ea.		15.95		15.95	Includes material and labor to install
Install	Ea.	234	44		278	good modular unit with melamine
Demolish and Install	Ea.	234	59.95		293.95	laminated to particle board, in
Reinstall	Ea.		35.26		35.26	textured colors or wood grain print
Clean	Ea.	1.07	11		12.07	finish, hinges and pulls.
Paint	Ea.	1.17	23.50		24.67	
Minimum Charge	Job		227		227	
24″ w / Blind Corner						
Demolish	Ea.		18.35		18.35	Includes material and labor to install
Install	Ea.	261	44.50		305.50	good modular unit with melamine
Demolish and Install	Ea.	261	62.85		323.85	laminated to particle board, in
Reinstall	Ea.		35.78		35.78	textured colors or wood grain print
Clean	Ea.	1.22	12.70		13.92	finish, hinges and pulls.
Paint	Ea.	1.33	27		28.33	
Minimum Charge	Job		227		227	

Cabinets and Countertops

Wall Cabinets	Unit	Material	Labor	Equip.	Total	Specification
27" Wide, 2 Door						
Demolish	Ea.		20.50		20.50	Includes material and labor to install
Install	Ea.	293	46		339	good modular unit with melamine
Demolish and Install	Ea.	293	66.50		359.50	laminated to particle board, in
Reinstall	Ea.		36.69		36.69	textured colors or wood grain print
Clean	Ea.	1.38	14.35		15.73	finish, hinges and pulls.
Paint	Ea.	1.51	31.50		33.01	
Minimum Charge	Job		227		227	
30" Wide, 2 Door						
Demolish	Ea.		23		23	Includes material and labor to install
Install	Ea.	305	47		352	good modular unit with melamine
Demolish and Install	Ea.	305	70		375	laminated to particle board, in
Reinstall	Ea.		37.64		37.64	textured colors or wood grain print
Clean	Ea.	1.53	15.70		17.23	finish, hinges and pulls.
Paint	Ea.	1.67	34.50		36.17	
Minimum Charge	Job		227		227	
36" Wide, 2 Door						
Demolish	Ea.		27		27	Includes material and labor to install
Install	Ea.	345	48.50		393.50	good modular unit with melamine
Demolish and Install	Ea.	345	75.50		420.50	laminated to particle board, in
Reinstall	Ea.		38.64		38.64	textured colors or wood grain print
Clean	Ea.	1.84	19.40		21.24	finish, hinges and pulls.
Paint	Ea.	2.01	42		44.01	
Minimum Charge	Job		227		227	
48" Wide, 2 Door						
Demolish	Ea.		36.50		36.50	Includes material and labor to install
Install	Ea.	425	49.50		474.50	good modular unit with melamine
Demolish and Install	Ea.	425	86		511	laminated to particle board, in
Reinstall	Ea.		39.48		39.48	textured colors or wood grain print
Clean	Ea.	2.44	25.50		27.94	finish, hinges and pulls.
Paint	Ea.	2.67	54		56.67	
Minimum Charge	Job		227		227	
Above-Appliance						
Demolish	Ea.		23		23	Includes material and labor to install
Install	Ea.	202	36.50		238.50	good modular unit with melamine
Demolish and Install	Ea.	202	59.50		261.50	laminated to particle board, in
Reinstall	Ea.		29.29		29.29	textured colors or wood grain print
Clean	Ea.	1.53	15.70		17.23	finish, hinges and pulls.
Paint	Ea.	1.67	34.50		36.17	
Minimum Charge	Job		227		227	
Better Grade Wood						
12" Wide, 1 Door						
Demolish	Ea.		9.15		9.15	Includes material and labor to install
Install	Ea.	257	41.50		298.50	custom modular unit with solid
Demolish and Install	Ea.	257	50.65		307.65	hardwood faced frames, hardwood
Reinstall	Ea.		33.02		33.02	door frames and drawer fronts,
Clean	Ea.	.62	6.35		6.97	hardwood veneer on raised door
Paint	Ea.	.67	14.50		15.17	panels, hinges and pulls.
Minimum Charge	Job		227		227	
15" Wide, 1 Door						
Demolish	Ea.		11.45		11.45	Includes material and labor to install
Install	Ea.	271	42.50		313.50	custom modular unit with solid
Demolish and Install	Ea.	271	53.95		324.95	hardwood faced frames, hardwood
Reinstall	Ea.		33.94		33.94	door frames and drawer fronts,
Clean	Ea.	.76	7.85		8.61	hardwood veneer on raised door
Paint	Ea.	.84	18		18.84	panels, hinges and pulls.
Minimum Charge	Job		227		227	

Cabinets and Countertops

Wall Cabinets	Unit	Material	Labor	Equip.	Total	Specification
18″ Wide, 1 Door						
Demolish	Ea.		13.85		13.85	Includes material and labor to install
Install	Ea.	297	43.50		340.50	custom modular unit with solid
Demolish and Install	Ea.	297	57.35		354.35	hardwood faced frames, hardwood
Reinstall	Ea.		34.76		34.76	door frames and drawer fronts,
Clean	Ea.	.91	9.45		10.36	hardwood veneer on raised door
Paint	Ea.	.99	22		22.99	panels, hinges and pulls.
Minimum Charge	Job		227		227	
21″ Wide, 1 Door						
Demolish	Ea.		15.95		15.95	Includes material and labor to install
Install	Ea.	310	44		354	custom modular unit with solid
Demolish and Install	Ea.	310	59.95		369.95	hardwood faced frames, hardwood
Reinstall	Ea.		35.26		35.26	door frames and drawer fronts,
Clean	Ea.	1.07	11		12.07	hardwood veneer on raised door
Paint	Ea.	1.17	25		26.17	panels, hinges and pulls.
Minimum Charge	Job		227		227	
24″ w / Blind Corner						
Demolish	Ea.		18.35		18.35	Includes material and labor to install
Install	Ea.	345	44.50		389.50	custom modular unit with solid
Demolish and Install	Ea.	345	62.85		407.85	hardwood faced frames, hardwood
Reinstall	Ea.		35.78		35.78	door frames and drawer fronts,
Clean	Ea.	1.22	12.70		13.92	hardwood veneer on raised door
Paint	Ea.	1.33	29		30.33	panels, hinges and pulls.
Minimum Charge	Job		227		227	
27″ Wide, 2 Door						
Demolish	Ea.		20.50		20.50	Includes material and labor to install
Install	Ea.	390	46		436	custom modular unit with solid
Demolish and Install	Ea.	390	66.50		456.50	hardwood faced frames, hardwood
Reinstall	Ea.		36.69		36.69	door frames and drawer fronts,
Clean	Ea.	1.38	14.35		15.73	hardwood veneer on raised door
Paint	Ea.	1.51	31.50		33.01	panels, hinges and pulls.
Minimum Charge	Job		227		227	
30″ Wide, 2 Door						
Demolish	Ea.		23		23	Includes material and labor to install
Install	Ea.	405	47		452	custom modular unit with solid
Demolish and Install	Ea.	405	70		475	hardwood faced frames, hardwood
Reinstall	Ea.		37.64		37.64	door frames and drawer fronts,
Clean	Ea.	1.53	15.70		17.23	hardwood veneer on raised door
Paint	Ea.	1.67	38		39.67	panels, hinges and pulls.
Minimum Charge	Job		227		227	
36″ Wide, 2 Door						
Demolish	Ea.		27		27	Includes material and labor to install
Install	Ea.	460	48.50		508.50	custom modular unit with solid
Demolish and Install	Ea.	460	75.50		535.50	hardwood faced frames, hardwood
Reinstall	Ea.		38.64		38.64	door frames and drawer fronts,
Clean	Ea.	1.84	19.40		21.24	hardwood veneer on raised door
Paint	Ea.	2.01	42		44.01	panels, hinges and pulls.
Minimum Charge	Job		227		227	
48″ Wide, 2 Door						
Demolish	Ea.		36.50		36.50	Includes material and labor to install
Install	Ea.	560	49.50		609.50	custom modular unit with solid
Demolish and Install	Ea.	560	86		646	hardwood faced frames, hardwood
Reinstall	Ea.		39.48		39.48	door frames and drawer fronts,
Clean	Ea.	2.44	25.50		27.94	hardwood veneer on raised door
Paint	Ea.	2.67	54		56.67	panels, hinges and pulls.
Minimum Charge	Job		227		227	

213

Cabinets and Countertops

Wall Cabinets	Unit	Material	Labor	Equip.	Total	Specification
Above-appliance						
Demolish	Ea.		23		23	Includes material and labor to install
Install	Ea.	270	36.50		306.50	custom modular unit with solid
Demolish and Install	Ea.	270	59.50		329.50	hardwood faced frames, hardwood
Reinstall	Ea.		29.29		29.29	door frames and drawer fronts,
Clean	Ea.	1.53	15.70		17.23	hardwood veneer on raised door
Paint	Ea.	1.67	38		39.67	panels, hinges and pulls.
Minimum Charge	Job		227		227	
Premium Grade Wood						
12" Wide, 1 Door						
Demolish	Ea.		9.15		9.15	Includes material and labor to install
Install	Ea.	310	41.50		351.50	premium modular unit with solid
Demolish and Install	Ea.	310	50.65		360.65	hardwood faced frames, hardwood
Reinstall	Ea.		33.02		33.02	door frames and drawer fronts,
Clean	Ea.	.62	6.35		6.97	hardwood veneer on raised door
Paint	Ea.	.67	15.75		16.42	panels, hinges and pulls.
Minimum Charge	Job		227		227	
15" Wide, 1 Door						
Demolish	Ea.		11.45		11.45	Includes material and labor to install
Install	Ea.	360	42.50		402.50	premium modular unit with solid
Demolish and Install	Ea.	360	53.95		413.95	hardwood faced frames, hardwood
Reinstall	Ea.		33.94		33.94	door frames and drawer fronts,
Clean	Ea.	.76	7.85		8.61	hardwood veneer on raised door
Paint	Ea.	.84	19.85		20.69	panels, hinges and pulls.
Minimum Charge	Job		227		227	
18" Wide, 1 Door						
Demolish	Ea.		13.85		13.85	Includes material and labor to install
Install	Ea.	360	43.50		403.50	premium modular unit with solid
Demolish and Install	Ea.	360	57.35		417.35	hardwood faced frames, hardwood
Reinstall	Ea.		34.76		34.76	door frames and drawer fronts,
Clean	Ea.	.91	9.45		10.36	hardwood veneer on raised door
Paint	Ea.	.99	23.50		24.49	panels, hinges and pulls.
Minimum Charge	Job		227		227	
21" Wide, 1 Door						
Demolish	Ea.		15.95		15.95	Includes material and labor to install
Install	Ea.	375	44		419	premium modular unit with solid
Demolish and Install	Ea.	375	59.95		434.95	hardwood faced frames, hardwood
Reinstall	Ea.		35.26		35.26	door frames and drawer fronts,
Clean	Ea.	1.07	11		12.07	hardwood veneer on raised door
Paint	Ea.	1.17	27		28.17	panels, hinges and pulls.
Minimum Charge	Job		227		227	
24" w / Blind Corner						
Demolish	Ea.		18.35		18.35	Includes material and labor to install
Install	Ea.	420	44.50		464.50	premium modular unit with solid
Demolish and Install	Ea.	420	62.85		482.85	hardwood faced frames, hardwood
Reinstall	Ea.		35.78		35.78	door frames and drawer fronts,
Clean	Ea.	1.22	12.70		13.92	hardwood veneer on raised door
Paint	Ea.	1.33	31.50		32.83	panels, hinges and pulls.
Minimum Charge	Job		227		227	
27" Wide, 2 Door						
Demolish	Ea.		20.50		20.50	Includes material and labor to install
Install	Ea.	385	46		431	premium modular unit with solid
Demolish and Install	Ea.	385	66.50		451.50	hardwood faced frames, hardwood
Reinstall	Ea.		36.69		36.69	door frames and drawer fronts,
Clean	Ea.	1.38	14.35		15.73	hardwood veneer on raised door
Paint	Ea.	1.51	34.50		36.01	panels, hinges and pulls.
Minimum Charge	Job		227		227	

For customer support on your Contractor's Pricing Guide: Residential Repair & Remodeling, call 888.606.7279.

Cabinets and Countertops

Wall Cabinets		Unit	Material	Labor	Equip.	Total	Specification
30" Wide, 2 Door							
	Demolish	Ea.		23		23	Includes material and labor to install
	Install	Ea.	485	47		532	premium modular unit with solid
	Demolish and Install	Ea.	485	70		555	hardwood faced frames, hardwood
	Reinstall	Ea.		37.64		37.64	door frames and drawer fronts,
	Clean	Ea.	1.53	15.70		17.23	hardwood veneer on raised door
	Paint	Ea.	1.67	38		39.67	panels, hinges and pulls.
	Minimum Charge	Job		227		227	
36" Wide, 2 Door							
	Demolish	Ea.		27		27	Includes material and labor to install
	Install	Ea.	555	48.50		603.50	premium modular unit with solid
	Demolish and Install	Ea.	555	75.50		630.50	hardwood faced frames, hardwood
	Reinstall	Ea.		38.64		38.64	door frames and drawer fronts,
	Clean	Ea.	1.84	19.40		21.24	hardwood veneer on raised door
	Paint	Ea.	2.01	47		49.01	panels, hinges and pulls.
	Minimum Charge	Job		227		227	
48" Wide, 2 Door							
	Demolish	Ea.		36.50		36.50	Includes material and labor to install
	Install	Ea.	675	49.50		724.50	premium modular unit with solid
	Demolish and Install	Ea.	675	86		761	hardwood faced frames, hardwood
	Reinstall	Ea.		39.48		39.48	door frames and drawer fronts,
	Clean	Ea.	2.44	25.50		27.94	hardwood veneer on raised door
	Paint	Ea.	2.67	63		65.67	panels, hinges and pulls.
	Minimum Charge	Job		227		227	
Above-appliance							
	Demolish	Ea.		23		23	Includes material and labor to install
	Install	Ea.	325	36.50		361.50	premium modular unit with solid
	Demolish and Install	Ea.	325	59.50		384.50	hardwood faced frames, hardwood
	Reinstall	Ea.		29.29		29.29	door frames and drawer fronts,
	Clean	Ea.	1.53	15.70		17.23	hardwood veneer on raised door
	Paint	Ea.	1.67	38		39.67	panels, hinges, pulls.
	Minimum Charge	Job		227		227	
Premium Hardwood							
12" Wide, 1 Door							
	Demolish	Ea.		9.15		9.15	Includes material and labor to install
	Install	Ea.	335	41.50		376.50	premium hardwood modular unit with
	Demolish and Install	Ea.	335	50.65		385.65	solid hardwood faced frames, drawer
	Reinstall	Ea.		33.02		33.02	fronts, and door panels, steel drawer
	Clean	Ea.	.62	6.35		6.97	guides, hinges, pulls, laminated
	Paint	Ea.	.67	17.15		17.82	interior.
	Minimum Charge	Job		227		227	
15" Wide, 1 Door							
	Demolish	Ea.		11.45		11.45	Includes material and labor to install
	Install	Ea.	475	42.50		517.50	premium hardwood modular unit with
	Demolish and Install	Ea.	475	53.95		528.95	solid hardwood faced frames, drawer
	Reinstall	Ea.		33.94		33.94	fronts, and door panels, steel drawer
	Clean	Ea.	.76	7.85		8.61	guides, hinges, pulls, laminated
	Paint	Ea.	.84	21		21.84	interior.
	Minimum Charge	Job		227		227	
18" Wide, 1 Door							
	Demolish	Ea.		13.85		13.85	Includes material and labor to install
	Install	Ea.	390	43.50		433.50	premium hardwood modular unit with
	Demolish and Install	Ea.	390	57.35		447.35	solid hardwood faced frames, drawer
	Reinstall	Ea.		34.76		34.76	fronts, and door panels, steel drawer
	Clean	Ea.	.91	9.45		10.36	guides, hinges, pulls, laminated
	Paint	Ea.	.99	25		25.99	interior.
	Minimum Charge	Job		227		227	

215

Cabinets and Countertops

Wall Cabinets	Unit	Material	Labor	Equip.	Total	Specification
21" Wide, 1 Door						
Demolish	Ea.		15.95		15.95	Includes material and labor to install
Install	Ea.	405	44		449	premium hardwood modular unit with
Demolish and Install	Ea.	405	59.95		464.95	solid hardwood faced frames, drawer
Reinstall	Ea.		35.26		35.26	fronts, and door panels, steel drawer
Clean	Ea.	1.07	11		12.07	guides, hinges, pulls, laminated
Paint	Ea.	1.17	29		30.17	interior.
Minimum Charge	Job		227		227	
24" w / Blind Corner						
Demolish	Ea.		18.35		18.35	Includes material and labor to install
Install	Ea.	455	44.50		499.50	premium hardwood modular unit with
Demolish and Install	Ea.	455	62.85		517.85	solid hardwood faced frames, drawer
Reinstall	Ea.		35.78		35.78	fronts, and door panels, steel drawer
Clean	Ea.	1.22	12.70		13.92	guides, hinges, pulls, laminated
Paint	Ea.	1.33	34.50		35.83	interior.
Minimum Charge	Job		227		227	
27" Wide, 2 Door						
Demolish	Ea.		20.50		20.50	Includes material and labor to install
Install	Ea.	510	46		556	premium hardwood modular unit with
Demolish and Install	Ea.	510	66.50		576.50	solid hardwood faced frames, drawer
Reinstall	Ea.		36.69		36.69	fronts, and door panels, steel drawer
Clean	Ea.	1.38	14.35		15.73	guides, hinges, pulls, laminated
Paint	Ea.	1.51	38		39.51	interior.
Minimum Charge	Job		227		227	
30" Wide, 2 Door						
Demolish	Ea.		23		23	Includes material and labor to install
Install	Ea.	530	47		577	premium hardwood modular unit with
Demolish and Install	Ea.	530	70		600	solid hardwood faced frames, drawer
Reinstall	Ea.		37.64		37.64	fronts, and door panels, steel drawer
Clean	Ea.	1.53	15.70		17.23	guides, hinges, pulls, laminated
Paint	Ea.	1.67	42		43.67	interior.
Minimum Charge	Job		227		227	
36" Wide, 2 Door						
Demolish	Ea.		27		27	Includes material and labor to install
Install	Ea.	605	48.50		653.50	premium hardwood modular unit with
Demolish and Install	Ea.	605	75.50		680.50	solid hardwood faced frames, drawer
Reinstall	Ea.		38.64		38.64	fronts, and door panels, steel drawer
Clean	Ea.	1.84	19.40		21.24	guides, hinges, pulls, laminated
Paint	Ea.	2.01	54		56.01	interior.
Minimum Charge	Job		227		227	
48" Wide, 2 Door						
Demolish	Ea.		36.50		36.50	Includes material and labor to install
Install	Ea.	735	49.50		784.50	premium hardwood modular unit with
Demolish and Install	Ea.	735	86		821	solid hardwood faced frames, drawer
Reinstall	Ea.		39.48		39.48	fronts, and door panels, steel drawer
Clean	Ea.	2.44	25.50		27.94	guides, hinges, pulls, laminated
Paint	Ea.	2.67	63		65.67	interior.
Minimum Charge	Job		227		227	
Above-appliance						
Demolish	Ea.		23		23	Includes material and labor to install
Install	Ea.	350	36.50		386.50	premium hardwood modular unit with
Demolish and Install	Ea.	350	59.50		409.50	solid hardwood faced frames, drawer
Reinstall	Ea.		29.29		29.29	fronts, and door panels, steel drawer
Clean	Ea.	1.53	15.70		17.23	guides, hinges, pulls, laminated
Paint	Ea.	1.67	42		43.67	interior.
Minimum Charge	Job		227		227	

Wall Cabinets

Wall Cabinets	Unit	Material	Labor	Equip.	Total	Specification
Medicine Cabinet						
Good Grade						
Demolish	Ea.		15.30		15.30	Cost includes material and labor to
Install	Ea.	165	32.50		197.50	install 14" x 18" recessed medicine
Demolish and Install	Ea.	165	47.80		212.80	cabinet with stainless steel frame,
Reinstall	Ea.		25.94		25.94	mirror. Light not included.
Clean	Ea.	.03	15.20		15.23	
Paint	Ea.	2.05	23.50		25.55	
Minimum Charge	Job		227		227	
Good Grade Lighted						
Demolish	Ea.		15.30		15.30	Cost includes material and labor to
Install	Ea.	420	38		458	install 14" x 18" recessed medicine
Demolish and Install	Ea.	420	53.30		473.30	cabinet with stainless steel frame,
Reinstall	Ea.		30.27		30.27	mirror and light.
Clean	Ea.	.03	15.20		15.23	
Paint	Ea.	2.05	23.50		25.55	
Minimum Charge	Job		227		227	
Better Grade						
Demolish	Ea.		15.30		15.30	Includes material and labor to install
Install	Ea.	275	32.50		307.50	recessed medicine cabinet with
Demolish and Install	Ea.	275	47.80		322.80	polished chrome frame and mirror.
Reinstall	Ea.		25.94		25.94	Light not included.
Clean	Ea.	.03	15.20		15.23	
Paint	Ea.	2.05	23.50		25.55	
Minimum Charge	Job		227		227	
Better Grade Lighted						
Demolish	Ea.		15.30		15.30	Includes material and labor to install
Install	Ea.	530	38		568	recessed medicine cabinet with
Demolish and Install	Ea.	530	53.30		583.30	polished chrome frame, mirror and
Reinstall	Ea.		30.27		30.27	light.
Clean	Ea.	.03	15.20		15.23	
Paint	Ea.	2.05	23.50		25.55	
Minimum Charge	Job		227		227	
Premium Grade						
Demolish	Ea.		15.30		15.30	Cost includes material and labor to
Install	Ea.	330	32.50		362.50	install 14" x 24" recessed medicine
Demolish and Install	Ea.	330	47.80		377.80	cabinet with beveled mirror, frameless
Reinstall	Ea.		25.94		25.94	swing door. Light not included.
Clean	Ea.	.03	15.20		15.23	
Paint	Ea.	2.05	23.50		25.55	
Minimum Charge	Job		227		227	
Premium Grade Lighted						
Demolish	Ea.		15.30		15.30	Cost includes material and labor to
Install	Ea.	585	38		623	install 36" x 30" surface mounted
Demolish and Install	Ea.	585	53.30		638.30	medicine cabinet with beveled tri-view
Reinstall	Ea.		30.27		30.27	mirror, frameless swing door and light.
Clean	Ea.	.03	15.20		15.23	
Paint	Ea.	2.05	23.50		25.55	
Minimum Charge	Job		227		227	
Bath Mirror						
Demolish	Ea.		11.45		11.45	Cost includes material and labor to
Install	Ea.	54.50	22.50		77	install 18" x 24" surface mount mirror
Demolish and Install	Ea.	54.50	33.95		88.45	with stainless steel frame.
Reinstall	Ea.		18.16		18.16	
Clean	Ea.	.03	1.01		1.04	
Minimum Charge	Job		227		227	

Cabinets and Countertops

Wall Cabinets

Wall Cabinets	Unit	Material	Labor	Equip.	Total	Specification
Garage						Includes material and labor to install
Demolish	L.F.		9.15		9.15	good modular unit with melamine
Install	L.F.	98.50	22.50		121	laminated to particle board, in
Demolish and Install	L.F.	98.50	31.65		130.15	textured colors or wood grain print
Reinstall	L.F.		18.16		18.16	finish, hinges and pulls.
Clean	L.F.	.62	6.35		6.97	
Paint	L.F.	.67	13.50		14.17	
Minimum Charge	Job		227		227	
Strip and Refinish						
Install	L.F.	4.87	47		51.87	Includes labor and material to strip,
Minimum Charge	Job		189		189	prep and refinish exterior of cabinets.

Tall Cabinets

Tall Cabinets	Unit	Material	Labor	Equip.	Total	Specification
Good Grade Laminated						Includes material and labor to install
Demolish	L.F.		16.30		16.30	custom modular unit with solid
Install	L.F.	243	50.50		293.50	hardwood faced frames, hardwood
Demolish and Install	L.F.	243	66.80		309.80	door frames and drawer fronts,
Reinstall	L.F.		40.36		40.36	hardwood veneer on raised door
Clean	L.F.	1.63	11.80		13.43	panels, hinges and pulls.
Paint	L.F.	1.55	27		28.55	
Minimum Charge	Job		227		227	
Better Grade Wood						Includes material and labor to install
Demolish	L.F.		16.30		16.30	custom modular unit with solid
Install	L.F.	345	50.50		395.50	hardwood faced frames, hardwood
Demolish and Install	L.F.	345	66.80		411.80	door frames and drawer fronts,
Reinstall	L.F.		40.36		40.36	hardwood veneer on raised door
Clean	L.F.	1.63	11.80		13.43	panels, hinges and pulls.
Paint	L.F.	1.55	29		30.55	
Minimum Charge	Job		227		227	
Premium Grade Wood						Includes material and labor to install
Demolish	L.F.		16.30		16.30	premium modular unit with solid
Install	L.F.	495	50.50		545.50	hardwood faced frames, hardwood
Demolish and Install	L.F.	495	66.80		561.80	door frames and drawer fronts,
Reinstall	L.F.		40.36		40.36	hardwood veneer on raised door
Clean	L.F.	1.63	11.80		13.43	panels, hinges and pulls.
Paint	L.F.	1.55	31.50		33.05	
Minimum Charge	Job		227		227	
Premium Hardwood						Includes material and labor to install
Demolish	L.F.		16.30		16.30	premium hardwood modular unit with
Install	L.F.	700	50.50		750.50	solid hardwood faced frames, drawer
Demolish and Install	L.F.	700	66.80		766.80	fronts, and door panels, steel drawer
Reinstall	L.F.		40.36		40.36	guides, hinges and pulls, laminated
Clean	L.F.	1.63	11.80		13.43	interior.
Paint	L.F.	1.55	34.50		36.05	
Minimum Charge	Job		227		227	

Cabinets and Countertops

Tall Cabinets

	Unit	Material	Labor	Equip.	Total	Specification
Single Oven, 27″ Wide						
Good Grade Laminated						Includes material and labor to install
Demolish	Ea.		36.50		36.50	good modular unit with melamine
Install	Ea.	840	114		954	laminated to particle board, in
Demolish and Install	Ea.	840	150.50		990.50	textured colors or wood grain print
Reinstall	Ea.		90.80		90.80	finish, hinges and pulls.
Clean	Ea.	5.10	20.50		25.60	
Paint	Ea.	2.61	60.50		63.11	
Minimum Charge	Job		227		227	
Better Grade Veneer						Includes material and labor to install
Demolish	Ea.		36.50		36.50	custom modular unit with solid
Install	Ea.	1475	114		1589	hardwood faced frames, hardwood
Demolish and Install	Ea.	1475	150.50		1625.50	door frames and drawer fronts,
Reinstall	Ea.		90.80		90.80	hardwood veneer on raised door
Clean	Ea.	5.10	20.50		25.60	panels, hinges and pulls.
Paint	Ea.	2.61	65		67.61	
Minimum Charge	Job		227		227	
Premium Grade Veneer						Includes material and labor to install
Demolish	Ea.		36.50		36.50	premium modular unit with solid
Install	Ea.	1350	114		1464	hardwood faced frames, hardwood
Demolish and Install	Ea.	1350	150.50		1500.50	door frames and drawer fronts,
Reinstall	Ea.		90.80		90.80	hardwood veneer on raised door
Clean	Ea.	5.10	20.50		25.60	panels, hinges and pulls.
Paint	Ea.	2.61	71.50		74.11	
Minimum Charge	Job		227		227	
Premium Hardwood						Includes material and labor to install
Demolish	Ea.		36.50		36.50	premium hardwood modular unit with
Install	Ea.	1450	114		1564	solid hardwood faced frames, drawer
Demolish and Install	Ea.	1450	150.50		1600.50	fronts, and door panels, steel drawer
Reinstall	Ea.		90.80		90.80	guides, hinges and pulls, laminated
Clean	Ea.	5.10	20.50		25.60	interior.
Paint	Ea.	2.61	77		79.61	
Minimum Charge	Job		227		227	
Single Oven, 30″ Wide						
Good Grade Laminated						Includes material and labor to install
Demolish	Ea.		36.50		36.50	good modular unit with melamine
Install	Ea.	980	114		1094	laminated to particle board, in
Demolish and Install	Ea.	980	150.50		1130.50	textured colors or wood grain print
Reinstall	Ea.		90.80		90.80	finish, hinges and pulls.
Clean	Ea.	5.10	20.50		25.60	
Paint	Ea.	2.61	60.50		63.11	
Minimum Charge	Job		227		227	
Better Grade Veneer						Includes material and labor to install
Demolish	Ea.		36.50		36.50	custom modular unit with solid
Install	Ea.	1100	114		1214	hardwood faced frames, hardwood
Demolish and Install	Ea.	1100	150.50		1250.50	door frames and drawer fronts,
Reinstall	Ea.		90.80		90.80	hardwood veneer on raised door
Clean	Ea.	5.10	20.50		25.60	panels, hinges and pulls.
Paint	Ea.	2.61	60.50		63.11	
Minimum Charge	Job		227		227	

219

Cabinets and Countertops

Tall Cabinets		Unit	Material	Labor	Equip.	Total	Specification
Premium Grade Veneer							
	Demolish	Ea.		36.50		36.50	Includes material and labor to install
	Install	Ea.	1250	114		1364	premium modular unit with solid
	Demolish and Install	Ea.	1250	150.50		1400.50	hardwood faced frames, hardwood
	Reinstall	Ea.		90.80		90.80	door frames and drawer fronts,
	Clean	Ea.	5.10	20.50		25.60	hardwood veneer on raised door
	Paint	Ea.	2.61	60.50		63.11	panels, hinges and pulls.
	Minimum Charge	Job		227		227	
Premium Hardwood							
	Demolish	Ea.		36.50		36.50	Includes material and labor to install
	Install	Ea.	1375	114		1489	premium hardwood modular unit with
	Demolish and Install	Ea.	1375	150.50		1525.50	solid hardwood faced frames, drawer
	Reinstall	Ea.		90.80		90.80	fronts, and door panels, steel drawer
	Clean	Ea.	5.10	20.50		25.60	guides, hinges and pulls, laminated
	Paint	Ea.	2.61	60.50		63.11	interior.
	Minimum Charge	Job		227		227	

Utility 18″ W x 12″ D

		Unit	Material	Labor	Equip.	Total	Specification
Good Grade Laminated							
	Demolish	Ea.		36.50		36.50	Includes material and labor to install
	Install	Ea.	595	91		686	good modular unit with melamine
	Demolish and Install	Ea.	595	127.50		722.50	laminated to particle board, in
	Reinstall	Ea.		72.64		72.64	textured colors or wood grain print
	Clean	Ea.	1.20	10		11.20	finish, hinges and pulls.
	Paint	Ea.	.99	23.50		24.49	
	Minimum Charge	Job		227		227	
Better Grade Veneer							
	Demolish	Ea.		36.50		36.50	Includes material and labor to install
	Install	Ea.	800	91		891	custom modular unit with solid
	Demolish and Install	Ea.	800	127.50		927.50	hardwood faced frames, hardwood
	Reinstall	Ea.		72.64		72.64	door frames and drawer fronts,
	Clean	Ea.	1.20	10		11.20	hardwood veneer on raised door
	Paint	Ea.	.99	25		25.99	panels, hinges and pulls.
	Minimum Charge	Job		227		227	
Premium Grade Veneer							
	Demolish	Ea.		36.50		36.50	Includes material and labor to install
	Install	Ea.	955	91		1046	premium modular unit with solid
	Demolish and Install	Ea.	955	127.50		1082.50	hardwood faced frames, hardwood
	Reinstall	Ea.		72.64		72.64	door frames and drawer fronts,
	Clean	Ea.	1.20	10		11.20	hardwood veneer on raised door
	Paint	Ea.	.99	29		29.99	panels, hinges and pulls.
	Minimum Charge	Job		227		227	
Premium Hardwood							
	Demolish	Ea.		36.50		36.50	Includes material and labor to install
	Install	Ea.	1050	91		1141	premium hardwood modular unit with
	Demolish and Install	Ea.	1050	127.50		1177.50	solid hardwood faced frames, drawer
	Reinstall	Ea.		72.64		72.64	fronts, and door panels, steel drawer
	Clean	Ea.	1.20	10		11.20	guides, hinges and pulls, laminated
	Paint	Ea.	.99	31.50		32.49	interior.
	Minimum Charge	Job		227		227	

Cabinets and Countertops

Tall Cabinets	Unit	Material	Labor	Equip.	Total	Specification
Utility 24″ W x 24″ D						
Good Grade Laminated						
Demolish	Ea.		36.50		36.50	Includes material and labor to install
Install	Ea.	705	114		819	good modular unit with melamine
Demolish and Install	Ea.	705	150.50		855.50	laminated to particle board, in
Reinstall	Ea.		90.80		90.80	textured colors or wood grain print
Clean	Ea.	1.56	13.20		14.76	finish, hinges and pulls.
Paint	Ea.	1.33	31.50		32.83	
Minimum Charge	Job		227		227	
Better Grade Veneer						
Demolish	Ea.		36.50		36.50	Includes material and labor to install
Install	Ea.	940	114		1054	custom modular unit with solid
Demolish and Install	Ea.	940	150.50		1090.50	hardwood faced frames, hardwood
Reinstall	Ea.		90.80		90.80	door frames and drawer fronts,
Clean	Ea.	1.56	13.20		14.76	hardwood veneer on raised door
Paint	Ea.	1.33	34.50		35.83	panels, hinges and pulls.
Minimum Charge	Job		227		227	
Premium Grade Veneer						
Demolish	Ea.		36.50		36.50	Includes material and labor to install
Install	Ea.	1125	114		1239	premium modular unit with solid
Demolish and Install	Ea.	1125	150.50		1275.50	hardwood faced frames, hardwood
Reinstall	Ea.		90.80		90.80	door frames and drawer fronts,
Clean	Ea.	1.56	13.20		14.76	hardwood veneer on raised door
Paint	Ea.	1.33	38		39.33	panels, hinges and pulls.
Minimum Charge	Job		227		227	
Premium Hardwood						
Demolish	Ea.		36.50		36.50	Includes material and labor to install
Install	Ea.	1250	114		1364	premium hardwood modular unit with
Demolish and Install	Ea.	1250	150.50		1400.50	solid hardwood faced frames, drawer
Reinstall	Ea.		90.80		90.80	fronts, and door panels, steel drawer
Clean	Ea.	1.56	13.20		14.76	guides, hinges and pulls, laminated
Paint	Ea.	1.33	42		43.33	interior.
Minimum Charge	Job		227		227	
Plain Shelves						
Install	Ea.	33.50	25		58.50	Includes labor and material to install
Rotating Shelves						tall cabinet shelving, per set.
Install	Ea.	99.50	22.50		122	Includes labor and material to install
Strip and Refinish						lazy susan shelving for wall cabinet.
Install	L.F.	9.75	75.50		85.25	Includes labor and material to strip,
Minimum Charge	Job		189		189	prep and refinish exterior of cabinets.
Stain						
Install	L.F.	.29	3.43		3.72	Includes labor and material to replace
Minimum Charge	Job		189		189	normal prep and stain on doors and
						exterior cabinets.

Cabinets and Countertops

Base Cabinets	Unit	Material	Labor	Equip.	Total	Specification
Good Grade Laminated						
12″ w, 1 Door, 1 Drawer						
Demolish	Ea.		9.15		9.15	Includes material and labor to install
Install	Ea.	223	36.50		259.50	good modular unit with melamine
Demolish and Install	Ea.	223	45.65		268.65	laminate, textured colors or wood
Reinstall	Ea.		29.29		29.29	grain print finish including hinges and
Clean	Ea.	.81	6.60		7.41	pulls.
Paint	Ea.	.67	15.75		16.42	
Minimum Charge	Job		227		227	
15″ w, 1 Door, 1 Drawer						
Demolish	Ea.		11.45		11.45	Includes material and labor to install
Install	Ea.	233	38		271	good modular unit with melamine
Demolish and Install	Ea.	233	49.45		282.45	laminate, textured colors or wood
Reinstall	Ea.		30.27		30.27	grain print finish including hinges and
Clean	Ea.	1.02	8.25		9.27	pulls.
Paint	Ea.	.84	19.85		20.69	
Minimum Charge	Job		227		227	
18″ w, 1 Door, 1 Drawer						
Demolish	Ea.		13.60		13.60	Includes material and labor to install
Install	Ea.	253	39		292	good modular unit with melamine
Demolish and Install	Ea.	253	52.60		305.60	laminate, textured colors or wood
Reinstall	Ea.		31.18		31.18	grain print finish including hinges and
Clean	Ea.	1.20	10		11.20	pulls.
Paint	Ea.	.99	23.50		24.49	
Minimum Charge	Job		227		227	
21″ w, 1 Door, 1 Drawer						
Demolish	Ea.		15.95		15.95	Includes material and labor to install
Install	Ea.	262	40		302	good modular unit with melamine
Demolish and Install	Ea.	262	55.95		317.95	laminate, textured colors or wood
Reinstall	Ea.		32		32	grain print finish including hinges and
Clean	Ea.	1.45	11.80		13.25	pulls.
Paint	Ea.	1.17	27		28.17	
Minimum Charge	Job		227		227	
24″ w, 1 Door, 1 Drawer						
Demolish	Ea.		18.35		18.35	Includes material and labor to install
Install	Ea.	305	40.50		345.50	good modular unit with melamine
Demolish and Install	Ea.	305	58.85		363.85	laminate, textured colors or wood
Reinstall	Ea.		32.57		32.57	grain print finish including hinges and
Clean	Ea.	1.56	13.20		14.76	pulls.
Paint	Ea.	1.33	31.50		32.83	
Minimum Charge	Job		227		227	
30″ Wide Sink Base						
Demolish	Ea.		23		23	Includes material and labor to install
Install	Ea.	299	42.50		341.50	good modular unit with melamine
Demolish and Install	Ea.	299	65.50		364.50	laminate, textured colors or wood
Reinstall	Ea.		33.94		33.94	grain print finish including hinges and
Clean	Ea.	2.04	16.50		18.54	pulls.
Paint	Ea.	1.67	38		39.67	
Minimum Charge	Job		227		227	
36″ Wide Sink Base						
Demolish	Ea.		28		28	Includes material and labor to install
Install	Ea.	335	44.50		379.50	good modular unit with melamine
Demolish and Install	Ea.	335	72.50		407.50	laminate, textured colors or wood
Reinstall	Ea.		35.78		35.78	grain print finish including hinges and
Clean	Ea.	2.27	19.40		21.67	pulls.
Paint	Ea.	2.01	47		49.01	
Minimum Charge	Job		227		227	

Cabinets and Countertops

Base Cabinets	Unit	Material	Labor	Equip.	Total	Specification
36″ Blind Corner						
Demolish	Ea.		28		28	Includes material and labor to install
Install	Ea.	555	50.50		605.50	good modular unit with melamine
Demolish and Install	Ea.	555	78.50		633.50	laminate, textured colors or wood
Reinstall	Ea.		40.36		40.36	grain print finish including hinges and
Clean	Ea.	2.27	19.40		21.67	pulls.
Paint	Ea.	2.01	47		49.01	
Minimum Charge	Job		227		227	
42″ w, 2 Door, 2 Drawer						
Demolish	Ea.		33.50		33.50	Includes material and labor to install
Install	Ea.	405	46		451	good modular unit with melamine
Demolish and Install	Ea.	405	79.50		484.50	laminate, textured colors or wood
Reinstall	Ea.		36.69		36.69	grain print finish including hinges and
Clean	Ea.	2.92	23.50		26.42	pulls.
Paint	Ea.	2.31	54		56.31	
Minimum Charge	Job		227		227	
Clean Interior						
Clean	L.F.	.81	5		5.81	Includes labor and materials to clean
Minimum Charge	Job		165		165	the interior of a 27″ cabinet.
Better Grade Wood						
12″ w, 1 Door, 1 Drawer						
Demolish	Ea.		9.15		9.15	Includes material and labor to install
Install	Ea.	298	36.50		334.50	custom modular unit including solid
Demolish and Install	Ea.	298	45.65		343.65	hardwood faced frames, hardwood
Reinstall	Ea.		29.29		29.29	door frames and drawer fronts,
Clean	Ea.	.81	6.60		7.41	hardwood veneer on raised door
Paint	Ea.	.67	17.15		17.82	panels, hinges and pulls.
Minimum Charge	Job		227		227	
15″ w, 1 Door, 1 Drawer						
Demolish	Ea.		11.45		11.45	Includes material and labor to install
Install	Ea.	310	38		348	custom modular unit including solid
Demolish and Install	Ea.	310	49.45		359.45	hardwood faced frames, hardwood
Reinstall	Ea.		30.27		30.27	door frames and drawer fronts,
Clean	Ea.	1.02	8.25		9.27	hardwood veneer on raised door
Paint	Ea.	.84	22		22.84	panels, hinges and pulls.
Minimum Charge	Job		227		227	
18″ w, 1 Door, 1 Drawer						
Demolish	Ea.		13.60		13.60	Includes material and labor to install
Install	Ea.	335	39		374	custom modular unit including solid
Demolish and Install	Ea.	335	52.60		387.60	hardwood faced frames, hardwood
Reinstall	Ea.		31.18		31.18	door frames and drawer fronts,
Clean	Ea.	1.20	10		11.20	hardwood veneer on raised door
Paint	Ea.	.99	25		25.99	panels, hinges and pulls.
Minimum Charge	Job		227		227	
21″ w, 1 Door, 1 Drawer						
Demolish	Ea.		15.95		15.95	Includes material and labor to install
Install	Ea.	345	40		385	custom modular unit including solid
Demolish and Install	Ea.	345	55.95		400.95	hardwood faced frames, hardwood
Reinstall	Ea.		32		32	door frames and drawer fronts,
Clean	Ea.	1.45	11.80		13.25	hardwood veneer on raised door
Paint	Ea.	1.17	29		30.17	panels, hinges and pulls.
Minimum Charge	Job		227		227	

223

Cabinets and Countertops

Base Cabinets	Unit	Material	Labor	Equip.	Total	Specification
24" w, 1 Door, 1 Drawer						
Demolish	Ea.		18.35		18.35	Includes material and labor to install
Install	Ea.	405	40.50		445.50	custom modular unit including solid
Demolish and Install	Ea.	405	58.85		463.85	hardwood faced frames, hardwood
Reinstall	Ea.		32.57		32.57	door frames and drawer fronts,
Clean	Ea.	1.56	13.20		14.76	hardwood veneer on raised door
Paint	Ea.	1.33	34.50		35.83	panels, hinges and pulls.
Minimum Charge	Job		227		227	
30" Wide Sink Base						
Demolish	Ea.		23		23	Includes material and labor to install
Install	Ea.	400	42.50		442.50	custom modular unit including solid
Demolish and Install	Ea.	400	65.50		465.50	hardwood faced frames, hardwood
Reinstall	Ea.		33.94		33.94	door frames and drawer fronts,
Clean	Ea.	2.04	16.50		18.54	hardwood veneer on raised door
Paint	Ea.	1.67	42		43.67	panels, hinges and pulls.
Minimum Charge	Job		227		227	
36" Wide Sink Base						
Demolish	Ea.		28		28	Includes material and labor to install
Install	Ea.	445	44.50		489.50	custom modular unit including solid
Demolish and Install	Ea.	445	72.50		517.50	hardwood faced frames, hardwood
Reinstall	Ea.		35.78		35.78	door frames and drawer fronts,
Clean	Ea.	2.27	19.40		21.67	hardwood veneer on raised door
Paint	Ea.	2.01	54		56.01	panels, hinges and pulls.
Minimum Charge	Job		227		227	
36" Blind Corner						
Demolish	Ea.		28		28	Includes material and labor to install
Install	Ea.	735	50.50		785.50	custom modular unit including solid
Demolish and Install	Ea.	735	78.50		813.50	hardwood faced frames, hardwood
Reinstall	Ea.		40.36		40.36	door frames and drawer fronts,
Clean	Ea.	2.27	19.40		21.67	hardwood veneer on raised door
Paint	Ea.	2.01	54		56.01	panels, hinges and pulls.
Minimum Charge	Job		227		227	
42" w, 2 Door, 2 Drawer						
Demolish	Ea.		33.50		33.50	Includes material and labor to install
Install	Ea.	545	46		591	modular unit including solid hardwood
Demolish and Install	Ea.	545	79.50		624.50	faced frames, hardwood door frames
Reinstall	Ea.		36.69		36.69	and drawer fronts, hardwood veneer
Clean	Ea.	2.92	23.50		26.42	on raised door panels, hinges and
Paint	Ea.	2.31	63		65.31	pulls.
Minimum Charge	Job		227		227	
Clean Interior						
Clean	L.F.	.81	5		5.81	Includes labor and materials to clean
Minimum Charge	Job		165		165	the interior of a 27" cabinet.
Premium Grade Wood						
12" w, 1 Door, 1 Drawer						
Demolish	Ea.		9.15		9.15	Includes material and labor to install
Install	Ea.	360	18.30		378.30	premium hardwood modular unit with
Demolish and Install	Ea.	360	27.45		387.45	solid hardwood faced frames, drawer
Reinstall	Ea.		14.65		14.65	fronts, and door panels, steel drawer
Clean	Ea.	.81	6.60		7.41	guides, hinges, pulls and laminated
Paint	Ea.	.67	18.90		19.57	interior.
Minimum Charge	Job		227		227	

Cabinets and Countertops

Base Cabinets		Unit	Material	Labor	Equip.	Total	Specification
15" w, 1 Door, 1 Drawer							
	Demolish	Ea.		11.45		11.45	Includes material and labor to install
	Install	Ea.	375	38		413	premium modular unit including solid
	Demolish and Install	Ea.	375	49.45		424.45	hardwood faced frames, hardwood
	Reinstall	Ea.		30.27		30.27	door frames and drawer fronts,
	Clean	Ea.	1.02	8.25		9.27	hardwood veneer on raised door
	Paint	Ea.	.84	23.50		24.34	panels, hinges and pulls.
	Minimum Charge	Job		227		227	
18" w, 1 Door, 1 Drawer							
	Demolish	Ea.		13.60		13.60	Includes material and labor to install
	Install	Ea.	405	39		444	premium modular unit including solid
	Demolish and Install	Ea.	405	52.60		457.60	hardwood faced frames, hardwood
	Reinstall	Ea.		31.18		31.18	door frames and drawer fronts,
	Clean	Ea.	1.20	10		11.20	hardwood veneer on raised door
	Paint	Ea.	.99	29		29.99	panels, hinges and pulls.
	Minimum Charge	Job		227		227	
21" w, 1 Door, 1 Drawer							
	Demolish	Ea.		15.95		15.95	Includes material and labor to install
	Install	Ea.	425	40		465	premium modular unit including solid
	Demolish and Install	Ea.	425	55.95		480.95	hardwood faced frames, hardwood
	Reinstall	Ea.		32		32	door frames and drawer fronts,
	Clean	Ea.	1.45	11.80		13.25	hardwood veneer on raised door
	Paint	Ea.	1.17	31.50		32.67	panels, hinges and pulls.
	Minimum Charge	Job		227		227	
24" w, 1 Door, 1 Drawer							
	Demolish	Ea.		18.35		18.35	Includes material and labor to install
	Install	Ea.	490	40.50		530.50	premium modular unit including solid
	Demolish and Install	Ea.	490	58.85		548.85	hardwood faced frames, hardwood
	Reinstall	Ea.		32.57		32.57	door frames and drawer fronts,
	Clean	Ea.	1.56	13.20		14.76	hardwood veneer on raised door
	Paint	Ea.	1.33	38		39.33	panels, hinges and pulls.
	Minimum Charge	Job		227		227	
30" Wide Sink Base							
	Demolish	Ea.		23		23	Includes material and labor to install
	Install	Ea.	480	42.50		522.50	premium modular unit including solid
	Demolish and Install	Ea.	480	65.50		545.50	hardwood faced frames, hardwood
	Reinstall	Ea.		33.94		33.94	door frames and drawer fronts,
	Clean	Ea.	2.04	16.50		18.54	hardwood veneer on raised door
	Paint	Ea.	1.67	47		48.67	panels, hinges and pulls.
	Minimum Charge	Job		227		227	
36" Wide Sink Base							
	Demolish	Ea.		28		28	Includes material and labor to install
	Install	Ea.	540	44.50		584.50	premium modular unit including solid
	Demolish and Install	Ea.	540	72.50		612.50	hardwood faced frames, hardwood
	Reinstall	Ea.		35.78		35.78	door frames and drawer fronts,
	Clean	Ea.	2.27	19.40		21.67	hardwood veneer on raised door
	Paint	Ea.	2.01	54		56.01	panels, hinges and pulls.
	Minimum Charge	Job		227		227	
36" Blind Corner							
	Demolish	Ea.		28		28	Includes material and labor to install
	Install	Ea.	885	50.50		935.50	premium modular unit including solid
	Demolish and Install	Ea.	885	78.50		963.50	hardwood faced frames, hardwood
	Reinstall	Ea.		40.36		40.36	door frames and drawer fronts,
	Clean	Ea.	2.27	19.40		21.67	hardwood veneer on raised door
	Paint	Ea.	2.01	54		56.01	panels, hinges and pulls.
	Minimum Charge	Job		227		227	

225

Base Cabinets	Unit	Material	Labor	Equip.	Total	Specification
42" w, 2 Door, 2 Drawer						
Demolish	Ea.		33.50		33.50	Includes material and labor to install
Install	Ea.	725	46		771	premium modular unit including solid
Demolish and Install	Ea.	725	79.50		804.50	hardwood faced frames, hardwood
Reinstall	Ea.		36.69		36.69	door frames and drawer fronts,
Clean	Ea.	2.92	23.50		26.42	hardwood veneer on raised door
Paint	Ea.	2.31	63		65.31	panels, hinges and pulls.
Minimum Charge	Job		227		227	
Clean Interior						
Clean	L.F.	.81	5		5.81	Includes labor and materials to clean
Minimum Charge	Job		165		165	the interior of a 27" cabinet.
Premium Hardwood						
12" w, 1 Door, 1 Drawer						
Demolish	Ea.		9.15		9.15	Includes material and labor to install
Install	Ea.	390	36.50		426.50	premium hardwood modular unit with
Demolish and Install	Ea.	390	45.65		435.65	solid hardwood faced frames, drawer
Reinstall	Ea.		29.29		29.29	fronts, and door panels, steel drawer
Clean	Ea.	.81	6.60		7.41	guides, hinges, pulls and laminated
Paint	Ea.	.67	21		21.67	interior.
Minimum Charge	Job		227		227	
15" w, 1 Door, 1 Drawer						
Demolish	Ea.		11.45		11.45	Includes material and labor to install
Install	Ea.	405	38		443	premium hardwood modular unit with
Demolish and Install	Ea.	405	49.45		454.45	solid hardwood faced frames, drawer
Reinstall	Ea.		30.27		30.27	fronts, and door panels, steel drawer
Clean	Ea.	1.02	8.25		9.27	guides, hinges, pulls and laminated
Paint	Ea.	.84	27		27.84	interior.
Minimum Charge	Job		227		227	
18" w, 1 Door, 1 Drawer						
Demolish	Ea.		13.60		13.60	Includes material and labor to install
Install	Ea.	440	39		479	premium hardwood modular unit with
Demolish and Install	Ea.	440	52.60		492.60	solid hardwood faced frames, drawer
Reinstall	Ea.		31.18		31.18	fronts, and door panels, steel drawer
Clean	Ea.	1.20	10		11.20	guides, hinges, pulls and laminated
Paint	Ea.	.99	31.50		32.49	interior.
Minimum Charge	Job		227		227	
21" w, 1 Door, 1 Drawer						
Demolish	Ea.		15.95		15.95	Includes material and labor to install
Install	Ea.	455	40		495	premium hardwood modular unit with
Demolish and Install	Ea.	455	55.95		510.95	solid hardwood faced frames, drawer
Reinstall	Ea.		32		32	fronts, and door panels, steel drawer
Clean	Ea.	1.45	11.80		13.25	guides, hinges, pulls and laminated
Paint	Ea.	1.17	34.50		35.67	interior.
Minimum Charge	Job		227		227	
24" w, 1 Door, 1 Drawer						
Demolish	Ea.		18.35		18.35	Includes material and labor to install
Install	Ea.	535	40.50		575.50	premium hardwood modular unit with
Demolish and Install	Ea.	535	58.85		593.85	solid hardwood faced frames, drawer
Reinstall	Ea.		32.57		32.57	fronts, and door panels, steel drawer
Clean	Ea.	1.56	13.20		14.76	guides, hinges and pulls, laminated
Paint	Ea.	1.33	42		43.33	interior.
Minimum Charge	Job		227		227	

Cabinets and Countertops

Base Cabinets		Unit	Material	Labor	Equip.	Total	Specification
30" Wide Sink Base							
	Demolish	Ea.		23		23	Includes material and labor to install
	Install	Ea.	525	42.50		567.50	premium hardwood modular unit with
	Demolish and Install	Ea.	525	65.50		590.50	solid hardwood faced frames, drawer
	Reinstall	Ea.		33.94		33.94	fronts, and door panels, steel drawer
	Clean	Ea.	2.04	16.50		18.54	guides, hinges, pulls and laminated
	Paint	Ea.	1.67	47		48.67	interior.
	Minimum Charge	Job		227		227	
36" Wide Sink Base							
	Demolish	Ea.		28		28	Includes material and labor to install
	Install	Ea.	590	44.50		634.50	premium hardwood modular unit with
	Demolish and Install	Ea.	590	72.50		662.50	solid hardwood faced frames, drawer
	Reinstall	Ea.		35.78		35.78	fronts, and door panels, steel drawer
	Clean	Ea.	2.27	19.40		21.67	guides, hinges, pulls and laminated
	Paint	Ea.	2.01	63		65.01	interior.
	Minimum Charge	Job		227		227	
36" Blind Corner							
	Demolish	Ea.		28		28	Includes material and labor to install
	Install	Ea.	970	50.50		1020.50	premium hardwood modular unit with
	Demolish and Install	Ea.	970	78.50		1048.50	solid hardwood faced frames, drawer
	Reinstall	Ea.		40.36		40.36	fronts, and door panels, steel drawer
	Clean	Ea.	2.27	19.40		21.67	guides, hinges, pulls and laminated
	Paint	Ea.	2.01	63		65.01	interior.
	Minimum Charge	Job		227		227	
42" w, 2 Door, 2 Drawer							
	Demolish	Ea.		33.50		33.50	Includes material and labor to install
	Install	Ea.	955	46		1001	premium hardwood modular unit with
	Demolish and Install	Ea.	955	79.50		1034.50	solid hardwood faced frames, drawer
	Reinstall	Ea.		36.69		36.69	fronts, and door panels, steel drawer
	Clean	Ea.	2.92	23.50		26.42	guides, hinges, pulls and laminated
	Paint	Ea.	2.31	75.50		77.81	interior.
	Minimum Charge	Job		227		227	
Clean Interior							
	Clean	L.F.	.81	5		5.81	Includes labor and materials to clean
	Minimum Charge	Job		165		165	the interior of a 27" cabinet.
Island							
24" Wide							
	Demolish	Ea.		18.35		18.35	Includes material and labor to install
	Install	Ea.	635	38		673	custom modular unit with solid
	Demolish and Install	Ea.	635	56.35		691.35	hardwood faced frames, hardwood
	Reinstall	Ea.		30.27		30.27	door frames and drawer fronts,
	Clean	Ea.	2.27	19.40		21.67	hardwood veneer on raised door
	Paint	Ea.	1.99	54		55.99	panels, hinges and pulls.
	Minimum Charge	Job		227		227	
30" Wide							
	Demolish	Ea.		23		23	Includes material and labor to install
	Install	Ea.	705	47.50		752.50	custom modular unit with solid
	Demolish and Install	Ea.	705	70.50		775.50	hardwood faced frames, hardwood
	Reinstall	Ea.		37.83		37.83	door frames and drawer fronts,
	Clean	Ea.	2.92	23.50		26.42	hardwood veneer on raised door
	Paint	Ea.	2.50	63		65.50	panels, hinges and pulls.
	Minimum Charge	Job		227		227	

Cabinets and Countertops

Base Cabinets		Unit	Material	Labor	Equip.	Total	Specification
36" Wide							
	Demolish	Ea.		28		28	Includes material and labor to install
	Install	Ea.	810	57		867	custom modular unit with solid
	Demolish and Install	Ea.	810	85		895	hardwood faced frames, hardwood
	Reinstall	Ea.		45.40		45.40	door frames and drawer fronts,
	Clean	Ea.	3.40	30		33.40	hardwood veneer on raised door
	Paint	Ea.	3.03	75.50		78.53	panels, hinges and pulls.
	Minimum Charge	Job		227		227	
48" Wide							
	Demolish	Ea.		36.50		36.50	Includes material and labor to install
	Install	Ea.	1275	75.50		1350.50	custom modular unit with solid
	Demolish and Install	Ea.	1275	112		1387	hardwood faced frames, hardwood
	Reinstall	Ea.		60.53		60.53	door frames and drawer fronts,
	Clean	Ea.	4.07	36.50		40.57	hardwood veneer on raised door
	Paint	Ea.	3.47	94.50		97.97	panels, hinges and pulls.
	Minimum Charge	Job		227		227	
60" Wide							
	Demolish	Ea.		40.50		40.50	Includes material and labor to install
	Install	Ea.	1400	94.50		1494.50	custom modular unit with solid
	Demolish and Install	Ea.	1400	135		1535	hardwood faced frames, hardwood
	Reinstall	Ea.		75.67		75.67	door frames and drawer fronts,
	Clean	Ea.	5.10	47		52.10	hardwood veneer on raised door
	Paint	Ea.	3.47	126		129.47	panels, hinges and pulls.
	Minimum Charge	Job		227		227	
Clean Interior							
	Clean	L.F.	.81	5		5.81	Includes labor and materials to clean
	Minimum Charge	Job		165		165	the interior of a 27" cabinet.
Liquor Bar							
	Demolish	L.F.		9.15		9.15	Includes material and labor to install
	Install	L.F.	193	45.50		238.50	custom modular unit with solid
	Demolish and Install	L.F.	193	54.65		247.65	hardwood faced frames, hardwood
	Reinstall	L.F.		36.32		36.32	door frames and drawer fronts,
	Clean	L.F.	.81	6.90		7.71	hardwood veneer on raised door
	Paint	L.F.	5.30	17.15		22.45	panels, hinges and pulls.
	Minimum Charge	Job		227		227	
Clean Interior							
	Clean	L.F.	.81	5		5.81	Includes labor and materials to clean
	Minimum Charge	Job		165		165	the interior of a 27" cabinet.
Seal Interior							
	Install	L.F.	.57	6.20		6.77	Includes labor and materials to seal a
	Minimum Charge	Job		189		189	cabinet interior.
Back Bar							
	Demolish	L.F.		9.15		9.15	Includes material and labor to install
	Install	L.F.	193	45.50		238.50	custom modular unit with solid
	Demolish and Install	L.F.	193	54.65		247.65	hardwood faced frames, hardwood
	Reinstall	L.F.		36.32		36.32	door frames and drawer fronts,
	Clean	L.F.	.81	6.90		7.71	hardwood veneer on raised door
	Paint	L.F.	5.30	17.15		22.45	panels, hinges and pulls.
	Minimum Charge	Job		227		227	
Clean Interior							
	Clean	L.F.	.81	5		5.81	Includes labor and materials to clean
	Minimum Charge	Job		165		165	the interior of a 27" cabinet.

Cabinets and Countertops

Base Cabinets		Unit	Material	Labor	Equip.	Total	Specification
Seal Interior							
	Install	L.F.	.57	6.20		6.77	Includes labor and materials to seal a
	Minimum Charge	Job		189		189	cabinet interior.
Storage (garage)							
	Demolish	L.F.		9.15		9.15	Includes material and labor to install
	Install	L.F.	144	45.50		189.50	good modular unit with melamine
	Demolish and Install	L.F.	144	54.65		198.65	laminate, textured colors or wood
	Reinstall	L.F.		36.32		36.32	grain print finish including hinges and
	Clean	L.F.	.81	6.90		7.71	pulls.
	Paint	L.F.	5.30	17.15		22.45	
	Minimum Charge	Job		227		227	
Clean Interior							
	Clean	L.F.	.81	5		5.81	Includes labor and materials to clean
	Minimum Charge	Job		165		165	the interior of a 27" cabinet.
Seal Interior							
	Install	L.F.	.57	6.20		6.77	Includes labor and materials to seal a
	Minimum Charge	Job		189		189	cabinet interior.
Built-in Desk							
	Demolish	L.F.		9.15		9.15	Includes material and labor to install
	Install	L.F.	193	45.50		238.50	custom modular unit with solid
	Demolish and Install	L.F.	193	54.65		247.65	hardwood faced frames, hardwood
	Reinstall	L.F.		36.32		36.32	door frames and drawer fronts,
	Clean	L.F.	.81	6.90		7.71	hardwood veneer on raised door
	Paint	L.F.	5.30	17.15		22.45	panels, hinges and pulls.
	Minimum Charge	Job		227		227	
Built-in Bookcase							
	Demolish	L.F.		12.20		12.20	Includes material and labor to install
	Install	L.F.	44	45.50		89.50	custom modular bookcase unit with
	Demolish and Install	L.F.	44	57.70		101.70	clear pine faced frames, shelves 12"
	Reinstall	L.F.		36.32		36.32	O.C., 7' high, 8" deep.
	Clean	L.F.	1.63	11.80		13.43	
	Paint	L.F.	1.55	29		30.55	
	Minimum Charge	Job		227		227	
Strip and Refinish							
	Install	L.F.	4.87	47		51.87	Includes labor and material to strip,
	Minimum Charge	Job		189		189	prep and refinish exterior of cabinets.
Stain							
	Install	L.F.	.64	1.76		2.40	Includes labor and material to stain
	Minimum Charge	Job		227		227	door and exterior of cabinets.

Countertop		Unit	Material	Labor	Equip.	Total	Specification
Laminated							
With Splash							
	Demolish	L.F.		6.10		6.10	Includes material and labor to install
	Install	L.F.	47.50	16.20		63.70	one piece laminated top with 4"
	Demolish and Install	L.F.	47.50	22.30		69.80	backsplash.
	Reinstall	L.F.		12.97		12.97	
	Clean	L.F.	.10	.83		.93	
	Minimum Charge	Job		227		227	

229

Cabinets and Countertops

Countertop		Unit	Material	Labor	Equip.	Total	Specification
Roll Top							
	Demolish	L.F.		6.10		6.10	Includes material and labor to install
	Install	L.F.	11.40	15.15		26.55	one piece laminated top with rolled
	Demolish and Install	L.F.	11.40	21.25		32.65	drip edge (post formed) and
	Reinstall	L.F.		12.11		12.11	backsplash.
	Clean	L.F.	.14	1.25		1.39	
	Minimum Charge	Job		227		227	
Ceramic Tile							
	Demolish	S.F.		3.06		3.06	Cost includes material and labor to
	Install	S.F.	18.80	9.10		27.90	install 4-1/4" x 4-1/4" to 6" x 6"
	Demolish and Install	S.F.	18.80	12.16		30.96	glazed tile set in mortar bed and grout
	Clean	S.F.	.08	.63		.71	on particle board substrate.
	Minimum Charge	Job		202		202	
Cultured Marble							
	Demolish	L.F.		6.10		6.10	Includes material and labor to install
	Install	L.F.	51	43		94	marble countertop, 24" wide, no
	Demolish and Install	L.F.	51	49.10		100.10	backsplash.
	Reinstall	L.F.		34.56		34.56	
	Clean	L.F.	.14	1.25		1.39	
	Minimum Charge	Job		202		202	
With Splash							
	Demolish	L.F.		6.10		6.10	Includes material and labor to install
	Install	L.F.	59.50	43		102.50	marble countertop, 24" wide, with
	Demolish and Install	L.F.	59.50	49.10		108.60	backsplash.
	Reinstall	L.F.		34.56		34.56	
	Clean	L.F.	.14	1.25		1.39	
	Minimum Charge	Job		202		202	
Quarry Tile							
	Demolish	S.F.		3.06		3.06	Includes material and labor to install
	Install	S.F.	9.15	9.10		18.25	quarry tile countertop with no
	Demolish and Install	S.F.	9.15	12.16		21.31	backsplash.
	Clean	S.F.	.08	.63		.71	
	Minimum Charge	Job		202		202	
With Splash							
	Demolish	S.F.		3.06		3.06	Includes material and labor to install
	Install	S.F.	12.20	10.10		22.30	quarry tile countertop with backsplash.
	Demolish and Install	S.F.	12.20	13.16		25.36	
	Clean	S.F.	.08	.63		.71	
	Minimum Charge	Job		202		202	
Butcher Block							
	Demolish	S.F.		3.06		3.06	Includes material and labor to install
	Install	S.F.	42.50	8.10		50.60	solid laminated maple countertop with
	Demolish and Install	S.F.	42.50	11.16		53.66	no backsplash.
	Reinstall	S.F.		6.49		6.49	
	Clean	S.F.	.04	.26		.30	
	Minimum Charge	Job		227		227	
Solid Surface							
	Demolish	L.F.		6.10		6.10	Includes material and labor to install
	Install	L.F.	60	32.50		92.50	solid surface counter top 22" deep.
	Demolish and Install	L.F.	60	38.60		98.60	
	Reinstall	L.F.		25.94		25.94	
	Clean	L.F.	.14	1.25		1.39	
	Minimum Charge	Job		227		227	

Cabinets and Countertops

Vanity Cabinets	Unit	Material	Labor	Equip.	Total	Specification
Good Grade Laminated						
24" Wide, 2 Door						
Demolish	Ea.		18.35		18.35	Includes material and labor to install
Install	Ea.	260	45.50		305.50	modular unit with melamine laminated
Demolish and Install	Ea.	260	63.85		323.85	to particle board, textured colors or
Reinstall	Ea.		36.32		36.32	wood grain print finish, hinges and
Clean	Ea.	1.56	13.20		14.76	pulls.
Paint	Ea.	1.33	31.50		32.83	
Minimum Charge	Job		227		227	
30" Wide, 2 Door						
Demolish	Ea.		23		23	Includes material and labor to install
Install	Ea.	310	57		367	modular unit with melamine laminated
Demolish and Install	Ea.	310	80		390	to particle board, textured colors or
Reinstall	Ea.		45.40		45.40	wood grain print finish, hinges and
Clean	Ea.	2.04	16.50		18.54	pulls.
Paint	Ea.	1.67	38		39.67	
Minimum Charge	Job		227		227	
36" Wide, 2 Door						
Demolish	Ea.		28		28	Includes material and labor to install
Install	Ea.	300	68		368	modular unit with melamine laminated
Demolish and Install	Ea.	300	96		396	to particle board, textured colors or
Reinstall	Ea.		54.49		54.49	wood grain print finish, hinges and
Clean	Ea.	2.27	19.40		21.67	pulls.
Paint	Ea.	2.01	47		49.01	
Minimum Charge	Job		227		227	
42" w, 3 Doors / Drawer						
Demolish	Ea.		33.50		33.50	Includes material and labor to install
Install	Ea.	500	73.50		573.50	modular unit with melamine laminated
Demolish and Install	Ea.	500	107		607	to particle board, textured colors or
Reinstall	Ea.		58.68		58.68	wood grain print finish, hinges and
Clean	Ea.	2.92	23.50		26.42	pulls.
Paint	Ea.	2.31	54		56.31	
Minimum Charge	Job		227		227	
48" w, 3 Doors / Drawer						
Demolish	Ea.		36.50		36.50	Includes material and labor to install
Install	Ea.	395	79.50		474.50	modular unit with melamine laminated
Demolish and Install	Ea.	395	116		511	to particle board, textured colors or
Reinstall	Ea.		63.55		63.55	wood grain print finish, hinges and
Clean	Ea.	4.07	36.50		40.57	pulls.
Paint	Ea.	3.47	94.50		97.97	
Minimum Charge	Job		227		227	
Clean Interior						
Install	L.F.	.81	6.90		7.71	Includes labor and materials to clean
Minimum Charge	Job		165		165	base cabinetry.
Better Grade Wood						
24" Wide, 2 Door						
Demolish	Ea.		18.35		18.35	Includes material and labor to install
Install	Ea.	345	45.50		390.50	custom modular unit with solid
Demolish and Install	Ea.	345	63.85		408.85	hardwood faced frames, hardwood
Reinstall	Ea.		36.32		36.32	door frames and drawer fronts,
Clean	Ea.	1.56	13.20		14.76	hardwood veneer on raised door
Paint	Ea.	1.33	34.50		35.83	panels, hinges and pulls.
Minimum Charge	Job		227		227	

For customer support on your Contractor's Pricing Guide: Residential Repair & Remodeling, call 888.606.7279.

Cabinets and Countertops

Vanity Cabinets	Unit	Material	Labor	Equip.	Total	Specification
30" Wide, 2 Door						
Demolish	Ea.		23		23	Includes material and labor to install
Install	Ea.	415	57		472	custom modular unit with solid
Demolish and Install	Ea.	415	80		495	hardwood faced frames, hardwood
Reinstall	Ea.		45.40		45.40	door frames and drawer fronts,
Clean	Ea.	2.04	16.50		18.54	hardwood veneer on raised door
Paint	Ea.	1.67	42		43.67	panels, hinges and pulls.
Minimum Charge	Job		227		227	
36" Wide, 2 Door						
Demolish	Ea.		28		28	Includes material and labor to install
Install	Ea.	400	68		468	custom modular unit with solid
Demolish and Install	Ea.	400	96		496	hardwood faced frames, hardwood
Reinstall	Ea.		54.49		54.49	door frames and drawer fronts,
Clean	Ea.	2.27	19.40		21.67	hardwood veneer on raised door
Paint	Ea.	2.01	54		56.01	panels, hinges and pulls.
Minimum Charge	Job		227		227	
42" w, 3 Doors / Drawer						
Demolish	Ea.		33.50		33.50	Includes material and labor to install
Install	Ea.	500	73.50		573.50	custom modular unit with solid
Demolish and Install	Ea.	500	107		607	hardwood faced frames, hardwood
Reinstall	Ea.		58.68		58.68	door frames and drawer fronts,
Clean	Ea.	2.92	23.50		26.42	hardwood veneer on raised door
Paint	Ea.	2.31	63		65.31	panels, hinges and pulls.
Minimum Charge	Job		227		227	
48" w, 3 Doors / Drawer						
Demolish	Ea.		36.50		36.50	Includes material and labor to install
Install	Ea.	530	79.50		609.50	custom modular unit with solid
Demolish and Install	Ea.	530	116		646	hardwood faced frames, hardwood
Reinstall	Ea.		63.55		63.55	door frames and drawer fronts,
Clean	Ea.	4.07	36.50		40.57	hardwood veneer on raised door
Paint	Ea.	3.47	94.50		97.97	panels, hinges and pulls.
Minimum Charge	Job		227		227	
Clean Interior						
Install	L.F.	.81	6.90		7.71	Includes labor and materials to clean
Minimum Charge	Job		165		165	base cabinetry.
Premium Grade Wood						
24" Wide, 2 Door						
Demolish	Ea.		18.35		18.35	Includes material and labor to install
Install	Ea.	370	45.50		415.50	premium modular unit with solid
Demolish and Install	Ea.	370	63.85		433.85	hardwood faced frames, hardwood
Reinstall	Ea.		36.32		36.32	door frames and drawer fronts,
Clean	Ea.	1.56	13.20		14.76	hardwood veneer on raised door
Paint	Ea.	1.33	38		39.33	panels, hinges and pulls.
Minimum Charge	Job		227		227	
30" Wide, 2 Door						
Demolish	Ea.		23		23	Includes material and labor to install
Install	Ea.	510	57		567	premium modular unit with solid
Demolish and Install	Ea.	510	80		590	hardwood faced frames, hardwood
Reinstall	Ea.		45.40		45.40	door frames and drawer fronts,
Clean	Ea.	2.04	16.50		18.54	hardwood veneer on raised door
Paint	Ea.	1.67	47		48.67	panels, hinges and pulls.
Minimum Charge	Job		227		227	

For customer support on your Contractor's Pricing Guide: Residential Repair & Remodeling, call 888.606.7279.

Cabinets and Countertops

Vanity Cabinets	Unit	Material	Labor	Equip.	Total	Specification
36" Wide, 2 Door						
Demolish	Ea.		28		28	Includes material and labor to install
Install	Ea.	495	68		563	premium modular unit with solid
Demolish and Install	Ea.	495	96		591	hardwood faced frames, hardwood
Reinstall	Ea.		54.49		54.49	door frames and drawer fronts,
Clean	Ea.	2.27	19.40		21.67	hardwood veneer on raised door
Paint	Ea.	2.01	54		56.01	panels, hinges and pulls.
Minimum Charge	Job		227		227	
42" w, 3 Doors / Drawer						
Demolish	Ea.		33.50		33.50	Includes material and labor to install
Install	Ea.	580	73.50		653.50	premium modular unit with solid
Demolish and Install	Ea.	580	107		687	hardwood faced frames, hardwood
Reinstall	Ea.		58.68		58.68	door frames and drawer fronts,
Clean	Ea.	2.92	23.50		26.42	hardwood veneer on raised door
Paint	Ea.	2.31	63		65.31	panels, hinges and pulls.
Minimum Charge	Job		227		227	
48" w, 3 Doors / Drawer						
Demolish	Ea.		36.50		36.50	Includes material and labor to install
Install	Ea.	605	79.50		684.50	premium modular unit with solid
Demolish and Install	Ea.	605	116		721	hardwood faced frames, hardwood
Reinstall	Ea.		63.55		63.55	door frames and drawer fronts,
Clean	Ea.	4.07	36.50		40.57	hardwood veneer on raised door
Paint	Ea.	3.47	94.50		97.97	panels, hinges and pulls.
Minimum Charge	Job		227		227	
Clean Interior						
Install	L.F.	.81	6.90		7.71	Includes labor and materials to clean
Minimum Charge	Job		165		165	base cabinetry.
Premium Hardwood						
24" Wide, 2 Door						
Demolish	Ea.		18.35		18.35	Includes material and labor to install
Install	Ea.	430	45.50		475.50	premium modular unit with solid
Demolish and Install	Ea.	430	63.85		493.85	hardwood faced frames, hardwood
Reinstall	Ea.		36.32		36.32	door frames and drawer fronts,
Clean	Ea.	1.56	13.20		14.76	hardwood veneer on raised door
Paint	Ea.	1.33	42		43.33	panels, hinges and pulls.
Minimum Charge	Job		227		227	
30" Wide, 2 Door						
Demolish	Ea.		23		23	Includes material and labor to install
Install	Ea.	650	57		707	premium modular unit with solid
Demolish and Install	Ea.	650	80		730	hardwood faced frames, hardwood
Reinstall	Ea.		45.40		45.40	door frames and drawer fronts,
Clean	Ea.	2.04	16.50		18.54	hardwood veneer on raised door
Paint	Ea.	1.67	47		48.67	panels, hinges and pulls.
Minimum Charge	Job		227		227	
36" Wide, 2 Door						
Demolish	Ea.		28		28	Includes material and labor to install
Install	Ea.	900	68		968	premium modular unit with solid
Demolish and Install	Ea.	900	96		996	hardwood faced frames, hardwood
Reinstall	Ea.		54.49		54.49	door frames and drawer fronts,
Clean	Ea.	2.27	19.40		21.67	hardwood veneer on raised door
Paint	Ea.	2.01	63		65.01	panels, hinges and pulls.
Minimum Charge	Job		227		227	

Cabinets and Countertops

Vanity Cabinets

Vanity Cabinets	Unit	Material	Labor	Equip.	Total	Specification
42″ w, 3 Doors / Drawer						Includes material and labor to install
Demolish	Ea.		33.50		33.50	premium modular unit with solid
Install	Ea.	585	73.50		658.50	hardwood faced frames, hardwood
Demolish and Install	Ea.	585	107		692	door frames and drawer fronts,
Reinstall	Ea.		58.68		58.68	hardwood veneer on raised door
Clean	Ea.	2.92	23.50		26.42	panels, hinges and pulls.
Paint	Ea.	2.31	75.50		77.81	
Minimum Charge	Job		227		227	
48″ w, 3 Doors / Drawer						Includes material and labor to install
Demolish	Ea.		36.50		36.50	premium modular unit with solid
Install	Ea.	650	79.50		729.50	hardwood faced frames, hardwood
Demolish and Install	Ea.	650	116		766	door frames and drawer fronts,
Reinstall	Ea.		63.55		63.55	hardwood veneer on raised door
Clean	Ea.	4.07	36.50		40.57	panels, hinges and pulls.
Paint	Ea.	3.47	94.50		97.97	
Minimum Charge	Job		227		227	
Clean Interior						Includes labor and materials to clean
Install	L.F.	.81	6.90		7.71	base cabinetry.
Minimum Charge	Job		165		165	
Strip and Refinish						Includes labor and material to strip,
Install	L.F.	4.87	47		51.87	prep and refinish exterior of cabinets.
Minimum Charge	Job		189		189	

Vanity Cabinet Tops

Vanity Cabinet Tops	Unit	Material	Labor	Equip.	Total	Specification
Cultured Marble Top						Includes material and labor to install
Demolish	L.F.		6.10		6.10	cultured marble vanity top with
Install	L.F.	47	25.50		72.50	integral sink, 22″ deep.
Demolish and Install	L.F.	47	31.60		78.60	
Reinstall	L.F.		20.33		20.33	
Clean	L.F.	.14	1.25		1.39	
Minimum Charge	Job		202		202	
Laminated Top						Includes material and labor to install
Demolish	L.F.		3.67		3.67	one piece laminated top with 4″
Install	L.F.	47.50	16.20		63.70	backsplash.
Demolish and Install	L.F.	47.50	19.87		67.37	
Reinstall	L.F.		12.97		12.97	
Clean	L.F.	.10	.83		.93	
Minimum Charge	Job		227		227	
Solid Surface						Includes material and labor to install
Demolish	L.F.		6.10		6.10	solid surface vanity top with integral
Install	L.F.	83.50	33.50		117	sink installed, 22″ deep.
Demolish and Install	L.F.	83.50	39.60		123.10	
Reinstall	L.F.		26.90		26.90	
Clean	L.F.	.14	1.25		1.39	
Minimum Charge	Job		227		227	

Painting

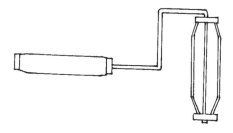

Roller Handle

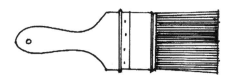

Paint Brush

Paint Preparation, Walls	Unit	Material	Labor	Equip.	Total	Specification
Cover / Protect Floors						
Install	S.F.	.03	.15		.18	Includes material and labor to install
Minimum Charge	Job		190		190	plastic masking sheet on large uninterrupted areas.
Cover / Protect Walls						
Install	S.F.	.03	.15		.18	Includes material and labor to install
Minimum Charge	Job		190		190	plastic masking sheet on walls.
Clean Walls						
Light						
Clean	S.F.		.12		.12	Includes labor to wash gypsum
Minimum Charge	Job		190		190	drywall or plaster wall surfaces.
Heavy						
Clean	S.F.		.15		.15	Includes labor and material for heavy
Minimum Charge	Job		190		190	cleaning (multiple applications) with detergent and solvent.
Prep Walls						
Paint	S.F.	.04	.19		.23	Includes labor and material to prepare
Minimum Charge	Job		190		190	for painting including scraping, patching and puttying.
Sand Walls						
Paint	S.F.		.12		.12	Includes labor to sand gypsum drywall
Minimum Charge	Job		190		190	wall surfaces.
Seal Walls						
Large Area						
Paint	S.F.	.07	.28		.35	Includes labor and material to paint
Minimum Charge	Job		190		190	with primer / sealer by roller.
Spot						
Paint	S.F.	.07	.33		.40	Includes labor and material to seal
Minimum Charge	Job		190		190	existing drywall with shellac-based material.

235

Painting

Paint Preparation, Ceilings

Paint Preparation, Ceilings		Unit	Material	Labor	Equip.	Total	Specification
Cover / Protect Floors							
	Install	S.F.	.03	.15		.18	Includes material and labor to install
	Minimum Charge	Job		190		190	plastic masking sheet on large uninterrupted areas.
Cover / Protect Walls							
	Install	S.F.	.03	.15		.18	Includes material and labor to install
	Minimum Charge	Job		190		190	plastic masking sheet on walls.
Prep Ceiling							
	Paint	S.F.	.07	.75		.82	Includes labor and material to fill in
	Minimum Charge	Job		190		190	hairline cracks up to 1/8" with filler, sand and prep for paint.
Sand Ceiling							
	Paint	S.F.		.17		.17	Includes labor to sand gypsum drywall
	Minimum Charge	Job		190		190	ceiling surfaces.
Seal Ceiling							
Large Area							
	Paint	S.F.	.08	.29		.37	Includes labor and material to paint
	Minimum Charge	Job		189		189	ceiling, one coat flat latex.
Spot							
	Paint	S.F.	.07	.33		.40	Includes labor and material to seal
	Minimum Charge	Job		189		189	existing drywall with shellac-based material.

Paint / Texture, Walls

Paint / Texture, Walls		Unit	Material	Labor	Equip.	Total	Specification
Texture Walls							
Spray							
	Install	S.F.	.04	.54		.58	Includes labor and material to install
	Minimum Charge	Job		189		189	by spray, texture finish.
Trowel							
	Install	S.F.		.27		.27	Includes labor and material to apply
	Minimum Charge	Job		189		189	hand troweled texture.
Paint Walls 1 Coat							
	Paint	S.F.	.08	.29		.37	Includes labor and material to paint,
	Minimum Charge	Job		189		189	one coat flat latex.
Paint Walls 2 Coats							
	Paint	S.F.	.15	.47		.62	Includes labor and material to paint,
	Minimum Charge	Job		189		189	two coats flat latex.
Paint Walls 3 Coats							
	Paint	S.F.	.23	.58		.81	Includes labor and material to paint,
	Minimum Charge	Job		189		189	three coats flat latex.
Clean and Seal							
	Paint	S.F.	.43	1.01		1.44	Includes labor and material to do light
	Minimum Charge	Job		189		189	cleaning, sealing with 2 coats of latex paint.

Painting

Paint / Texture, Ceilings		Unit	Material	Labor	Equip.	Total	Specification
Texture Ceiling							
Spray							
	Install	S.F.	.04	.54		.58	Includes labor and material to install
	Minimum Charge	Job		189		189	by spray, texture finish.
Trowel							
	Install	S.F.		.27		.27	Includes labor and material to apply
	Minimum Charge	Job		189		189	hand troweled texture.
Paint Ceiling, 1 Coat							
	Paint	S.F.	.08	.29		.37	Includes labor and material to paint
	Minimum Charge	Job		189		189	ceiling, one coat flat latex.
Paint Ceiling, 2 Coats							
	Paint	S.F.	.23	.47		.70	Includes labor and material to paint
	Minimum Charge	Job		189		189	two coats flat latex paint.
Paint Ceiling, 3 Coats							
	Paint	S.F.	.34	.56		.90	Includes labor and material to paint
	Minimum Charge	Job		189		189	ceiling, three coats flat latex.
Clean and Seal							
	Paint	S.F.	.34	.79		1.13	Includes labor and material for light
	Minimum Charge	Job		189		189	cleaning, sealing and paint 3 coats of
							flat latex paint.

For customer support on your Contractor's Pricing Guide: Residential Repair & Remodeling, call 888.606.7279.

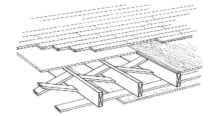

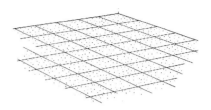

Plank Flooring		Wood Strip Flooring			Tile Flooring	

Wood Plank Flooring	Unit	Material	Labor	Equip.	Total	Specification
Maple						
2-1/4"						
Demolish	S.F.		1.47		1.47	Includes material and labor to install
Install	S.F.	4.42	2.26		6.68	unfinished T&G maple flooring,
Demolish and Install	S.F.	4.42	3.73		8.15	25/32" thick, 2-1/4" wide, random
Reinstall	S.F.		1.81		1.81	3' to 16' lengths including felt
Clean	S.F.		.41		.41	underlayment, nailed in place over
Paint	S.F.	1.47	2.54		4.01	prepared subfloor.
Minimum Charge	Job		227		227	
3-1/4"						
Demolish	S.F.		1.47		1.47	Includes material and labor to install
Install	S.F.	4.11	2.26		6.37	unfinished T&G maple flooring,
Demolish and Install	S.F.	4.11	3.73		7.84	33/32" thick, 3-1/4" wide, random
Reinstall	S.F.		1.81		1.81	3' to 16' lengths including felt
Clean	S.F.		.41		.41	underlayment, nailed in place over
Paint	S.F.	1.47	2.54		4.01	prepared subfloor.
Minimum Charge	Job		227		227	
Sand and Finish						
Install	S.F.	.99	.83		1.82	Includes labor and material to refinish
Minimum Charge	Job		227		227	wood floor.
Oak						
2-1/4"						
Demolish	S.F.		1.47		1.47	Includes material and labor to install
Install	S.F.	4.83	2.26		7.09	unfinished T&G oak flooring, 25/32"
Demolish and Install	S.F.	4.83	3.73		8.56	thick, 2-1/4" wide, random 3' to 16'
Reinstall	S.F.		1.81		1.81	lengths including felt underlayment,
Clean	S.F.		.41		.41	nailed in place over prepared
Paint	S.F.	1.47	2.54		4.01	subfloor.
Minimum Charge	Job		227		227	
3-1/4"						
Demolish	S.F.		1.47		1.47	Includes material and labor to install
Install	S.F.	4.36	1.60		5.96	unfinished T&G oak flooring, 25/32"
Demolish and Install	S.F.	4.36	3.07		7.43	thick, 3-1/4" wide, random 3' to 16'
Reinstall	S.F.		1.28		1.28	lengths including felt underlayment,
Clean	S.F.		.41		.41	nailed in place over prepared
Paint	S.F.	1.47	2.54		4.01	subfloor.
Minimum Charge	Job		227		227	
Sand and Finish						
Install	S.F.	1.47	2.54		4.01	Includes labor and material to refinish
Minimum Charge	Job		227		227	wood floor.

Flooring

Wood Plank Flooring		Unit	Material	Labor	Equip.	Total	Specification
Walnut							
3" to 7"							
	Demolish	S.F.		1.47		1.47	Includes material and labor to install
	Install	S.F.	14.45	1.60		16.05	unfinished T&G walnut flooring,
	Demolish and Install	S.F.	14.45	3.07		17.52	25/32" thick, random 3' to 16'
	Reinstall	S.F.		1.28		1.28	lengths including felt underlayment,
	Clean	S.F.		.41		.41	nailed in place over prepared
	Paint	S.F.	1.47	2.54		4.01	subfloor.
	Minimum Charge	Job		227		227	
Sand and Finish							
	Install	S.F.	1.47	2.54		4.01	Includes labor and material to refinish
	Minimum Charge	Job		227		227	wood floor.
Pine							
2-1/4"							
	Demolish	S.F.		1.47		1.47	Includes material and labor to install
	Install	S.F.	7.80	1.71		9.51	unfinished T&G pine flooring, 25/32"
	Demolish and Install	S.F.	7.80	3.18		10.98	thick, 2-1/4" wide, random 3' to 16'
	Reinstall	S.F.		1.37		1.37	lengths including felt underlayment,
	Clean	S.F.		.41		.41	nailed in place over prepared
	Paint	S.F.	.99	.83		1.82	subfloor.
	Minimum Charge	Job		227		227	
Sand and Finish							
	Install	S.F.	1.47	2.54		4.01	Includes labor and material to refinish
	Minimum Charge	Job		227		227	wood floor.
Prefinished Oak							
2-1/4"							
	Demolish	S.F.		1.47		1.47	Includes material and labor to install
	Install	S.F.	4.36	2.26		6.62	prefinished T&G oak flooring, 2-1/4"
	Demolish and Install	S.F.	4.36	3.73		8.09	wide, random 3' to 16' lengths
	Reinstall	S.F.		1.81		1.81	including felt underlayment, nailed in
	Clean	S.F.		.41		.41	place over prepared subfloor.
	Minimum Charge	Job		227		227	
3-1/4"							
	Demolish	S.F.		1.47		1.47	Includes material and labor to install
	Install	S.F.	4.74	2.08		6.82	prefinished T&G oak flooring, 25/32"
	Demolish and Install	S.F.	4.74	3.55		8.29	thick, 3" to 7" wide, random 3' to 16'
	Reinstall	S.F.		1.66		1.66	lengths including felt underlayment,
	Clean	S.F.		.41		.41	nailed in place over prepared
	Minimum Charge	Job		227		227	subfloor.
Sand and Finish New							
	Install	S.F.	.99	1.12		2.11	Includes labor and material to refinish
	Minimum Charge	Job		227		227	wood floor.
Refinish Existing							
	Install	S.F.	.99	.83		1.82	Includes labor and material to refinish
	Minimum Charge	Job		227		227	wood floor.

Flooring

Wood Plank Flooring		Unit	Material	Labor	Equip.	Total	Specification
Urethane Coat							
	Install	S.F.	.17	.20		.37	Includes labor and material to install
	Minimum Charge	Job		227		227	two coats of urethane, on existing floor.
Clean and Wax							
	Clean	S.F.	.04	.26		.30	Includes labor and material to wash
	Minimum Charge	Job		227		227	and wax hardwood.
Underlayment Per S.F.							
	Demolish	S.F.		.66		.66	Cost includes material and labor to
	Install	S.F.	.98	.61		1.59	install 1/4" lauan subfloor, standard
	Demolish and Install	S.F.	.98	1.27		2.25	interior grade, nailed every 6".
	Clean	S.F.		.32		.32	
	Minimum Charge	Job		227		227	

Wood Parquet Tile Floor		Unit	Material	Labor	Equip.	Total	Specification
Oak							
9" x 9"							
	Demolish	S.F.		1.12		1.12	Includes material and labor to install
	Install	S.F.	7.15	3.63		10.78	prefinished oak 9" x 9" parquet block
	Demolish and Install	S.F.	7.15	4.75		11.90	flooring, 5/16" thick, installed in
	Reinstall	S.F.		2.91		2.91	mastic.
	Clean	S.F.		.41		.41	
	Paint	S.F.	1.47	2.54		4.01	
	Minimum Charge	Job		227		227	
13" x 13"							
	Demolish	S.F.		1.12		1.12	Includes material and labor to install
	Install	S.F.	10.10	2.84		12.94	prefinished oak 13" x 13" parquet
	Demolish and Install	S.F.	10.10	3.96		14.06	block flooring, 5/16" thick, installed in
	Reinstall	S.F.		2.27		2.27	mastic.
	Clean	S.F.		.41		.41	
	Paint	S.F.	1.47	2.54		4.01	
	Minimum Charge	Job		227		227	
Cherry							
9" x 9"							
	Demolish	S.F.		1.12		1.12	Includes material and labor to install
	Install	S.F.	19.25	3.63		22.88	prefinished cherry 9" x 9" parquet
	Demolish and Install	S.F.	19.25	4.75		24	block flooring, 5/16" thick, installed in
	Reinstall	S.F.		3.63		3.63	mastic.
	Clean	S.F.		.41		.41	
	Paint	S.F.	1.47	2.54		4.01	
	Minimum Charge	Job		227		227	
13" x 13"							
	Demolish	S.F.		1.12		1.12	Includes material and labor to install
	Install	S.F.	19.25	2.84		22.09	prefinished cherry 13" x 13" parquet
	Demolish and Install	S.F.	19.25	3.96		23.21	block flooring, 5/16" thick, installed in
	Reinstall	S.F.		2.84		2.84	mastic.
	Clean	S.F.		.41		.41	
	Paint	S.F.	1.47	2.54		4.01	
	Minimum Charge	Job		227		227	

Wood Parquet Tile Floor	Unit	Material	Labor	Equip.	Total	Specification
Walnut						
9″ x 9″						
Demolish	S.F.		1.12		1.12	Includes material and labor to install
Install	S.F.	6.80	3.63		10.43	prefinished walnut 9″ x 9″ parquet
Demolish and Install	S.F.	6.80	4.75		11.55	flooring in mastic.
Reinstall	S.F.		2.91		2.91	
Clean	S.F.		.41		.41	
Paint	S.F.	1.47	2.54		4.01	
Minimum Charge	Job		227		227	
13″ x 13″						
Demolish	S.F.		1.12		1.12	Includes material and labor to install
Install	S.F.	12.10	2.84		14.94	prefinished walnut 13″ x 13″ parquet
Demolish and Install	S.F.	12.10	3.96		16.06	block flooring, 5/16″ thick, installed in
Reinstall	S.F.		2.84		2.84	mastic.
Clean	S.F.		.41		.41	
Paint	S.F.	1.47	2.54		4.01	
Minimum Charge	Job		227		227	
Acrylic Impregnated						
Oak 12″ x 12″						
Demolish	S.F.		1.12		1.12	Includes material and labor to install
Install	S.F.	11.10	3.24		14.34	red oak 12″ x 12″ acrylic
Demolish and Install	S.F.	11.10	4.36		15.46	impregnated parquet block flooring,
Reinstall	S.F.		3.24		3.24	5/16″ thick, installed in mastic.
Clean	S.F.		.41		.41	
Minimum Charge	Job		227		227	
Cherry 12″ x 12″						
Demolish	S.F.		1.12		1.12	Includes material and labor to install
Install	S.F.	22	3.24		25.24	prefinished cherry 12″ x 12″ acrylic
Demolish and Install	S.F.	22	4.36		26.36	impregnated parquet block flooring,
Reinstall	S.F.		3.24		3.24	5/16″ thick, installed in mastic.
Clean	S.F.		.41		.41	
Minimum Charge	Job		227		227	
Ash 12″ x 12″						
Demolish	S.F.		1.12		1.12	Includes material and labor to install
Install	S.F.	20.50	3.24		23.74	ash 12″ x 12″ acrylic impregnated
Demolish and Install	S.F.	20.50	4.36		24.86	parquet block flooring, 5/16″ thick in
Reinstall	S.F.		3.24		3.24	mastic.
Clean	S.F.		.41		.41	
Minimum Charge	Job		227		227	
Teak						
Demolish	S.F.		1.12		1.12	Includes material and labor to install
Install	S.F.	9.45	2.84		12.29	prefinished teak parquet block
Demolish and Install	S.F.	9.45	3.96		13.41	flooring, 5/16″ thick, and installed in
Reinstall	S.F.		2.84		2.84	mastic.
Clean	S.F.		.41		.41	
Paint	S.F.	1.47	2.54		4.01	
Minimum Charge	Job		227		227	
Clean and Wax						
Clean	S.F.	.04	.26		.30	Includes labor and material to wash
Minimum Charge	Job		227		227	and wax hardwood.

Flooring

Ceramic Tile Flooring		Unit	Material	Labor	Equip.	Total	Specification
Economy Grade							
	Demolish	S.F.		1.09		1.09	Cost includes material and labor to
	Install	S.F.	8.70	7.90		16.60	install 4-1/4" x 4-1/4" ceramic tile in
	Demolish and Install	S.F.	8.70	8.99		17.69	mortar bed including grout.
	Clean	S.F.		.63		.63	
	Minimum Charge	Job		202		202	
Average Grade							
	Demolish	S.F.		1.09		1.09	Cost includes material and labor to
	Install	S.F.	11.05	8.05		19.10	install 4-1/4" x 4-1/4" ceramic tile in
	Demolish and Install	S.F.	11.05	9.14		20.19	mortar bed including grout.
	Clean	S.F.		.63		.63	
	Minimum Charge	Job		202		202	
Premium Grade							
	Demolish	S.F.		1.09		1.09	Cost includes material and labor to
	Install	S.F.	12.15	8.05		20.20	install 4-1/4" x 4-1/4" ceramic tile in
	Demolish and Install	S.F.	12.15	9.14		21.29	mortar bed including grout.
	Clean	S.F.		.63		.63	
	Minimum Charge	Job		202		202	
Re-grout							
	Install	S.F.	.15	3.23		3.38	Includes material and labor to regrout
	Minimum Charge	Job		202		202	tile floors.

Hard Tile Flooring		Unit	Material	Labor	Equip.	Total	Specification
Thick Set Paver							
Brick							
	Demolish	S.F.		2.29		2.29	Cost includes material and labor to
	Install	S.F.	1.45	18.75		20.20	install 6" x 12" adobe brick paver with
	Demolish and Install	S.F.	1.45	21.04		22.49	1/2" mortar joints.
	Clean	S.F.		.63		.63	
	Minimum Charge	Job		202		202	
Mexican Red							
	Demolish	S.F.		2.29		2.29	Cost includes material and labor to
	Install	S.F.	1.89	8.40		10.29	install 12" x 12" red Mexican paver
	Demolish and Install	S.F.	1.89	10.69		12.58	tile in mortar bed with grout.
	Clean	S.F.		.63		.63	
	Minimum Charge	Job		202		202	
Saltillo							
	Demolish	S.F.		1.09		1.09	Cost includes material and labor to
	Install	S.F.	1.64	8.40		10.04	install 12" x 12" saltillo tile in mortar
	Demolish and Install	S.F.	1.64	9.49		11.13	bed with grout.
	Clean	S.F.		.63		.63	
	Minimum Charge	Job		202		202	
Marble							
Premium Grade							
	Demolish	S.F.		1.09		1.09	Cost includes material and labor to
	Install	S.F.	18.80	17.55		36.35	install 3/8" x 12" x 12" marble, in
	Demolish and Install	S.F.	18.80	18.64		37.44	mortar bed with grout.
	Clean	S.F.		.63		.63	
	Minimum Charge	Job		202		202	

Flooring

Hard Tile Flooring	Unit	Material	Labor	Equip.	Total	Specification
Terrazzo						
Gray Cement						
Demolish	S.F.		1.45		1.45	Cost includes material and labor to
Install	S.F.	3.69	9.90	4.54	18.13	install 1-3/4" terrazzo , #1 and #2
Demolish and Install	S.F.	3.69	11.35	4.54	19.58	chips in gray Portland cement.
Clean	S.F.		.63		.63	
Minimum Charge	Job		370	170	540	
White Cement						
Demolish	S.F.		1.45		1.45	Cost includes material and labor to
Install	S.F.	4.10	9.90	4.54	18.54	install 1-3/4" terrazzo, #1 and #2
Demolish and Install	S.F.	4.10	11.35	4.54	19.99	chips in white Portland cement.
Clean	S.F.		.63		.63	
Minimum Charge	Job		370	170	540	
Non-skid Gray						
Demolish	S.F.		1.45		1.45	Cost includes material and labor to
Install	S.F.	6.80	12.40	5.70	24.90	install 1-3/4" terrazzo, #1 and #2
Demolish and Install	S.F.	6.80	13.85	5.70	26.35	chips in gray Portland cement with
Clean	S.F.		.63		.63	light non-skid abrasive.
Minimum Charge	Job		370	170	540	
Non-skid White						
Demolish	S.F.		1.45		1.45	Cost includes material and labor to
Install	S.F.	7.15	12.40	5.70	25.25	install 1-3/4" terrazzo, #1 and #2
Demolish and Install	S.F.	7.15	13.85	5.70	26.70	chips in white Portland cement with
Clean	S.F.		.63		.63	light non-skid abrasive.
Minimum Charge	Job		370	170	540	
Gray w / Brass Divider						
Demolish	S.F.		1.47		1.47	Cost includes material and labor to
Install	S.F.	10.30	13.50	6.20	30	install 1-3/4" terrazzo, #1 and #2
Demolish and Install	S.F.	10.30	14.97	6.20	31.47	chips in gray Portland cement with
Clean	S.F.		.63		.63	brass strips 2' O.C. each way.
Minimum Charge	Job		370	170	540	
Clean						
Clean	S.F.	.18	.42		.60	Includes labor and material to clean
Minimum Charge	Job		370	170	540	terrazzo flooring.
Slate						
Demolish	S.F.		1.31		1.31	Includes material and labor to install
Install	S.F.	8.25	3.99		12.24	slate tile flooring in thin set with grout.
Demolish and Install	S.F.	8.25	5.30		13.55	
Clean	S.F.		.63		.63	
Minimum Charge	Job		202		202	
Granite						
Demolish	S.F.		1.08		1.08	Includes material and labor to install
Install	S.F.	28.50	4.03		32.53	granite tile flooring in thin set with
Demolish and Install	S.F.	28.50	5.11		33.61	grout.
Reinstall	S.F.		4.03		4.03	
Clean	S.F.		.63		.63	
Minimum Charge	Job		202		202	
Re-grout						
Install	S.F.	.15	3.23		3.38	Includes material and labor to regrout
Minimum Charge	Job		202		202	tile floors.

Flooring

Sheet Vinyl Per S.F.

	Unit	Material	Labor	Equip.	Total	Specification
Economy Grade						
Demolish	S.F.		.52		.52	Includes material and labor to install
Install	S.F.	2.49	1.26		3.75	resilient sheet vinyl flooring.
Demolish and Install	S.F.	2.49	1.78		4.27	
Clean	S.F.		.41		.41	
Minimum Charge	Job		202		202	
Average Grade						
Demolish	S.F.		.52		.52	Includes material and labor to install
Install	S.F.	3.73	1.26		4.99	resilient sheet vinyl flooring.
Demolish and Install	S.F.	3.73	1.78		5.51	
Clean	S.F.		.41		.41	
Minimum Charge	Job		202		202	
Premium Grade						
Demolish	S.F.		.52		.52	Includes material and labor to install
Install	S.F.	3.61	1.26		4.87	resilient sheet vinyl flooring.
Demolish and Install	S.F.	3.61	1.78		5.39	
Clean	S.F.		.41		.41	
Minimum Charge	Job		202		202	
Clean and Wax						
Clean	S.F.	.04	.46		.50	Includes labor and material to wash
Minimum Charge	Job		202		202	and wax a vinyl tile floor.

Sheet Vinyl

	Unit	Material	Labor	Equip.	Total	Specification
Economy Grade						
Demolish	S.Y.		4.73		4.73	Includes material and labor to install
Install	S.Y.	22.50	11.35		33.85	no wax sheet vinyl flooring.
Demolish and Install	S.Y.	22.50	16.08		38.58	
Clean	S.Y.		3.71		3.71	
Minimum Charge	Job		202		202	
Average Grade						
Demolish	S.Y.		4.73		4.73	Includes material and labor to install
Install	S.Y.	33.50	11.35		44.85	no wax sheet vinyl flooring.
Demolish and Install	S.Y.	33.50	16.08		49.58	
Clean	S.Y.		3.71		3.71	
Minimum Charge	Job		202		202	
Premium Grade						
Demolish	S.Y.		4.73		4.73	Includes material and labor to install
Install	S.Y.	32.50	11.35		43.85	no wax sheet vinyl flooring.
Demolish and Install	S.Y.	32.50	16.08		48.58	
Clean	S.Y.		3.71		3.71	
Minimum Charge	Job		202		202	
Clean and Wax						
Clean	S.F.	.04	.46		.50	Includes labor and material to wash
Minimum Charge	Job		202		202	and wax a vinyl tile floor.

Vinyl Tile

	Unit	Material	Labor	Equip.	Total	Specification
Economy Grade						
Demolish	S.F.		.73		.73	Includes material and labor to install
Install	S.F.	4.08	.81		4.89	no wax 12" x 12" vinyl tile.
Demolish and Install	S.F.	4.08	1.54		5.62	
Clean	S.F.		.41		.41	
Minimum Charge	Job		202		202	

245

Flooring

Vinyl Tile		Unit	Material	Labor	Equip.	Total	Specification
Average Grade							
	Demolish	S.F.		.73		.73	Includes material and labor to install
	Install	S.F.	3.58	.81		4.39	no wax 12" x 12" vinyl tile.
	Demolish and Install	S.F.	3.58	1.54		5.12	
	Clean	S.F.		.41		.41	
	Minimum Charge	Job		202		202	
Premium Grade							
	Demolish	S.F.		.73		.73	Includes material and labor to install
	Install	S.F.	6.90	.81		7.71	no wax 12" x 12" vinyl tile.
	Demolish and Install	S.F.	6.90	1.54		8.44	
	Clean	S.F.		.41		.41	
	Minimum Charge	Job		202		202	
Clean and Wax							
	Clean	S.F.	.04	.46		.50	Includes labor and material to wash
	Minimum Charge	Job		202		202	and wax a vinyl tile floor.

Carpeting Per S.F.		Unit	Material	Labor	Equip.	Total	Specification
Economy Grade							
	Demolish	S.F.		.08		.08	Includes material and labor to install
	Install	S.F.	1.29	.60		1.89	carpet including tack strips and hot
	Demolish and Install	S.F.	1.29	.68		1.97	melt tape on seams.
	Clean	S.F.		.57		.57	
	Minimum Charge	Job		202		202	
Average Grade							
	Demolish	S.F.		.08		.08	Includes material and labor to install
	Install	S.F.	1.64	.60		2.24	carpet including tack strips and hot
	Demolish and Install	S.F.	1.64	.68		2.32	melt tape on seams.
	Clean	S.F.		.57		.57	
	Minimum Charge	Job		202		202	
Premium Grade							
	Demolish	S.F.		.08		.08	Includes material and labor to install
	Install	S.F.	2.45	.60		3.05	carpet including tack strips and hot
	Demolish and Install	S.F.	2.45	.68		3.13	melt tape on seams.
	Clean	S.F.		.57		.57	
	Minimum Charge	Job		202		202	
Indoor/Outdoor							
	Demolish	S.F.		.08		.08	Includes material and labor to install
	Install	S.F.	1.27	.60		1.87	indoor-outdoor carpet.
	Demolish and Install	S.F.	1.27	.68		1.95	
	Clean	S.F.		.57		.57	
	Minimum Charge	Job		202		202	
100% Wool							
Average Grade							
	Demolish	S.F.		.08		.08	Includes material and labor to install
	Install	S.F.	8.15	.80		8.95	carpet including tack strips and hot
	Demolish and Install	S.F.	8.15	.88		9.03	melt tape on seams.
	Clean	S.F.		.57		.57	
	Minimum Charge	Job		202		202	
Premium Grade							
	Demolish	S.F.		.08		.08	Includes material and labor to install
	Install	S.F.	12.25	.80		13.05	carpet including tack strips and hot
	Demolish and Install	S.F.	12.25	.88		13.13	melt tape on seams.
	Clean	S.F.		.57		.57	
	Minimum Charge	Job		202		202	

Flooring

Carpeting Per S.F.		Unit	Material	Labor	Equip.	Total	Specification
Luxury Grade							
	Demolish	S.F.		.08		.08	Includes material and labor to install
	Install	S.F.	12	.80		12.80	carpet including tack strips and hot
	Demolish and Install	S.F.	12	.88		12.88	melt tape on seams.
	Clean	S.F.		.57		.57	
	Minimum Charge	Job		202		202	
Berber							
Average Grade							
	Demolish	S.F.		.08		.08	Includes material and labor to install
	Install	S.F.	5.55	.95		6.50	carpet including tack strips and hot
	Demolish and Install	S.F.	5.55	1.03		6.58	melt tape on seams.
	Clean	S.F.		.57		.57	
	Minimum Charge	Job		202		202	
Premium Grade							
	Demolish	S.F.		.08		.08	Includes material and labor to install
	Install	S.F.	8.25	.95		9.20	carpet including tack strips and hot
	Demolish and Install	S.F.	8.25	1.03		9.28	melt tape on seams.
	Clean	S.F.		.57		.57	
	Minimum Charge	Job		202		202	

Carpeting		Unit	Material	Labor	Equip.	Total	Specification
Indoor/Outdoor							
	Demolish	S.Y.		1.22		1.22	Includes material and labor to install
	Install	S.Y.	11.35	5.40		16.75	indoor-outdoor carpet.
	Demolish and Install	S.Y.	11.35	6.62		17.97	
	Reinstall	S.Y.		4.30		4.30	
	Clean	S.Y.		5.15		5.15	
	Minimum Charge	Job		202		202	
Economy Grade							
	Demolish	S.Y.		1.22		1.22	Includes material and labor to install
	Install	S.Y.	11.55	5.40		16.95	carpet including tack strips and hot
	Demolish and Install	S.Y.	11.55	6.62		18.17	melt tape on seams.
	Reinstall	S.Y.		4.30		4.30	
	Clean	S.Y.		5.15		5.15	
	Minimum Charge	Job		202		202	
Average Grade							
	Demolish	S.Y.		1.22		1.22	Includes material and labor to install
	Install	S.Y.	14.70	5.40		20.10	carpet including tack strips and hot
	Demolish and Install	S.Y.	14.70	6.62		21.32	melt tape on seams.
	Reinstall	S.Y.		4.30		4.30	
	Clean	S.Y.		5.15		5.15	
	Minimum Charge	Job		202		202	
Premium Grade							
	Demolish	S.Y.		1.22		1.22	Includes material and labor to install
	Install	S.Y.	22	5.40		27.40	carpet including tack strips and hot
	Demolish and Install	S.Y.	22	6.62		28.62	melt tape on seams.
	Reinstall	S.Y.		4.30		4.30	
	Clean	S.Y.		5.15		5.15	
	Minimum Charge	Job		202		202	

Flooring

Carpeting

Carpeting	Unit	Material	Labor	Equip.	Total	Specification
100% Wool						
Average Grade						
Demolish	S.Y.		1.22		1.22	Includes material and labor to install
Install	S.Y.	73	7.20		80.20	wool carpet including tack strips and
Demolish and Install	S.Y.	73	8.42		81.42	hot melt tape on seams.
Reinstall	S.Y.		5.76		5.76	
Clean	S.Y.		5.15		5.15	
Minimum Charge	Job		202		202	
Premium Grade						
Demolish	S.Y.		1.22		1.22	Includes material and labor to install
Install	S.Y.	110	7.20		117.20	carpet including tack strips and hot
Demolish and Install	S.Y.	110	8.42		118.42	melt tape on seams.
Reinstall	S.Y.		5.76		5.76	
Clean	S.Y.		5.15		5.15	
Minimum Charge	Job		202		202	
Luxury Grade						
Demolish	S.Y.		1.22		1.22	Includes material and labor to install
Install	S.Y.	108	7.20		115.20	carpet including tack strips and hot
Demolish and Install	S.Y.	108	8.42		116.42	melt tape on seams.
Reinstall	S.Y.		5.76		5.76	
Clean	S.Y.		5.15		5.15	
Minimum Charge	Job		202		202	
Berber						
Average Grade						
Demolish	S.Y.		1.22		1.22	Includes material and labor to install
Install	S.Y.	50	8.60		58.60	berber carpet including tack strips and
Demolish and Install	S.Y.	50	9.82		59.82	hot melt tape on seams.
Reinstall	S.Y.		6.86		6.86	
Clean	S.Y.		5.15		5.15	
Minimum Charge	Job		202		202	
Premium Grade						
Demolish	S.Y.		1.22		1.22	Includes material and labor to install
Install	S.Y.	74.50	8.60		83.10	carpet including tack strips and hot
Demolish and Install	S.Y.	74.50	9.82		84.32	melt tape on seams.
Reinstall	S.Y.		6.86		6.86	
Clean	S.Y.		5.15		5.15	
Minimum Charge	Job		202		202	

Carpet Pad Per S.F.	Unit	Material	Labor	Equip.	Total	Specification
Urethane						
Demolish	S.F.		.02		.02	Includes material and labor to install
Install	S.F.	.73	.30		1.03	urethane carpet pad.
Demolish and Install	S.F.	.73	.32		1.05	
Minimum Charge	Job		202		202	
Foam Rubber Slab						
Demolish	S.F.		.02		.02	Includes material and labor to install
Install	S.F.	.70	.30		1	carpet pad, 3/8" thick foam rubber.
Demolish and Install	S.F.	.70	.32		1.02	
Minimum Charge	Job		202		202	
Waffle						
Demolish	S.F.		.02		.02	Includes material and labor to install
Install	S.F.	.87	.03		.90	rubber waffle carpet pad.
Demolish and Install	S.F.	.87	.05		.92	
Minimum Charge	Job		202		202	

Flooring

Carpet Pad Per S.F.

	Unit	Material	Labor	Equip.	Total	Specification
Jute						
Demolish	S.F.		.02		.02	Includes material and labor to install
Install	S.F.	.85	.26		1.11	jute hair carpet pad.
Demolish and Install	S.F.	.85	.28		1.13	
Minimum Charge	Job		202		202	
Rebound						
Demolish	S.F.		.02		.02	Includes material and labor to install
Install	S.F.	.46	.22		.68	rebound carpet pad.
Demolish and Install	S.F.	.46	.24		.70	
Minimum Charge	Job		202		202	

Carpet Pad

	Unit	Material	Labor	Equip.	Total	Specification
Urethane						
Demolish	S.Y.		.21		.21	Includes material and labor to install
Install	S.Y.	6.55	2.69		9.24	urethane carpet pad.
Demolish and Install	S.Y.	6.55	2.90		9.45	
Reinstall	S.Y.		2.15		2.15	
Minimum Charge	Job		202		202	
Foam Rubber Slab						
Demolish	S.Y.		.21		.21	Includes material and labor to install
Install	S.Y.	6.35	2.69		9.04	rubber slab carpet pad.
Demolish and Install	S.Y.	6.35	2.90		9.25	
Reinstall	S.Y.		2.15		2.15	
Minimum Charge	Job		202		202	
Waffle						
Demolish	S.Y.		.21		.21	Includes material and labor to install
Install	S.Y.	7.80	.30		8.10	rubber waffle carpet pad.
Demolish and Install	S.Y.	7.80	.51		8.31	
Reinstall	S.Y.		.24		.24	
Minimum Charge	Job		202		202	
Jute						
Demolish	S.Y.		.21		.21	Includes material and labor to install
Install	S.Y.	7.60	2.37		9.97	jute hair carpet pad.
Demolish and Install	S.Y.	7.60	2.58		10.18	
Reinstall	S.Y.		1.90		1.90	
Minimum Charge	Job		202		202	
Rebound						
Demolish	S.Y.		.21		.21	Includes material and labor to install
Install	S.Y.	4.15	2.01		6.16	rebound carpet pad.
Demolish and Install	S.Y.	4.15	2.22		6.37	
Reinstall	S.Y.		1.61		1.61	
Minimum Charge	Job		202		202	

Stair Components

	Unit	Material	Labor	Equip.	Total	Specification
Carpeting						
Demolish	Ea.		2.22		2.22	Includes material and labor to install
Install	Riser	58.50	10.60		69.10	carpet and pad on stairs.
Demolish and Install	Riser	58.50	12.82		71.32	
Reinstall	Riser		8.49		8.49	
Clean	Ea.		5.15		5.15	
Minimum Charge	Job		202		202	

249

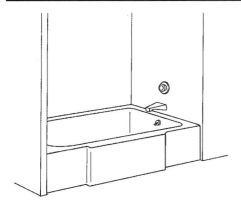

Tub/Shower

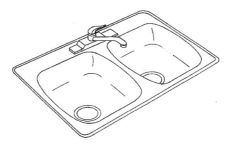

Stainless Steel Double Sink

Water Closet

Commode	Unit	Material	Labor	Equip.	Total	Specification
Floor Mounted w / Tank						
Demolish	Ea.		51.50		51.50	Includes material and labor to install a
Install	Ea.	253	175		428	toilet with valve, seat and cover.
Demolish and Install	Ea.	253	226.50		479.50	
Reinstall	Ea.		140.03		140.03	
Clean	Ea.	.41	16.50		16.91	
Minimum Charge	Job		258		258	
Wall w / Flush Valve						
Demolish	Ea.		51.50		51.50	Includes material and labor to install a
Install	Ea.	1050	160		1210	wall hung toilet with flush valve, seat
Demolish and Install	Ea.	1050	211.50		1261.50	and cover.
Reinstall	Ea.		127.96		127.96	
Clean	Ea.	.41	16.50		16.91	
Minimum Charge	Job		258		258	
Designer Floor Mount						
Demolish	Ea.		51.50		51.50	Includes material and labor to install a
Install	Ea.	840	175		1015	toilet with valve, seat and cover.
Demolish and Install	Ea.	840	226.50		1066.50	
Reinstall	Ea.		140.03		140.03	
Clean	Ea.	.41	16.50		16.91	
Minimum Charge	Job		258		258	
Urinal						
Demolish	Ea.		51.50		51.50	Includes material and labor to install a
Install	Ea.	305	310		615	wall hung urinal with flush valve.
Demolish and Install	Ea.	305	361.50		666.50	
Reinstall	Ea.		247.38		247.38	
Clean	Ea.	.41	16.50		16.91	
Minimum Charge	Job		258		258	
Bidet						
Vitreous China						
Demolish	Ea.		51.50		51.50	Includes material and labor to install a
Install	Ea.	660	186		846	vitreous china bidet complete with
Demolish and Install	Ea.	660	237.50		897.50	trim.
Reinstall	Ea.		148.43		148.43	
Clean	Ea.	.41	16.50		16.91	
Minimum Charge	Job		258		258	

Finish Mechanical

Commode		Unit	Material	Labor	Equip.	Total	Specification
Chrome Fittings							
	Install	Ea.	233	64.50		297.50	Includes labor and materials to install chrome fittings for a bidet.
Brass Fittings							
	Install	Ea.	257	64.50		321.50	Includes labor and materials to install brass fittings for a bidet.
Seat							
	Demolish	Ea.		21.50		21.50	Includes material and labor to install a toilet seat and cover.
	Install	Ea.	32	21.50		53.50	
	Demolish and Install	Ea.	32	43		75	
	Reinstall	Ea.		17.17		17.17	
	Clean	Ea.	.21	6.90		7.11	
	Minimum Charge	Job		258		258	
Rough-in							
	Install	Ea.	400	325		725	Includes labor and materials to install water supply pipe with valves and drain, waste and vent pipe with all couplings, hangers and fasteners necessary.

Sink (assembly)		Unit	Material	Labor	Equip.	Total	Specification
Single							
Porcelain							
	Demolish	Ea.		32		32	Includes material and labor to install a single bowl enamel finished cast iron kitchen sink with faucet and drain.
	Install	Ea.	605	166		771	
	Demolish and Install	Ea.	605	198		803	
	Reinstall	Ea.		132.53		132.53	
	Clean	Ea.	.26	11.80		12.06	
	Minimum Charge	Job		258		258	
Stainless Steel							
	Demolish	Ea.		32		32	Includes material and labor to install a single bowl stainless steel kitchen sink with faucet and drain.
	Install	Ea.	745	166		911	
	Demolish and Install	Ea.	745	198		943	
	Reinstall	Ea.		132.53		132.53	
	Clean	Ea.	.26	11.80		12.06	
	Minimum Charge	Job		258		258	
Double							
Porcelain							
	Demolish	Ea.		37		37	Includes material and labor to install a double bowl enamel finished cast iron kitchen sink with faucet and drain.
	Install	Ea.	405	193		598	
	Demolish and Install	Ea.	405	230		635	
	Reinstall	Ea.		154.61		154.61	
	Clean	Ea.	.41	16.50		16.91	
	Minimum Charge	Job		258		258	
Stainless Steel							
	Demolish	Ea.		37		37	Includes material and labor to install a double bowl stainless steel kitchen sink with faucet and drain.
	Install	Ea.	470	193		663	
	Demolish and Install	Ea.	470	230		700	
	Reinstall	Ea.		154.61		154.61	
	Clean	Ea.	.41	16.50		16.91	
	Minimum Charge	Job		258		258	

Finish Mechanical

Sink (assembly)

Sink (assembly)	Unit	Material	Labor	Equip.	Total	Specification
Bar						
Demolish	Ea.		32		32	Includes material and labor to install a
Install	Ea.	194	129		323	small single bowl stainless steel bar /
Demolish and Install	Ea.	194	161		355	vegetable / kitchen sink with faucet
Reinstall	Ea.		103.04		103.04	and drain.
Clean	Ea.	.26	11.80		12.06	
Minimum Charge	Job		258		258	
Floor						
Demolish	Ea.		103		103	Includes material and labor to install
Install	Ea.	1125	211		1336	an enameled cast iron floor mounted
Demolish and Install	Ea.	1125	314		1439	corner service sink with faucet and
Reinstall	Ea.		168.67		168.67	drain.
Clean	Ea.	.41	8.25		8.66	
Minimum Charge	Job		258		258	
Pedestal						
Demolish	Ea.		32		32	Includes material and labor to install a
Install	Ea.	370	145		515	vitreous china pedestal lavatory with
Demolish and Install	Ea.	370	177		547	faucet set and pop-up drain.
Reinstall	Ea.		115.96		115.96	
Clean	Ea.	.26	11.80		12.06	
Minimum Charge	Job		258		258	
Vanity Lavatory						
Demolish	Ea.		32		32	Includes material and labor to install a
Install	Ea.	267	172		439	vitreous china lavatory with faucet set
Demolish and Install	Ea.	267	204		471	and pop-up drain.
Reinstall	Ea.		137.43		137.43	
Clean	Ea.	.26	11.80		12.06	
Minimum Charge	Job		258		258	
Rough-in						
Install	Ea.	272	435		707	Includes labor and materials to install copper supply pipe with valves and drain, waste and vent pipe with all couplings, hangers and fasteners necessary.
Laundry						
Demolish	Ea.		33		33	Includes material and labor to install a
Install	Ea.	169	143		312	laundry sink (wall mounted or with
Demolish and Install	Ea.	169	176		345	legs) including faucet and pop up
Clean	Ea.	.26	11.80		12.06	drain.
Minimum Charge	Job		258		258	

Sink Only

Sink Only	Unit	Material	Labor	Equip.	Total	Specification
Single						
Porcelain						
Demolish	Ea.		32		32	Includes material and labor to install a
Install	Ea.	500	184		684	single bowl enamel finished cast iron
Demolish and Install	Ea.	500	216		716	kitchen sink.
Reinstall	Ea.		147.20		147.20	
Clean	Ea.	.26	11.80		12.06	
Minimum Charge	Job		258		258	

·Finish Mechanical

Sink Only		Unit	Material	Labor	Equip.	Total	Specification
Stainless Steel							
	Demolish	Ea.		32		32	Includes material and labor to install a
	Install	Ea.	640	129		769	single bowl stainless steel kitchen sink.
	Demolish and Install	Ea.	640	161		801	
	Reinstall	Ea.		103.04		103.04	
	Clean	Ea.	.26	11.80		12.06	
	Minimum Charge	Job		258		258	
Double							
Porcelain							
	Demolish	Ea.		37		37	Includes material and labor to install a
	Install	Ea.	289	215		504	double bowl enamel finished cast iron
	Demolish and Install	Ea.	289	252		541	kitchen sink.
	Reinstall	Ea.		171.73		171.73	
	Clean	Ea.	.41	16.50		16.91	
	Minimum Charge	Job		258		258	
Stainless Steel							
	Demolish	Ea.		37		37	Includes material and labor to install a
	Install	Ea.	955	166		1121	double bowl stainless steel kitchen
	Demolish and Install	Ea.	955	203		1158	sink.
	Reinstall	Ea.		132.95		132.95	
	Clean	Ea.	.41	16.50		16.91	
	Minimum Charge	Job		258		258	
Bar							
	Demolish	Ea.		32		32	Includes material and labor to install a
	Install	Ea.	85.50	129		214.50	small single bowl stainless steel bar /
	Demolish and Install	Ea.	85.50	161		246.50	vegetable / kitchen sink.
	Reinstall	Ea.		103.04		103.04	
	Clean	Ea.	.26	11.80		12.06	
	Minimum Charge	Job		258		258	
Floor							
	Demolish	Ea.		103		103	Includes material and labor to install
	Install	Ea.	995	234		1229	an enameled cast iron floor mounted
	Demolish and Install	Ea.	995	337		1332	corner service sink.
	Reinstall	Ea.		187.35		187.35	
	Clean	Ea.	.41	8.25		8.66	
	Minimum Charge	Job		258		258	
Pedestal							
	Demolish	Ea.		32		32	Includes material and labor to install a
	Install	Ea.	275	161		436	vitreous china pedestal lavatory.
	Demolish and Install	Ea.	275	193		468	
	Reinstall	Ea.		128.80		128.80	
	Clean	Ea.	.26	11.80		12.06	
	Minimum Charge	Job		258		258	
Vanity Lavatory							
	Demolish	Ea.		32		32	Includes material and labor to install a
	Install	Ea.	168	161		329	vitreous china lavatory.
	Demolish and Install	Ea.	168	193		361	
	Reinstall	Ea.		128.80		128.80	
	Clean	Ea.	.26	11.80		12.06	
	Minimum Charge	Job		258		258	
Rough-in							
	Install	Ea.	240	405		645	Includes labor and materials to install copper supply pipe with valves and drain, waste and vent pipe with all couplings, hangers and fasteners necessary.

Finish Mechanical

Sink Only

	Unit	Material	Labor	Equip.	Total	Specification
Faucet Sink						
Good Grade						
Demolish	Ea.		26		26	Includes material and labor to install a
Install	Ea.	85.50	35.50		121	sink faucet and fittings.
Demolish and Install	Ea.	85.50	61.50		147	
Reinstall	Ea.		28.44		28.44	
Clean	Ea.	.17	10.30		10.47	
Minimum Charge	Job		258		258	
Better Grade						
Demolish	Ea.		26		26	Includes material and labor to install a
Install	Ea.	143	35.50		178.50	sink faucet and fittings.
Demolish and Install	Ea.	143	61.50		204.50	
Reinstall	Ea.		28.44		28.44	
Clean	Ea.	.17	10.30		10.47	
Minimum Charge	Job		258		258	
Premium Grade						
Demolish	Ea.		26		26	Includes material and labor to install a
Install	Ea.	280	44.50		324.50	sink faucet and fittings.
Demolish and Install	Ea.	280	70.50		350.50	
Reinstall	Ea.		35.47		35.47	
Clean	Ea.	.17	10.30		10.47	
Minimum Charge	Job		258		258	
Drain & Basket						
Demolish	Ea.		16.10		16.10	Includes material and labor to install a
Install	Ea.	16.35	32		48.35	sink drain and basket.
Demolish and Install	Ea.	16.35	48.10		64.45	
Reinstall	Ea.		25.76		25.76	
Clean	Ea.	.17	10.30		10.47	
Minimum Charge	Job		258		258	

Bathtub

	Unit	Material	Labor	Equip.	Total	Specification
Enameled						
Steel						
Demolish	Ea.		93		93	Includes labor and material to install a
Install	Ea.	550	169		719	formed steel bathtub with spout,
Demolish and Install	Ea.	550	262		812	mixing valve, shower head and
Reinstall	Ea.		134.94		134.94	pop-up drain.
Clean	Ea.	.81	27.50		28.31	
Minimum Charge	Job		258		258	
Cast Iron						
Demolish	Ea.		93		93	Includes material and labor to install a
Install	Ea.	1275	211		1486	cast iron bathtub with spout, mixing
Demolish and Install	Ea.	1275	304		1579	valve, shower head and pop-up drain.
Reinstall	Ea.		168.67		168.67	
Clean	Ea.	.81	27.50		28.31	
Minimum Charge	Job		258		258	
Fiberglass						
Demolish	Ea.		93		93	Includes material and labor to install
Install	Ea.	1800	169		1969	an acrylic soaking tub with spout,
Demolish and Install	Ea.	1800	262		2062	mixing valve, shower head and
Reinstall	Ea.		134.94		134.94	pop-up drain.
Clean	Ea.	.81	27.50		28.31	
Minimum Charge	Job		258		258	

Bathtub	Unit	Material	Labor	Equip.	Total	Specification
Institutional						
Demolish	Ea.		93		93	Includes material and labor to install a
Install	Ea.	1425	310		1735	hospital / institutional type bathtub
Demolish and Install	Ea.	1425	403		1828	with spout, mixing valve, shower head
Reinstall	Ea.		247.38		247.38	and pop-up drain.
Clean	Ea.	.81	27.50		28.31	
Minimum Charge	Job		258		258	
Whirlpool (Acrylic)						
Demolish	Ea.		286		286	Includes material and labor to install a
Install	Ea.	1975	930		2905	molded fiberglass whirlpool tub.
Demolish and Install	Ea.	1975	1216		3191	
Reinstall	Ea.		742.14		742.14	
Clean	Ea.	.81	27.50		28.31	
Minimum Charge	Job		258		258	
Pump / Motor						
Demolish	Ea.		71.50		71.50	Includes material and labor to install a
Install	Ea.	288	155		443	whirlpool tub pump / motor.
Demolish and Install	Ea.	288	226.50		514.50	
Reinstall	Ea.		123.77		123.77	
Clean	Ea.	.21	20.50		20.71	
Minimum Charge	Job		258		258	
Heater / Motor						
Demolish	Ea.		71.50		71.50	Includes material and labor to install a
Install	Ea.	197	229		426	whirlpool tub heater.
Demolish and Install	Ea.	197	300.50		497.50	
Reinstall	Ea.		183.18		183.18	
Clean	Ea.	.21	20.50		20.71	
Minimum Charge	Job		258		258	
Thermal Cover						
Demolish	Ea.		11.45		11.45	Includes material and labor to install a
Install	Ea.	395	32		427	whirlpool tub thermal cover.
Demolish and Install	Ea.	395	43.45		438.45	
Reinstall	Ea.		25.76		25.76	
Clean	Ea.	2.04	20.50		22.54	
Minimum Charge	Job		258		258	
Tub / Shower Combination						
Demolish	Ea.		82		82	Includes material and labor to install a
Install	Ea.	910	232		1142	fiberglass module tub with shower
Demolish and Install	Ea.	910	314		1224	surround with spout, mixing valve,
Reinstall	Ea.		185.54		185.54	shower head and pop-up drain.
Clean	Ea.	1.02	41.50		42.52	
Minimum Charge	Job		258		258	
Accessories						
Sliding Door						
Demolish	Ea.		23		23	Includes material and labor to install a
Install	Ea.	635	106		741	48" aluminum framed tempered glass
Demolish and Install	Ea.	635	129		764	shower door.
Reinstall	Ea.		84.95		84.95	
Clean	Ea.	1.69	13.75		15.44	
Minimum Charge	Job		258		258	
Shower Head						
Demolish	Ea.		6.45		6.45	Includes material and labor to install a
Install	Ea.	49.50	21.50		71	water saving shower head.
Demolish and Install	Ea.	49.50	27.95		77.45	
Reinstall	Ea.		17.17		17.17	
Clean	Ea.	.02	4.13		4.15	
Minimum Charge	Job		258		258	

Finish Mechanical

Bathtub

	Unit	Material	Labor	Equip.	Total	Specification
Faucet Set						
Demolish	Ea.		32		32	Includes material and labor to install a
Install	Ea.	94.50	64.50		159	combination spout / diverter for a
Demolish and Install	Ea.	94.50	96.50		191	bathtub.
Reinstall	Ea.		51.52		51.52	
Clean	Ea.	.17	10.30		10.47	
Minimum Charge	Job		258		258	
Shower Rod						
Demolish	Ea.		3.82		3.82	Includes material and labor to install a
Install	Ea.	36	35		71	chrome plated shower rod.
Demolish and Install	Ea.	36	38.82		74.82	
Reinstall	Ea.		27.94		27.94	
Clean	Ea.		4.13		4.13	
Minimum Charge	Job		258		258	
Rough-in						
Install	Ea.	390	450		840	Includes labor and materials to install copper supply pipe with valves and drain, waste and vent pipe with all couplings, hangers and fasteners necessary.

Shower

	Unit	Material	Labor	Equip.	Total	Specification
Fiberglass						
32" x 32"						
Demolish	Ea.		193		193	Includes material and labor to install a
Install	Ea.	370	169		539	fiberglass shower stall with door,
Demolish and Install	Ea.	370	362		732	mixing valve and shower head and
Reinstall	Ea.		134.94		134.94	drain fitting.
Clean	Ea.	.81	27.50		28.31	
Minimum Charge	Job		258		258	
36" x 36"						
Demolish	Ea.		193		193	Includes material and labor to install a
Install	Ea.	425	169		594	fiberglass shower stall with door,
Demolish and Install	Ea.	425	362		787	mixing valve and shower head and
Reinstall	Ea.		134.94		134.94	drain fitting.
Clean	Ea.	.81	27.50		28.31	
Minimum Charge	Job		258		258	
35" x 60"						
Demolish	Ea.		193		193	Includes material and labor to install a
Install	Ea.	805	232		1037	handicap fiberglass shower stall with
Demolish and Install	Ea.	805	425		1230	door & seat, mixing valve and shower
Reinstall	Ea.		185.54		185.54	head and drain fitting.
Clean	Ea.	.81	27.50		28.31	
Minimum Charge	Job		258		258	
Accessories						
Single Door						
Demolish	Ea.		23		23	Includes material and labor to install a
Install	Ea.	158	56		214	shower door.
Demolish and Install	Ea.	158	79		237	
Reinstall	Ea.		44.84		44.84	
Clean	Ea.	1.69	13.75		15.44	
Minimum Charge	Job		258		258	

257

Finish Mechanical

Shower		Unit	Material	Labor	Equip.	Total	Specification
Shower Head							
	Demolish	Ea.		6.45		6.45	Includes material and labor to install a
	Install	Ea.	49.50	21.50		71	water saving shower head.
	Demolish and Install	Ea.	49.50	27.95		77.45	
	Reinstall	Ea.		17.17		17.17	
	Clean	Ea.	.02	4.13		4.15	
	Minimum Charge	Job		258		258	
Faucet Set							
	Demolish	Ea.		32		32	Includes material and labor to install a
	Install	Ea.	94.50	64.50		159	combination spout / diverter for a
	Demolish and Install	Ea.	94.50	96.50		191	bathtub.
	Reinstall	Ea.		51.52		51.52	
	Clean	Ea.	.17	10.30		10.47	
	Minimum Charge	Job		258		258	
Shower Pan							
	Demolish	Ea.		28.50		28.50	Includes material and labor to install
	Install	Ea.	109	64.50		173.50	the base portion of a fiberglass
	Demolish and Install	Ea.	109	93		202	shower unit.
	Reinstall	Ea.		51.52		51.52	
	Clean	Ea.	.81	27.50		28.31	
	Minimum Charge	Job		258		258	
Rough-in							
	Install	Ea.	380	455		835	Includes labor and materials to install copper supply pipe with valves and drain, waste and vent pipe with all couplings, hangers and fasteners necessary.

Ductwork		Unit	Material	Labor	Equip.	Total	Specification
Return Grill							
Under 10" Wide							
	Demolish	Ea.		11.20		11.20	Includes material and labor to install a
	Install	Ea.	18.90	28		46.90	10" x 10" air return register.
	Demolish and Install	Ea.	18.90	39.20		58.10	
	Reinstall	Ea.		22.42		22.42	
	Clean	Ea.	.21	9.15		9.36	
	Minimum Charge	Job		252		252	
12" to 20" Wide							
	Demolish	Ea.		11.20		11.20	Includes material and labor to install a
	Install	Ea.	36.50	29.50		66	16" x 16" air return register.
	Demolish and Install	Ea.	36.50	40.70		77.20	
	Reinstall	Ea.		23.74		23.74	
	Clean	Ea.	.21	9.15		9.36	
	Minimum Charge	Job		252		252	
Over 20" Wide							
	Demolish	Ea.		17		17	Includes material and labor to install a
	Install	Ea.	74.50	46		120.50	24" x 24" air return register.
	Demolish and Install	Ea.	74.50	63		137.50	
	Reinstall	Ea.		36.68		36.68	
	Clean	Ea.	.21	10.30		10.51	
	Minimum Charge	Job		252		252	

Finish Mechanical

Ductwork

	Unit	Material	Labor	Equip.	Total	Specification
Supply Grill						
Baseboard Type						
Demolish	Ea.		8		8	Includes material and labor to install a
Install	Ea.	7.15	25		32.15	14″ x 6″ baseboard air supply
Demolish and Install	Ea.	7.15	33		40.15	register.
Reinstall	Ea.		20.18		20.18	
Clean	Ea.	.21	9.15		9.36	
Minimum Charge	Job		252		252	
10″ Wide						
Demolish	Ea.		9.35		9.35	Includes material and labor to install a
Install	Ea.	15.20	25		40.20	10″ x 6″ air supply register.
Demolish and Install	Ea.	15.20	34.35		49.55	
Reinstall	Ea.		20.18		20.18	
Clean	Ea.	.21	9.15		9.36	
Minimum Charge	Job		252		252	
12″ to 15″ Wide						
Demolish	Ea.		9.35		9.35	Includes material and labor to install a
Install	Ea.	21	29.50		50.50	14″ x 8″ air supply register.
Demolish and Install	Ea.	21	38.85		59.85	
Reinstall	Ea.		23.74		23.74	
Clean	Ea.	.21	9.15		9.36	
Minimum Charge	Job		252		252	
18″ to 24″ Wide						
Demolish	Ea.		11.20		11.20	Includes material and labor to install a
Install	Ea.	29	39		68	24″ x 8″ air supply register.
Demolish and Install	Ea.	29	50.20		79.20	
Reinstall	Ea.		31.04		31.04	
Clean	Ea.	.21	10.30		10.51	
Minimum Charge	Job		252		252	
30″ to 36″ Wide						
Demolish	Ea.		14		14	Includes material and labor to install a
Install	Ea.	38	36		74	30″ x 8″ air supply register.
Demolish and Install	Ea.	38	50		88	
Reinstall	Ea.		28.82		28.82	
Clean	Ea.	.21	20.50		20.71	
Minimum Charge	Job		252		252	
Diffuser						
12″ x 12″						
Demolish	Ea.		11.20		11.20	Includes material and labor to install a
Install	Ea.	93	42		135	12″ x 12″ aluminum diffuser.
Demolish and Install	Ea.	93	53.20		146.20	
Reinstall	Ea.		33.63		33.63	
Clean	Ea.	.21	9.15		9.36	
Minimum Charge	Job		252		252	
14″ x 14″						
Demolish	Ea.		11.20		11.20	Includes material and labor to install a
Install	Ea.	141	42		183	14″ x 14″ aluminum diffuser.
Demolish and Install	Ea.	141	53.20		194.20	
Reinstall	Ea.		33.63		33.63	
Clean	Ea.	.21	10.30		10.51	
Minimum Charge	Job		252		252	
18″ x 18″						
Demolish	Ea.		17		17	Includes material and labor to install a
Install	Ea.	152	56		208	18″ x 18″ aluminum diffuser.
Demolish and Install	Ea.	152	73		225	
Reinstall	Ea.		44.84		44.84	
Clean	Ea.	.21	13.75		13.96	
Minimum Charge	Job		252		252	

259

Finish Mechanical

Ductwork		Unit	Material	Labor	Equip.	Total	Specification
12" Round							
	Demolish	Ea.		15.55		15.55	Includes material and labor to install a
	Install	Ea.	18	42		60	12" diameter aluminum diffuser with a
	Demolish and Install	Ea.	18	57.55		75.55	butterfly damper.
	Reinstall	Ea.		33.63		33.63	
	Clean	Ea.	.21	9.15		9.36	
	Minimum Charge	Job		252		252	
14" to 20" Round							
	Demolish	Ea.		21		21	Includes material and labor to install a
	Install	Ea.	166	56		222	20" diameter aluminum diffuser with a
	Demolish and Install	Ea.	166	77		243	butterfly damper.
	Reinstall	Ea.		44.84		44.84	
	Clean	Ea.	.21	10.30		10.51	
	Minimum Charge	Job		252		252	
Flush-out / Sanitize							
	Install	L.F.	.81	4.13		4.94	Includes labor and material costs to
	Minimum Charge	Job		252		252	clean ductwork or flue.

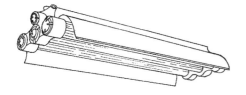

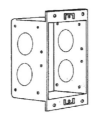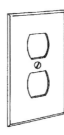

Lighting Fixture   Duplex Receptacle

Electrical Per S.F.	Unit	Material	Labor	Equip.	Total	Specification
Residential						
Light Fixtures						
Install	S.F.	.24	.24		.48	Includes labor and material to replace house light fixtures. Apply S.F. cost to floor area including garages but not basements.

Circuits	Unit	Material	Labor	Equip.	Total	Specification
120V						
Outlet with Wiring						
Demolish	Ea.		27		27	Includes material and labor to install general purpose outlet with cover, up to 20' of #12/2 wire.
Install	Ea.	11.35	33		44.35	
Demolish and Install	Ea.	11.35	60		71.35	
Reinstall	Ea.		26.55		26.55	
Clean	Ea.		2.06		2.06	
Minimum Charge	Job		241		241	
GFCI Outlet w / Wiring						
Demolish	Ea.		27		27	Cost includes material and labor to install 120V GFCI receptacle and wall plate, with up to 20' of #12/2 wire.
Install	Ea.	25.50	39		64.50	
Demolish and Install	Ea.	25.50	66		91.50	
Reinstall	Ea.		31.38		31.38	
Clean	Ea.		2.06		2.06	
Minimum Charge	Job		241		241	
Exterior Outlet w / Wiring						
Demolish	Ea.		27		27	Cost includes material and labor to install 110V waterproof external receptacle and cover plate with up to 20' of #12/2 wire.
Install	Ea.	43.50	40		83.50	
Demolish and Install	Ea.	43.50	67		110.50	
Reinstall	Ea.		32.19		32.19	
Clean	Ea.		2.06		2.06	
Minimum Charge	Job		241		241	
240V Outlet w / Wiring						
Demolish	Ea.		27		27	Cost includes material and labor to install 240V receptacle and cover plate with up to 20' of #12/2 wire.
Install	Ea.	64	75.50		139.50	
Demolish and Install	Ea.	64	102.50		166.50	
Reinstall	Ea.		60.26		60.26	
Clean	Ea.		2.06		2.06	
Minimum Charge	Job		241		241	

Finish Electrical

Circuits	Unit	Material	Labor	Equip.	Total	Specification
TV / Cable Jack w / Wiring						
Demolish	Ea.		27		27	Includes material and labor to install
Install	Ea.	17.60	30		47.60	TV/cable outlet box and cover plate
Demolish and Install	Ea.	17.60	57		74.60	including 20' of wiring.
Reinstall	Ea.		24.14		24.14	
Clean	Ea.		2.06		2.06	
Minimum Charge	Job		241		241	
Phone Jack with Wiring						
Demolish	Ea.		27		27	Includes material and labor to install
Install	Ea.	9.25	18.55		27.80	phone jack and cover plate including
Demolish and Install	Ea.	9.25	45.55		54.80	up to 20' of telephone cable.
Reinstall	Ea.		14.86		14.86	
Clean	Ea.		1.29		1.29	
Minimum Charge	Job		241		241	

Light Switch	Unit	Material	Labor	Equip.	Total	Specification
Single						
Switch						
Demolish	Ea.		5.35		5.35	Includes material and labor to install
Install	Ea.	15.85	23		38.85	single switch and wall plate. Wiring is
Demolish and Install	Ea.	15.85	28.35		44.20	not included.
Reinstall	Ea.		18.39		18.39	
Clean	Ea.		2.06		2.06	
Minimum Charge	Job		241		241	
Plate						
Demolish	Ea.		2.68		2.68	Includes material and labor to install a
Install	Ea.	.42	6.05		6.47	one gang brown wall switch plate.
Demolish and Install	Ea.	.42	8.73		9.15	
Reinstall	Ea.		4.83		4.83	
Clean	Ea.	.08	.41		.49	
Minimum Charge	Job		241		241	
Double						
Switch						
Demolish	Ea.		10.75		10.75	Includes material and labor to install
Install	Ea.	15.05	34.50		49.55	double switch and wall plate. Wiring
Demolish and Install	Ea.	15.05	45.25		60.30	is not included.
Reinstall	Ea.		27.59		27.59	
Clean	Ea.	.08	6.60		6.68	
Minimum Charge	Job		241		241	
Plate						
Demolish	Ea.		2.68		2.68	Includes material and labor to install a
Install	Ea.	.83	9.10		9.93	two gang brown wall switch plate.
Demolish and Install	Ea.	.83	11.78		12.61	
Reinstall	Ea.		7.29		7.29	
Clean	Ea.	.08	.41		.49	
Minimum Charge	Job		241		241	
Triple						
Switch						
Demolish	Ea.		16.25		16.25	Includes material and labor to install
Install	Ea.	37.50	40		77.50	triple switch and wall plate. Wiring is
Demolish and Install	Ea.	37.50	56.25		93.75	not included.
Reinstall	Ea.		32.19		32.19	
Clean	Ea.	.08	9.90		9.98	
Minimum Charge	Job		241		241	

Finish Electrical

Light Switch	Unit	Material	Labor	Equip.	Total	Specification
Plate						
Demolish	Ea.		2.68		2.68	Includes material and labor to install a
Install	Ea.	1.22	15.10		16.32	three gang brown wall switch plate.
Demolish and Install	Ea.	1.22	17.78		19	
Reinstall	Ea.		12.07		12.07	
Clean	Ea.	.08	.41		.49	
Minimum Charge	Job		241		241	
Dimmer						
Switch						
Demolish	Ea.		5.35		5.35	Includes material and labor to install
Install	Ea.	34.50	28.50		63	dimmer switch. Wiring is not included.
Demolish and Install	Ea.	34.50	33.85		68.35	
Reinstall	Ea.		22.72		22.72	
Clean	Ea.		2.06		2.06	
Minimum Charge	Job		241		241	
Plate						
Demolish	Ea.		2.68		2.68	Includes material and labor to install a
Install	Ea.	.42	6.05		6.47	one gang brown wall switch plate.
Demolish and Install	Ea.	.42	8.73		9.15	
Reinstall	Ea.		4.83		4.83	
Clean	Ea.	.08	.41		.49	
Minimum Charge	Job		241		241	
With Wiring						
Single						
Demolish	Ea.		27		27	Includes material and labor to install
Install	Ea.	20.50	28		48.50	single switch and plate including up to
Demolish and Install	Ea.	20.50	55		75.50	20' of romex cable.
Reinstall	Ea.		22.59		22.59	
Clean	Ea.		2.06		2.06	
Minimum Charge	Job		241		241	
Double						
Demolish	Ea.		27		27	Includes material and labor to install
Install	Ea.	29	48.50		77.50	double switch and plate including up to
Demolish and Install	Ea.	29	75.50		104.50	to 20' of romex cable.
Reinstall	Ea.		38.62		38.62	
Clean	Ea.	.08	6.60		6.68	
Minimum Charge	Job		241		241	
Triple						
Demolish	Ea.		40		40	Includes material and labor to install
Install	Ea.	44	54.50		98.50	triple switch and plate including up to
Demolish and Install	Ea.	44	94.50		138.50	20' of romex cable.
Reinstall	Ea.		43.45		43.45	
Clean	Ea.	.08	9.90		9.98	
Minimum Charge	Job		241		241	
Dimmer						
Demolish	Ea.		27		27	Includes material and labor to install
Install	Ea.	39	33		72	dimmer switch and cover plate
Demolish and Install	Ea.	39	60		99	including up to 20' of romex cable.
Reinstall	Ea.		26.55		26.55	
Clean	Ea.		2.06		2.06	
Minimum Charge	Job		241		241	

Finish Electrical

Cover for Outlet / Switch

	Unit	Material	Labor	Equip.	Total	Specification
High Grade						
Demolish	Ea.		2.68		2.68	Includes material and labor to install
Install	Ea.	2.82	6.05		8.87	wall plate, stainless steel, 1 gang.
Demolish and Install	Ea.	2.82	8.73		11.55	
Reinstall	Ea.		4.83		4.83	
Clean	Ea.	.08	.41		.49	
Minimum Charge	Job		241		241	
Deluxe Grade						
Demolish	Ea.		2.68		2.68	Includes material and labor to install
Install	Ea.	4.69	6.05		10.74	wall plate, brushed brass, 1 gang.
Demolish and Install	Ea.	4.69	8.73		13.42	
Reinstall	Ea.		4.83		4.83	
Clean	Ea.	.08	.41		.49	
Minimum Charge	Job		241		241	

Residential Light Fixture

	Unit	Material	Labor	Equip.	Total	Specification
Ceiling						
Good Quality						
Demolish	Ea.		11.20		11.20	Includes material and labor to install
Install	Ea.	29.50	12.05		41.55	standard quality incandescent ceiling
Demolish and Install	Ea.	29.50	23.25		52.75	light fixture. Wiring is not included.
Reinstall	Ea.		9.66		9.66	
Clean	Ea.	.10	27.50		27.60	
Minimum Charge	Job		241		241	
Better Quality						
Demolish	Ea.		11.20		11.20	Includes material and labor to install
Install	Ea.	74.50	25.50		100	custom quality incandescent ceiling
Demolish and Install	Ea.	74.50	36.70		111.20	light fixture. Wiring is not included.
Reinstall	Ea.		20.33		20.33	
Clean	Ea.	.10	27.50		27.60	
Minimum Charge	Job		241		241	
Premium Quality						
Demolish	Ea.		11.20		11.20	Includes material and labor to install
Install	Ea.	345	25.50		370.50	deluxe quality incandescent ceiling
Demolish and Install	Ea.	345	36.70		381.70	light fixture. Wiring is not included.
Reinstall	Ea.		20.33		20.33	
Clean	Ea.	.10	27.50		27.60	
Minimum Charge	Job		241		241	
Recessed Ceiling						
Demolish	Ea.		22.50		22.50	Includes material and labor to install
Install	Ea.	75	16.10		91.10	standard quality recessed eyeball
Demolish and Install	Ea.	75	38.60		113.60	spotlight with housing. Wiring is not
Reinstall	Ea.		12.87		12.87	included.
Clean	Ea.	.04	3.44		3.48	
Minimum Charge	Job		241		241	
Recessed Eyeball						
Demolish	Ea.		22.50		22.50	Includes material and labor to install
Install	Ea.	93	17.25		110.25	custom quality recessed eyeball
Demolish and Install	Ea.	93	39.75		132.75	spotlight with housing. Wiring is not
Reinstall	Ea.		13.79		13.79	included.
Clean	Ea.	.04	3.44		3.48	
Minimum Charge	Job		241		241	

Finish Electrical

Residential Light Fixture		Unit	Material	Labor	Equip.	Total	Specification
Recessed Shower Type							Includes material and labor to install deluxe quality recessed eyeball spotlight with housing. Wiring is not included.
	Demolish	Ea.		22.50		22.50	
	Install	Ea.	91	16.10		107.10	
	Demolish and Install	Ea.	91	38.60		129.60	
	Reinstall	Ea.		12.87		12.87	
	Clean	Ea.	.04	3.44		3.48	
	Minimum Charge	Job		241		241	
Run Wiring							Includes labor and material to install an electric fixture utility box with non-metallic cable.
	Install	Ea.	15.30	19.30		34.60	
	Minimum Charge	Job		241		241	
Fluorescent							
Good Quality							Includes material and labor to install standard quality 12" wide by 48" long decorative surface mounted fluorescent fixture w/acrylic diffuser.
	Demolish	Ea.		24.50		24.50	
	Install	Ea.	42	69		111	
	Demolish and Install	Ea.	42	93.50		135.50	
	Reinstall	Ea.		55.18		55.18	
	Clean	Ea.	.04	3.44		3.48	
	Minimum Charge	Job		241		241	
Better Quality							Includes material and labor to install an interior fluorescent lighting fixture.
	Demolish	Ea.		24.50		24.50	
	Install	Ea.	86	78		164	
	Demolish and Install	Ea.	86	102.50		188.50	
	Reinstall	Ea.		62.30		62.30	
	Clean	Ea.	.04	3.44		3.48	
	Minimum Charge	Job		241		241	
Premium Quality							Includes material and labor to install deluxe quality 24" wide by 48" long decorative surface mounted fluorescent fixture w/acrylic diffuser.
	Demolish	Ea.		24.50		24.50	
	Install	Ea.	89	91		180	
	Demolish and Install	Ea.	89	115.50		204.50	
	Reinstall	Ea.		72.88		72.88	
	Clean	Ea.	.04	3.44		3.48	
	Minimum Charge	Job		241		241	
Run Wiring							Includes labor and material to install an electric fixture utility box with non-metallic cable.
	Install	Ea.	15.30	19.30		34.60	
	Minimum Charge	Job		241		241	
Exterior							
Good Quality							Includes material and labor to install standard quality incandescent outdoor light fixture. Wiring is not included.
	Demolish	Ea.		21.50		21.50	
	Install	Ea.	66	30		96	
	Demolish and Install	Ea.	66	51.50		117.50	
	Reinstall	Ea.		24.14		24.14	
	Clean	Ea.	.21	13.75		13.96	
	Minimum Charge	Job		241		241	
Better Quality							Includes material and labor to install custom quality incandescent outdoor light fixture. Wiring is not included.
	Demolish	Ea.		21.50		21.50	
	Install	Ea.	104	30		134	
	Demolish and Install	Ea.	104	51.50		155.50	
	Reinstall	Ea.		24.14		24.14	
	Clean	Ea.	.21	13.75		13.96	
	Minimum Charge	Job		241		241	
Premium Quality							Includes material and labor to install deluxe quality incandescent outdoor light fixture. Wiring is not included.
	Demolish	Ea.		21.50		21.50	
	Install	Ea.	259	121		380	
	Demolish and Install	Ea.	259	142.50		401.50	
	Reinstall	Ea.		96.56		96.56	
	Clean	Ea.	.21	13.75		13.96	
	Minimum Charge	Job		241		241	

Finish Electrical

Residential Light Fixture

	Unit	Material	Labor	Equip.	Total	Specification
Outdoor w / Pole						Includes material and labor to install post lantern type incandescent outdoor light fixture. Wiring is not included.
Demolish	Ea.		21.50		21.50	
Install	Ea.	297	121		418	
Demolish and Install	Ea.	297	142.50		439.50	
Reinstall	Ea.		96.56		96.56	
Clean	Ea.	.21	13.75		13.96	
Minimum Charge	Job		241		241	
Run Wiring						Includes labor and material to install an electric fixture utility box with non-metallic cable.
Install	Ea.	15.30	19.30		34.60	
Minimum Charge	Job		241		241	
Track Lighting						Includes material and labor to install track lighting.
Demolish	L.F.		7.45		7.45	
Install	L.F.	18.35	20		38.35	
Demolish and Install	L.F.	18.35	27.45		45.80	
Reinstall	L.F.		16.09		16.09	
Clean	L.F.	.21	6.60		6.81	
Minimum Charge	Job		241		241	

Commercial Light Fixture

	Unit	Material	Labor	Equip.	Total	Specification
Fluorescent						
2 Bulb, 2'						Includes material and labor to install a 2' long surface mounted fluorescent light fixture with two 40 watt bulbs.
Demolish	Ea.		21.50		21.50	
Install	Ea.	37	60.50		97.50	
Demolish and Install	Ea.	37	82		119	
Reinstall	Ea.		48.28		48.28	
Clean	Ea.	1.02	23.50		24.52	
Minimum Charge	Job		241		241	
2 Bulb, 4'						Includes material and labor to install a 4' long surface mounted fluorescent light fixture with two 40 watt bulbs.
Demolish	Ea.		21.50		21.50	
Install	Ea.	73	69		142	
Demolish and Install	Ea.	73	90.50		163.50	
Reinstall	Ea.		55.18		55.18	
Clean	Ea.	1.02	23.50		24.52	
Minimum Charge	Job		241		241	
3 Bulb, 4'						Includes material and labor to install a 4' long pendant mounted fluorescent light fixture with three 34 watt bulbs.
Demolish	Ea.		21.50		21.50	
Install	Ea.	106	69		175	
Demolish and Install	Ea.	106	90.50		196.50	
Reinstall	Ea.		55.18		55.18	
Clean	Ea.	1.02	23.50		24.52	
Minimum Charge	Job		241		241	
4 Bulb, 4'						Includes material and labor to install a 4' long pendant mounted fluorescent light fixture with four 34 watt bulbs.
Demolish	Ea.		21.50		21.50	
Install	Ea.	96.50	74.50		171	
Demolish and Install	Ea.	96.50	96		192.50	
Reinstall	Ea.		59.42		59.42	
Clean	Ea.	1.02	23.50		24.52	
Minimum Charge	Job		241		241	
2 Bulb, 8'						Includes material and labor to install an 8' long surface mounted fluorescent light fixture with two 110 watt bulbs.
Demolish	Ea.		27		27	
Install	Ea.	108	91		199	
Demolish and Install	Ea.	108	118		226	
Reinstall	Ea.		72.88		72.88	
Clean	Ea.	1.02	23.50		24.52	
Minimum Charge	Job		241		241	

Finish Electrical

Commercial Light Fixture		Unit	Material	Labor	Equip.	Total	Specification
Run Wiring							
	Install	Ea.	15.30	19.30		34.60	Includes labor and material to install
	Minimum Charge	Job		241		241	an electric fixture utility box with
							non-metallic cable.
Ceiling, Recessed							
Eyeball Type							
	Demolish	Ea.		22.50		22.50	Includes material and labor to install
	Install	Ea.	93	17.25		110.25	custom quality recessed eyeball
	Demolish and Install	Ea.	93	39.75		132.75	spotlight with housing. Wiring is not
	Reinstall	Ea.		13.79		13.79	included.
	Clean	Ea.	.04	3.44		3.48	
	Minimum Charge	Job		241		241	
Shower Type							
	Demolish	Ea.		22.50		22.50	Includes material and labor to install
	Install	Ea.	91	16.10		107.10	deluxe quality recessed eyeball
	Demolish and Install	Ea.	91	38.60		129.60	spotlight with housing. Wiring is not
	Reinstall	Ea.		12.87		12.87	included.
	Clean	Ea.	.04	3.44		3.48	
	Minimum Charge	Job		241		241	
Run Wiring							
	Install	Ea.	15.30	19.30		34.60	Includes labor and material to install
	Minimum Charge	Job		241		241	an electric fixture utility box with
							non-metallic cable.
Ceiling Grid Type							
2 Tube							
	Demolish	Ea.		31.50		31.50	Cost includes material and labor to
	Install	Ea.	58	91		149	install 2 lamp fluorescent light fixture
	Demolish and Install	Ea.	58	122.50		180.50	in 2' x 4' ceiling grid system including
	Reinstall	Ea.		72.88		72.88	lens and tubes.
	Clean	Ea.	1.02	23.50		24.52	
	Minimum Charge	Job		241		241	
4 Tube							
	Demolish	Ea.		36		36	Cost includes material and labor to
	Install	Ea.	66.50	103		169.50	install 4 lamp fluorescent light fixture
	Demolish and Install	Ea.	66.50	139		205.50	in 2' x 4' ceiling grid system including
	Reinstall	Ea.		82.18		82.18	lens and tubes.
	Clean	Ea.	1.02	23.50		24.52	
	Minimum Charge	Job		241		241	
Acrylic Diffuser							
	Demolish	Ea.		5.60		5.60	Includes material and labor to install
	Install	Ea.	95.50	14.65		110.15	fluorescent light fixture w/acrylic
	Demolish and Install	Ea.	95.50	20.25		115.75	diffuser.
	Reinstall	Ea.		11.70		11.70	
	Clean	Ea.	.65	4.13		4.78	
	Minimum Charge	Job		241		241	
Louver Diffuser							
	Demolish	Ea.		6.70		6.70	Includes material and labor to install
	Install	Ea.	107	14.65		121.65	fluorescent light fixture w/louver
	Demolish and Install	Ea.	107	21.35		128.35	diffuser.
	Reinstall	Ea.		11.70		11.70	
	Clean	Ea.	.65	4.13		4.78	
	Minimum Charge	Job		241		241	
Run Wiring							
	Install	Ea.	15.30	19.30		34.60	Includes labor and material to install
	Minimum Charge	Job		241		241	an electric fixture utility box with
							non-metallic cable.

Finish Electrical

Commercial Light Fixture

Commercial Light Fixture	Unit	Material	Labor	Equip.	Total	Specification
Track Lighting						
Demolish	L.F.		7.45		7.45	Includes material and labor to install
Install	L.F.	18.35	20		38.35	track lighting.
Demolish and Install	L.F.	18.35	27.45		45.80	
Reinstall	L.F.		16.09		16.09	
Clean	L.F.	.21	6.60		6.81	
Minimum Charge	Job		241		241	
Emergency						
Demolish	Ea.		38.50		38.50	Includes material and labor to install
Install	Ea.	178	121		299	battery-powered emergency lighting
Demolish and Install	Ea.	178	159.50		337.50	unit with two (2) lights, including
Reinstall	Ea.		96.56		96.56	wiring.
Clean	Ea.	.51	23.50		24.01	
Minimum Charge	Job		241		241	
Run Wiring						
Install	Ea.	15.30	19.30		34.60	Includes labor and material to install
Minimum Charge	Job		241		241	an electric fixture utility box with
						non-metallic cable.

Fan

Fan	Unit	Material	Labor	Equip.	Total	Specification
Ceiling Paddle						
Good Quality						
Demolish	Ea.		30		30	Includes material and labor to install
Install	Ea.	90	48.50		138.50	standard ceiling fan.
Demolish and Install	Ea.	90	78.50		168.50	
Reinstall	Ea.		38.62		38.62	
Clean	Ea.	.41	20.50		20.91	
Minimum Charge	Job		241		241	
Better Quality						
Demolish	Ea.		30		30	Includes material and labor to install
Install	Ea.	208	48.50		256.50	better ceiling fan.
Demolish and Install	Ea.	208	78.50		286.50	
Reinstall	Ea.		38.62		38.62	
Clean	Ea.	.41	20.50		20.91	
Minimum Charge	Job		241		241	
Premium Quality						
Demolish	Ea.		30		30	Includes material and labor to install
Install	Ea.	272	60.50		332.50	deluxe ceiling fan.
Demolish and Install	Ea.	272	90.50		362.50	
Reinstall	Ea.		48.28		48.28	
Clean	Ea.	.41	20.50		20.91	
Minimum Charge	Job		241		241	
Light Kit						
Install	Ea.	45	60.50		105.50	Includes labor and material to install a
Minimum Charge	Job		241		241	ceiling fan light kit.
Bathroom Exhaust						
Ceiling Mounted						
Demolish	Ea.		25		25	Cost includes material and labor to
Install	Ea.	48.50	48		96.50	install 60 CFM ceiling mounted
Demolish and Install	Ea.	48.50	73		121.50	bathroom exhaust fan, up to 10' of
Reinstall	Ea.		38.51		38.51	wiring with one wall switch.
Clean	Ea.	.21	8.25		8.46	
Minimum Charge	Job		241		241	

Finish Electrical

Fan		Unit	Material	Labor	Equip.	Total	Specification
Wall Mounted							
	Demolish	Ea.		25		25	Cost includes material and labor to
	Install	Ea.	70.50	48		118.50	install 110 CFM wall mounted
	Demolish and Install	Ea.	70.50	73		143.50	bathroom exhaust fan, up to 10' of
	Reinstall	Ea.		38.51		38.51	wiring with one wall switch.
	Clean	Ea.	.21	8.25		8.46	
	Minimum Charge	Job		241		241	
Lighted							
	Demolish	Ea.		25		25	Cost includes material and labor to
	Install	Ea.	50.50	48		98.50	install 50 CFM ceiling mounted lighted
	Demolish and Install	Ea.	50.50	73		123.50	exhaust fan, up to 10' of wiring with
	Reinstall	Ea.		38.51		38.51	one wall switch..
	Clean	Ea.	.21	8.25		8.46	
	Minimum Charge	Job		241		241	
Blower Heater w / Light							
	Demolish	Ea.		25		25	Includes material and labor to install
	Install	Ea.	86.50	80		166.50	ceiling blower type ventilator with
	Demolish and Install	Ea.	86.50	105		191.50	switch, light and snap-on grill. Includes
	Reinstall	Ea.		64.18		64.18	1300 watt/120 volt heater.
	Clean	Ea.	.21	8.25		8.46	
	Minimum Charge	Job		241		241	
Fan Heater w / Light							
	Demolish	Ea.		25		25	Includes material and labor to install
	Install	Ea.	111	80		191	circular ceiling fan heater unit, 1500
	Demolish and Install	Ea.	111	105		216	watt, surface mounted.
	Reinstall	Ea.		64.18		64.18	
	Clean	Ea.	.21	8.25		8.46	
	Minimum Charge	Job		241		241	
Whole House Exhaust							
36" Ceiling Mounted							
	Install	Ea.	1100	120		1220	Includes material and labor to install a
	Clean	Ea.	.07	6.05		6.12	whole house exhaust fan.
	Minimum Charge	Job		241		241	
Kitchen Exhaust							
Standard Model							
	Install	Ea.	74.50	32		106.50	Includes labor and materials to install
	Clean	Ea.	.07	6.05		6.12	a bathroom or kitchen vent fan.
	Minimum Charge	Job		241		241	
Low Noise Model							
	Install	Ea.	101	32		133	Includes labor and materials to install
	Clean	Ea.	.07	6.05		6.12	a bathroom or kitchen vent fan, low
	Minimum Charge	Job		241		241	noise model.

Low Voltage Systems		Unit	Material	Labor	Equip.	Total	Specification
Smoke Detector							
Pre-wired							
	Demolish	Ea.		11.20		11.20	Includes material and labor to install
	Install	Ea.	22.50	90.50		113	AC type smoke detector including
	Demolish and Install	Ea.	22.50	101.70		124.20	wiring.
	Reinstall	Ea.		72.47		72.47	
	Clean	.Ea.	.08	10.30		10.38	
	Minimum Charge	Job		241		241	

For customer support on your Contractor's Pricing Guide: Residential Repair & Remodeling, call 888.606.7279.

Low Voltage Systems	Unit	Material	Labor	Equip.	Total	Specification
Run Wiring						
Install	Ea.	15.30	19.30		34.60	Includes labor and material to install
Minimum Charge	Job		241		241	an electric fixture utility box with
						non-metallic cable.
Door Bell / Chimes						
Good Quality						
Demolish	Ea.		33.50		33.50	Includes material and labor to install
Install	Ea.	49.50	121		170.50	surface mounted 2 note door bell.
Demolish and Install	Ea.	49.50	154.50		204	Wiring is not included.
Reinstall	Ea.		96.56		96.56	
Clean	Ea.	.21	10.30		10.51	
Minimum Charge	Job		241		241	
Better Quality						
Demolish	Ea.		33.50		33.50	Includes material and labor to install
Install	Ea.	74.50	121		195.50	standard quality surface mounted 2
Demolish and Install	Ea.	74.50	154.50		229	note door bell. Wiring is not included.
Reinstall	Ea.		96.56		96.56	
Clean	Ea.	.21	10.30		10.51	
Minimum Charge	Job		241		241	
Premium Quality						
Demolish	Ea.		33.50		33.50	Includes material and labor to install
Install	Ea.	103	121		224	deluxe quality surface mounted 2 note
Demolish and Install	Ea.	103	154.50		257.50	door bell. Wiring is not included.
Reinstall	Ea.		96.56		96.56	
Clean	Ea.	.21	10.30		10.51	
Minimum Charge	Job		241		241	
Run Wiring						
Install	Ea.	15.30	19.30		34.60	Includes labor and material to install
Minimum Charge	Job		241		241	an electric fixture utility box with
						non-metallic cable.
Intercom System						
Intercom (master)						
Demolish	Ea.		89.50		89.50	Includes material and labor to install
Install	Ea.	665	241		906	master intercom station with AM/FM
Demolish and Install	Ea.	665	330.50		995.50	music and room monitoring. Wiring is
Reinstall	Ea.		193.12		193.12	not included.
Clean	Ea.	.21	27.50		27.71	
Minimum Charge	Job		241		241	
Intercom (remote)						
Demolish	Ea.		53.50		53.50	Includes material and labor to install
Install	Ea.	60.50	60.50		121	remote station for master intercom
Demolish and Install	Ea.	60.50	114		174.50	system. Wiring is not included.
Reinstall	Ea.		48.28		48.28	
Clean	Ea.	.21	10.30		10.51	
Minimum Charge	Job		241		241	
Security System						
Alarm Panel						
Demolish	Ea.		107		107	Includes material and labor to install a
Install	Ea.	161	241		402	6 zone burglar alarm panel.
Demolish and Install	Ea.	161	348		509	
Reinstall	Ea.		193.12		193.12	
Clean	Ea.	.21	10.30		10.51	
Minimum Charge	Job		241		241	

Finish Electrical

Low Voltage Systems		Unit	Material	Labor	Equip.	Total	Specification
Motion Detector							
	Demolish	Ea.		76.50		76.50	Includes material and labor to install
	Install	Ea.	231	210		441	an ultrasonic motion detector for a
	Demolish and Install	Ea.	231	286.50		517.50	security system.
	Reinstall	Ea.		167.93		167.93	
	Clean	Ea.	.21	10.30		10.51	
	Minimum Charge	Job		241		241	
Door Switch							
	Demolish	Ea.		33.50		33.50	Includes material and labor to install a
	Install	Ea.	135	91		226	door / window contact for a security
	Demolish and Install	Ea.	135	124.50		259.50	system.
	Reinstall	Ea.		72.88		72.88	
	Clean	Ea.	.21	3.30		3.51	
	Minimum Charge	Job		241		241	
Window Switch							
	Demolish	Ea.		33.50		33.50	Includes material and labor to install a
	Install	Ea.	135	91		226	door / window contact for a security
	Demolish and Install	Ea.	135	124.50		259.50	system.
	Reinstall	Ea.		72.88		72.88	
	Clean	Ea.	.21	3.30		3.51	
	Minimum Charge	Job		241		241	
Satellite System							
Dish							
	Demolish	Ea.		149		149	Includes material, labor and
	Install	Ea.	1850	400		2250	equipment to install 10' mesh dish.
	Demolish and Install	Ea.	1850	549		2399	
	Reinstall	Ea.		321.87		321.87	
	Clean	Ea.	4.07	41.50		45.57	
	Minimum Charge	Job		241		241	
Motor							
	Demolish	Ea.		74.50		74.50	Includes material and labor to install a
	Install	Ea.	285	201		486	television satellite dish motor unit.
	Demolish and Install	Ea.	285	275.50		560.50	
	Reinstall	Ea.		160.93		160.93	
	Clean	Ea.	.21	10.30		10.51	
	Minimum Charge	Job		241		241	
Television Antenna							
40' Pole							
	Demolish	Ea.		84		84	Includes material and labor to install
	Install	Ea.	21.50	151		172.50	stand-alone 1"-2" metal pole, 10' long.
	Demolish and Install	Ea.	21.50	235		256.50	Cost does not include antenna.
	Reinstall	Ea.		120.70		120.70	
	Clean	Ea.	.81	41.50		42.31	
	Paint	Ea.	9	47		56	
	Minimum Charge	Job		241		241	
Rotor Unit							
	Demolish	Ea.		67		67	Includes material and labor to install a
	Install	Ea.	55	60.50		115.50	television antenna rotor.
	Demolish and Install	Ea.	55	127.50		182.50	
	Reinstall	Ea.		48.28		48.28	
	Clean	Ea.	.81	41.50		42.31	
	Minimum Charge	Job		241		241	
Single Booster							
	Demolish	Ea.		67		67	Includes material and labor to install
	Install	Ea.	27.50	60.50		88	standard grade VHF television signal
	Demolish and Install	Ea.	27.50	127.50		155	booster unit and all hardware required
	Reinstall	Ea.		48.28		48.28	for connection to TV cable.
	Clean	Ea.	.04	5.15		5.19	
	Minimum Charge	Job		241		241	

For customer support on your Contractor's Pricing Guide: Residential Repair & Remodeling, call 888.606.7279.

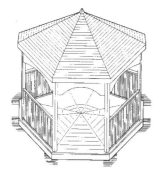

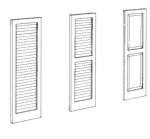

| Gazebo | Gutter | Shutter |

Window Treatment	Unit	Material	Labor	Equip.	Total	Specification
Drapery						
Good Grade						
Install	L.F.	46	6.60		52.60	Includes material and labor to install
Reinstall	L.F.		5.28		5.28	good quality drapery.
Clean	L.F.	1.10	.17		1.27	
Minimum Charge	Job		165		165	
Premium Grade						
Install	L.F.	76	6.60		82.60	Includes material and labor to install
Reinstall	L.F.		5.28		5.28	premium quality drapery.
Clean	L.F.	1.10	.18		1.28	
Minimum Charge	Job		165		165	
Lining						
Install	L.F.	25	6.60		31.60	Includes material and labor to install
Reinstall	L.F.		5.28		5.28	drapery lining.
Clean	L.F.	1.10	.09		1.19	
Minimum Charge	Job		165		165	
Valance Board						
Demolish	L.F.		1.83		1.83	Includes material and labor to install a
Install	L.F.	7.15	13.20		20.35	drapery valance board.
Demolish and Install	L.F.	7.15	15.03		22.18	
Reinstall	L.F.		10.56		10.56	
Clean	L.F.	1.10	.33		1.43	
Minimum Charge	Job		165		165	
Rod						
Demolish	L.F.		4.58		4.58	Includes material and labor to install
Install	L.F.	8.15	7.70		15.85	standard traverse drapery hardware.
Demolish and Install	L.F.	8.15	12.28		20.43	
Reinstall	L.F.		6.16		6.16	
Clean	L.F.	.08	.30		.38	
Paint	L.F.	.07	.73		.80	
Minimum Charge	Job		165		165	
Blinds						
Mini-type						
Demolish	S.F.		.99		.99	Includes material and labor to install
Install	S.F.	5.15	.77		5.92	horizontal 1" slat blinds.
Demolish and Install	S.F.	5.15	1.76		6.91	
Reinstall	S.F.		.62		.62	
Clean	S.F.	1.29	.03		1.32	
Minimum Charge	Job		165		165	

273

For customer support on your Contractor's Pricing Guide: Residential Repair & Remodeling, call 888.606.7279.

Window Treatment	Unit	Material	Labor	Equip.	Total	Specification
Vertical Type						
Demolish	S.F.		.99		.99	Includes material and labor to install
Install	S.F.	8.65	.99		9.64	vertical 3″ - 5″ cloth or PVC blinds.
Demolish and Install	S.F.	8.65	1.98		10.63	
Reinstall	S.F.		.79		.79	
Clean	S.F.	1.29	.03		1.32	
Minimum Charge	Job		165		165	
Vertical Type per L.F.						
Demolish	L.F.		1.53		1.53	Includes material and labor to install
Install	L.F.	51.50	4.40		55.90	vertical 3″ - 5″ cloth or PVC blinds x
Demolish and Install	L.F.	51.50	5.93		57.43	72″ long.
Reinstall	L.F.		3.52		3.52	
Clean	L.F.	7.70	.23		7.93	
Minimum Charge	Job		165		165	
Venetian Type						
Demolish	S.F.		.99		.99	Includes material and labor to install
Install	S.F.	6	.77		6.77	horizontal 2″ slat blinds.
Demolish and Install	S.F.	6	1.76		7.76	
Reinstall	S.F.		.62		.62	
Clean	S.F.	1.29	.03		1.32	
Minimum Charge	Job		165		165	

Wood Shutter

27″ x 36″, 4 Panel

	Unit	Material	Labor	Equip.	Total	Specification
Demolish	Ea.		31.50		31.50	Includes material and labor to install
Install	Pr.	105	26.50		131.50	movable louver, up to 33″ x 36″, 4
Demolish and Install	Pr.	105	58		163	panel shutter.
Reinstall	Pr.		21.36		21.36	
Clean	Ea.	.08	10.10		10.18	
Paint	Ea.	1.44	27		28.44	
Minimum Charge	Job		227		227	

33″ x 36″, 4 Panel

	Unit	Material	Labor	Equip.	Total	Specification
Demolish	Ea.		31.50		31.50	Includes material and labor to install
Install	Pr.	105	26.50		131.50	movable louver, up to 33″ x 36″, 4
Demolish and Install	Pr.	105	58		163	panel shutter.
Reinstall	Pr.		21.36		21.36	
Clean	Ea.	.08	10.10		10.18	
Paint	Ea.	1.50	29		30.50	
Minimum Charge	Job		227		227	

39″ x 36″, 4 Panel

	Unit	Material	Labor	Equip.	Total	Specification
Demolish	Ea.		31.50		31.50	Includes material and labor to install
Install	Pr.	113	26.50		139.50	movable louver, up to 47″ x 36″, 4
Demolish and Install	Pr.	113	58		171	panel shutter.
Reinstall	Pr.		21.36		21.36	
Clean	Ea.	.08	10.10		10.18	
Paint	Ea.	1.50	29		30.50	
Minimum Charge	Job		227		227	

47″ x 36″, 4 Panel

	Unit	Material	Labor	Equip.	Total	Specification
Demolish	Ea.		31.50		31.50	Includes material and labor to install
Install	Pr.	113	26.50		139.50	movable louver, up to 47″ x 36″, 4
Demolish and Install	Pr.	113	58		171	panel shutter.
Reinstall	Pr.		21.36		21.36	
Clean	Ea.	.08	10.10		10.18	
Paint	Ea.	1.56	31.50		33.06	
Minimum Charge	Job		227		227	

Improvements / Appliances / Treatments

Stairs

	Unit	Material	Labor	Equip.	Total	Specification
Concrete Steps						
Precast						
Demolish	Ea.		64.50	45.50	110	Includes material and labor to install
Install	Flight	470	156	42.50	668.50	precast concrete front entrance stairs,
Demolish and Install	Flight	470	220.50	88	778.50	4' wide, with 48" platform and 2
Clean	Ea.		4.58		4.58	risers.
Paint	Ea.	4.68	47		51.68	
Minimum Charge	Job		395		395	
Metal						
Concrete Filled						
Demolish	Ea.		18.35		18.35	Includes material and labor to install
Install	Riser	615	70.50	5.40	690.90	steel, cement filled metal pan stairs
Demolish and Install	Riser	615	88.85	5.40	709.25	with picket rail.
Clean	Ea.	.13	2.87		3	
Paint	Ea.	.70	2.80		3.50	
Minimum Charge	Job		262		262	
Landing						
Demolish	S.F.		4.58		4.58	Includes material and labor to install
Install	S.F.	130	8.30	.64	138.94	pre-erected conventional steel pan
Demolish and Install	S.F.	130	12.88	.64	143.52	landing.
Clean	S.F.	.13	.59		.72	
Paint	S.F.	.24	.80		1.04	
Minimum Charge	Job		262		262	
Railing						
Demolish	L.F.		2.53		2.53	Includes material and labor to install
Install	L.F.	17.05	12.05	.92	30.02	primed steel pipe.
Demolish and Install	L.F.	17.05	14.58	.92	32.55	
Clean	L.F.	.81	2.06		2.87	
Paint	L.F.	.87	6.30		7.17	
Minimum Charge	Job		262		262	

Elevator

	Unit	Material	Labor	Equip.	Total	Specification
2 Stop Residential						
Install	Ea.	12500	6925		19425	Includes labor, material and
Clean	Ea.	.98	20.50		21.48	equipment to install 2 stop elevator for
Minimum Charge	Job		690		690	residential use.
2 Stop Commercial						
Install	Ea.	43300	13800		57100	Includes labor, material and
Clean	Ea.	.98	20.50		21.48	equipment to install 2 stop elevator for
Minimum Charge	Job		690		690	apartment or commercial building with
3 Stop Commercial						2,000 lb/13 passenger capacity.
Install	Ea.	54000	21300		75300	Includes labor, material and
Clean	Ea.	.98	20.50		21.48	equipment to install 3 stop elevator for
Minimum Charge	Job		690		690	apartment or commercial building with
						2,000 lb/13 passenger capacity.
4 Stop Commercial						
Install	Ea.	64500	28800		93300	Includes labor, material and
Clean	Ea.	.98	20.50		21.48	equipment to install 4 stop elevator for
Minimum Charge	Job		690		690	apartment or commercial building with
						2,000 lb/13 passenger capacity.

Improvements / Appliances / Treatments

Elevator

	Unit	Material	Labor	Equip.	Total	Specification
5 Stop Commercial						
Install	Ea.	75000	36400		111400	Includes labor, material and
Clean	Ea.	.98	20.50		21.48	equipment to install 5 stop elevator for
Minimum Charge	Job		690		690	apartment or commercial building with
						2,000 lb/13 passenger capacity.

Rangetop

	Unit	Material	Labor	Equip.	Total	Specification
4-burner						
Demolish	Ea.		33.50		33.50	Cost includes material and labor to
Install	Ea.	385	80.50		465.50	install 4-burner electric surface unit.
Demolish and Install	Ea.	385	114		499	
Reinstall	Ea.		64.37		64.37	
Clean	Ea.	.07	30.50		30.57	
Minimum Charge	Job		82.50		82.50	
6-burner						
Demolish	Ea.		33.50		33.50	Includes material and labor to install a
Install	Ea.	2100	96.50		2196.50	6-burner cooktop unit.
Demolish and Install	Ea.	2100	130		2230	
Reinstall	Ea.		77.25		77.25	
Clean	Ea.	.07	30.50		30.57	
Minimum Charge	Job		82.50		82.50	
Premium Brand w / Grill						
Demolish	Ea.		33.50		33.50	Includes material and labor to install
Install	Ea.	1250	96.50		1346.50	electric downdraft cooktop with grill.
Demolish and Install	Ea.	1250	130		1380	
Reinstall	Ea.		77.25		77.25	
Clean	Ea.	.07	30.50		30.57	
Minimum Charge	Job		82.50		82.50	

Rangehood

	Unit	Material	Labor	Equip.	Total	Specification
Vented						
30″						
Demolish	Ea.		30.50		30.50	Includes material and labor to install a
Install	Ea.	91.50	96.50		188	vented range hood.
Demolish and Install	Ea.	91.50	127		218.50	
Reinstall	Ea.		77.25		77.25	
Clean	Ea.	.07	8.65		8.72	
Minimum Charge	Job		82.50		82.50	
36″						
Demolish	Ea.		30.50		30.50	Includes material and labor to install a
Install	Ea.	130	96.50		226.50	vented range hood.
Demolish and Install	Ea.	130	127		257	
Reinstall	Ea.		77.25		77.25	
Clean	Ea.	.07	8.65		8.72	
Minimum Charge	Job		82.50		82.50	
42″						
Demolish	Ea.		30.50		30.50	Includes material and labor to install a
Install	Ea.	183	115		298	vented range hood.
Demolish and Install	Ea.	183	145.50		328.50	
Reinstall	Ea.		91.95		91.95	
Clean	Ea.	.07	8.65		8.72	
Minimum Charge	Job		82.50		82.50	

For customer support on your Contractor's Pricing Guide: Residential Repair & Remodeling, call 888.606.7279.

Improvements / Appliances / Treatments

Rangehood

Rangehood	Unit	Material	Labor	Equip.	Total	Specification
Ventless						
30″						
Demolish	Ea.		30.50		30.50	Includes material and labor to install a
Install	Ea.	109	60.50		169.50	ventless range hood.
Demolish and Install	Ea.	109	91		200	
Reinstall	Ea.		48.28		48.28	
Clean	Ea.	.07	8.65		8.72	
Minimum Charge	Job		82.50		82.50	
36″						
Demolish	Ea.		30.50		30.50	Includes material and labor to install a
Install	Ea.	215	96.50		311.50	ventless range hood.
Demolish and Install	Ea.	215	127		342	
Reinstall	Ea.		77.25		77.25	
Clean	Ea.	.07	8.65		8.72	
Minimum Charge	Job		82.50		82.50	
42″						
Demolish	Ea.		30.50		30.50	Includes material and labor to install a
Install	Ea.	206	80.50		286.50	ventless range hood.
Demolish and Install	Ea.	206	111		317	
Reinstall	Ea.		64.37		64.37	
Clean	Ea.	.07	8.65		8.72	
Minimum Charge	Job		82.50		82.50	

Oven

Oven	Unit	Material	Labor	Equip.	Total	Specification
Free Standing						
Demolish	Ea.		33.50		33.50	Cost includes material and labor to
Install	Ea.	615	66		681	install 30″ free standing electric range
Demolish and Install	Ea.	615	99.50		714.50	with self-cleaning oven.
Reinstall	Ea.		52.80		52.80	
Clean	Ea.	.07	30.50		30.57	
Minimum Charge	Job		82.50		82.50	
Single Wall Oven						
Demolish	Ea.		33.50		33.50	Includes material and labor to install a
Install	Ea.	990	121		1111	built-in oven.
Demolish and Install	Ea.	990	154.50		1144.50	
Reinstall	Ea.		96.56		96.56	
Clean	Ea.	.07	30.50		30.57	
Minimum Charge	Job		82.50		82.50	
Double Wall Oven						
Demolish	Ea.		33.50		33.50	Includes material and labor to install a
Install	Ea.	1475	227		1702	conventional double oven.
Demolish and Install	Ea.	1475	260.50		1735.50	
Reinstall	Ea.		181.60		181.60	
Clean	Ea.	.07	30.50		30.57	
Minimum Charge	Job		82.50		82.50	
Drop-in Type w / Range						
Demolish	Ea.		33.50		33.50	Includes material and labor to install a
Install	Ea.	990	80.50		1070.50	built-in cooking range with oven.
Demolish and Install	Ea.	990	114		1104	
Reinstall	Ea.		64.37		64.37	
Clean	Ea.	.07	30.50		30.57	
Minimum Charge	Job		82.50		82.50	

Improvements / Appliances / Treatments

Oven

Oven	Unit	Material	Labor	Equip.	Total	Specification
Hi/Lo w / Range, Microwave						
Demolish	Ea.		33.50		33.50	Includes material and labor to install a
Install	Ea.	2775	455		3230	double oven, one conventional, one
Demolish and Install	Ea.	2775	488.50		3263.50	microwave.
Reinstall	Ea.		363.20		363.20	
Clean	Ea.	.07	30.50		30.57	
Minimum Charge	Job		82.50		82.50	
Hi/Lo w/ Range / Double Oven						
Demolish	Ea.		33.50		33.50	Includes material and labor to install a
Install	Ea.	1475	227		1702	conventional double oven.
Demolish and Install	Ea.	1475	260.50		1735.50	
Reinstall	Ea.		181.60		181.60	
Clean	Ea.	.07	30.50		30.57	
Minimum Charge	Job		82.50		82.50	
Countertop Microwave						
Demolish	Ea.		11.45		11.45	Includes material and labor to install a
Install	Ea.	164	41.50		205.50	1.0 cubic foot microwave oven.
Demolish and Install	Ea.	164	52.95		216.95	
Reinstall	Ea.		33		33	
Clean	Ea.	.07	15.20		15.27	
Minimum Charge	Job		82.50		82.50	
Cabinet / Wall Mounted Microwave						
Demolish	Ea.		23		23	Includes material and labor to install a
Install	Ea.	760	121		881	space saver microwave oven.
Demolish and Install	Ea.	760	144		904	
Reinstall	Ea.		96.56		96.56	
Clean	Ea.	.07	15.20		15.27	
Minimum Charge	Job		82.50		82.50	

Dryer

Dryer	Unit	Material	Labor	Equip.	Total	Specification
Basic Grade						
Demolish	Ea.		15.30		15.30	Includes material and labor to install
Install	Ea.	495	263		758	electric or gas dryer. Vent kit not
Demolish and Install	Ea.	495	278.30		773.30	included.
Reinstall	Ea.		210.26		210.26	
Clean	Ea.	.07	30.50		30.57	
Minimum Charge	Job		82.50		82.50	
Good Grade						
Demolish	Ea.		15.30		15.30	Includes material and labor to install
Install	Ea.	615	263		878	electric or gas dryer. Vent kit not
Demolish and Install	Ea.	615	278.30		893.30	included.
Reinstall	Ea.		210.26		210.26	
Clean	Ea.	.07	30.50		30.57	
Minimum Charge	Job		82.50		82.50	
Premium Grade						
Demolish	Ea.		15.30		15.30	Includes material and labor to install
Install	Ea.	1575	395		1970	electric or gas dryer. Vent kit not
Demolish and Install	Ea.	1575	410.30		1985.30	included.
Reinstall	Ea.		315.39		315.39	
Clean	Ea.	.07	30.50		30.57	
Minimum Charge	Job		82.50		82.50	

For customer support on your Contractor's Pricing Guide: Residential Repair & Remodeling, call 888.606.7279.

Improvements / Appliances / Treatments

Dryer

	Unit	Material	Labor	Equip.	Total	Specification
Washer / Dryer Combination Unit						
Demolish	Ea.		15.30		15.30	Includes material and labor to install
Install	Ea.	1550	66		1616	electric washer/dryer combination
Demolish and Install	Ea.	1550	81.30		1631.30	unit, 27" wide, stackable or unitized.
Reinstall	Ea.		52.80		52.80	
Clean	Ea.	.07	30.50		30.57	
Minimum Charge	Job		82.50		82.50	

Washer

	Unit	Material	Labor	Equip.	Total	Specification
Basic Grade						
Demolish	Ea.		15.30		15.30	Includes material and labor to install
Install	Ea.	600	172		772	washing machine. New hoses not
Demolish and Install	Ea.	600	187.30		787.30	included.
Reinstall	Ea.		137.39		137.39	
Clean	Ea.	.07	30.50		30.57	
Minimum Charge	Job		82.50		82.50	
Good Grade						
Demolish	Ea.		15.30		15.30	Includes material and labor to install
Install	Ea.	985	258		1243	washing machine. New hoses not
Demolish and Install	Ea.	985	273.30		1258.30	included.
Reinstall	Ea.		206.08		206.08	
Clean	Ea.	.07	30.50		30.57	
Minimum Charge	Job		82.50		82.50	
Premium Grade						
Demolish	Ea.		15.30		15.30	Includes material and labor to install
Install	Ea.	1825	515		2340	washing machine with digital
Demolish and Install	Ea.	1825	530.30		2355.30	computer readouts and large capacity.
Reinstall	Ea.		412.16		412.16	New hoses not included.
Clean	Ea.	.07	30.50		30.57	
Minimum Charge	Job		82.50		82.50	
Washer / Dryer Combination Unit						
Demolish	Ea.		15.30		15.30	Includes material and labor to install
Install	Ea.	1550	66		1616	electric washer/dryer combination
Demolish and Install	Ea.	1550	81.30		1631.30	unit, 27" wide, stackable or unitized.
Reinstall	Ea.		52.80		52.80	
Clean	Ea.	.07	30.50		30.57	
Minimum Charge	Job		82.50		82.50	

Refrigerator

	Unit	Material	Labor	Equip.	Total	Specification
12 Cubic Foot						
Demolish	Ea.		46		46	Includes material and labor to install a
Install	Ea.	525	66		591	refrigerator.
Demolish and Install	Ea.	525	112		637	
Reinstall	Ea.		52.80		52.80	
Clean	Ea.	.07	30.50		30.57	
Minimum Charge	Job		82.50		82.50	
16 Cubic Foot						
Demolish	Ea.		46		46	Includes material and labor to install a
Install	Ea.	640	73.50		713.50	refrigerator.
Demolish and Install	Ea.	640	119.50		759.50	
Reinstall	Ea.		58.67		58.67	
Clean	Ea.	.07	30.50		30.57	
Minimum Charge	Job		82.50		82.50	

For customer support on your Contractor's Pricing Guide: Residential Repair & Remodeling, call 888.606.7279.

Improvements / Appliances / Treatments

Refrigerator	Unit	Material	Labor	Equip.	Total	Specification
18 Cubic Foot						
Demolish	Ea.		46		46	Includes material and labor to install a refrigerator.
Install	Ea.	825	82.50		907.50	
Demolish and Install	Ea.	825	128.50		953.50	
Reinstall	Ea.		66		66	
Clean	Ea.	.07	30.50		30.57	
Minimum Charge	Job		82.50		82.50	
21 Cubic Foot						
Demolish	Ea.		46		46	Includes material and labor to install a refrigerator.
Install	Ea.	1275	94.50		1369.50	
Demolish and Install	Ea.	1275	140.50		1415.50	
Reinstall	Ea.		75.43		75.43	
Clean	Ea.	.07	30.50		30.57	
Minimum Charge	Job		82.50		82.50	
24 Cubic Foot						
Demolish	Ea.		46		46	Includes material and labor to install a refrigerator.
Install	Ea.	1675	94.50		1769.50	
Demolish and Install	Ea.	1675	140.50		1815.50	
Reinstall	Ea.		75.43		75.43	
Clean	Ea.	.07	30.50		30.57	
Minimum Charge	Job		82.50		82.50	
27 Cubic Foot						
Demolish	Ea.		46		46	Includes material and labor to install a refrigerator.
Install	Ea.	2750	66		2816	
Demolish and Install	Ea.	2750	112		2862	
Reinstall	Ea.		52.80		52.80	
Clean	Ea.	.07	30.50		30.57	
Minimum Charge	Job		82.50		82.50	
Ice Maker						
Install	Ea.	87	64.50		151.50	Includes labor and material to install an automatic ice maker in a refrigerator.
Minimum Charge	Job		82.50		82.50	

Freezer	Unit	Material	Labor	Equip.	Total	Specification
15 Cubic Foot						
Demolish	Ea.		46		46	Cost includes material and labor to install 15 cubic foot freezer.
Install	Ea.	625	66		691	
Demolish and Install	Ea.	625	112		737	
Reinstall	Ea.		52.80		52.80	
Clean	Ea.	.07	30.50		30.57	
Minimum Charge	Job		82.50		82.50	
18 Cubic Foot						
Demolish	Ea.		46		46	Cost includes material and labor to install 18 cubic foot freezer.
Install	Ea.	830	66		896	
Demolish and Install	Ea.	830	112		942	
Reinstall	Ea.		52.80		52.80	
Clean	Ea.	.07	30.50		30.57	
Minimum Charge	Job		82.50		82.50	
21 Cubic Foot						
Demolish	Ea.		46		46	Includes material and labor to install a 15 to 23 cubic foot deep freezer.
Install	Ea.	825	66		891	
Demolish and Install	Ea.	825	112		937	
Reinstall	Ea.		52.80		52.80	
Clean	Ea.	.07	30.50		30.57	
Minimum Charge	Job		82.50		82.50	

Improvements / Appliances / Treatments

Freezer

	Unit	Material	Labor	Equip.	Total	Specification
24 Cubic Foot						
Demolish	Ea.		46		46	Includes material and labor to install a
Install	Ea.	920	132		1052	24 cubic foot deep freezer.
Demolish and Install	Ea.	920	178		1098	
Reinstall	Ea.		105.60		105.60	
Clean	Ea.	.07	30.50		30.57	
Minimum Charge	Job		82.50		82.50	
Chest-type						
Demolish	Ea.		46		46	Cost includes material and labor to
Install	Ea.	625	66		691	install 15 cubic foot freezer.
Demolish and Install	Ea.	625	112		737	
Clean	Ea.	.07	30.50		30.57	
Minimum Charge	Job		82.50		82.50	

Dishwasher

	Unit	Material	Labor	Equip.	Total	Specification
4 Cycle						
Demolish	Ea.		36.50		36.50	Includes material and labor to install
Install	Ea.	470	159		629	good grade automatic dishwasher.
Demolish and Install	Ea.	470	195.50		665.50	
Reinstall	Ea.		127.18		127.18	
Clean	Ea.	.07	30.50		30.57	
Minimum Charge	Job		82.50		82.50	

Garbage Disposal

	Unit	Material	Labor	Equip.	Total	Specification
In Sink Unit						
Demolish	Ea.		36		36	Cost includes material and labor to
Install	Ea.	188	63.50		251.50	install 1/2 HP custom disposal unit.
Demolish and Install	Ea.	188	99.50		287.50	
Reinstall	Ea.		50.87		50.87	
Clean	Ea.	.03	8.65		8.68	
Minimum Charge	Job		82.50		82.50	

Trash Compactor

	Unit	Material	Labor	Equip.	Total	Specification
4 to 1 Compaction						
Demolish	Ea.		33.50		33.50	Includes material and labor to install a
Install	Ea.	1000	91		1091	trash compactor.
Demolish and Install	Ea.	1000	124.50		1124.50	
Reinstall	Ea.		72.64		72.64	
Clean	Ea.	.07	20.50		20.57	
Minimum Charge	Job		82.50		82.50	

Kitchenette Unit

	Unit	Material	Labor	Equip.	Total	Specification
Range, Refrigerator and Sink						
Demolish	Ea.		134		134	Includes material and labor to install
Install	Ea.	1650	530		2180	range/refrigerator/sink unit, 72"
Demolish and Install	Ea.	1650	664		2314	wide, base unit only.
Reinstall	Ea.		423.93		423.93	
Clean	Ea.	1.02	15.20		16.22	
Minimum Charge	Job		82.50		82.50	

For customer support on your Contractor's Pricing Guide: Residential Repair & Remodeling, call 888.606.7279.

Improvements / Appliances / Treatments

Bath Accessories		Unit	Material	Labor	Equip.	Total	Specification
Towel Bar							
	Demolish	Ea.		7.65		7.65	Includes material and labor to install
	Install	Ea.	54	18.90		72.90	towel bar up to 24" long.
	Demolish and Install	Ea.	54	26.55		80.55	
	Reinstall	Ea.		15.13		15.13	
	Clean	Ea.		4.13		4.13	
	Minimum Charge	Job		227		227	
Toothbrush Holder							
	Demolish	Ea.		7.65		7.65	Includes material and labor to install a
	Install	Ea.	14.85	22.50		37.35	surface mounted tumbler and
	Demolish and Install	Ea.	14.85	30.15		45	toothbrush holder.
	Reinstall	Ea.		18.16		18.16	
	Clean	Ea.		4.13		4.13	
	Minimum Charge	Job		227		227	
Grab Rail							
	Demolish	Ea.		11.45		11.45	Includes material and labor to install a
	Install	Ea.	36	19.75		55.75	24" long grab bar. Blocking is not
	Demolish and Install	Ea.	36	31.20		67.20	included.
	Reinstall	Ea.		15.79		15.79	
	Clean	Ea.	.41	6.90		7.31	
	Minimum Charge	Job		227		227	
Soap Dish							
	Demolish	Ea.		7.65		7.65	Includes material and labor to install a
	Install	Ea.	6.45	22.50		28.95	surface mounted soap dish.
	Demolish and Install	Ea.	6.45	30.15		36.60	
	Reinstall	Ea.		18.16		18.16	
	Clean	Ea.		4.13		4.13	
	Minimum Charge	Job		227		227	
Toilet Paper Roller							
	Demolish	Ea.		7.65		7.65	Includes material and labor to install a
	Install	Ea.	21	15.15		36.15	surface mounted toilet tissue dispenser.
	Demolish and Install	Ea.	21	22.80		43.80	
	Reinstall	Ea.		12.11		12.11	
	Clean	Ea.	.41	10.30		10.71	
	Minimum Charge	Job		227		227	
Dispenser							
Soap							
	Demolish	Ea.		10.20		10.20	Includes material and labor to install a
	Install	Ea.	55	22.50		77.50	surface mounted soap dispenser.
	Demolish and Install	Ea.	55	32.70		87.70	
	Reinstall	Ea.		18.16		18.16	
	Clean	Ea.	.41	10.30		10.71	
	Minimum Charge	Job		227		227	
Towel							
	Demolish	Ea.		11.45		11.45	Includes material and labor to install a
	Install	Ea.	52.50	28.50		81	surface mounted towel dispenser.
	Demolish and Install	Ea.	52.50	39.95		92.45	
	Reinstall	Ea.		22.70		22.70	
	Clean	Ea.	.41	13.75		14.16	
	Minimum Charge	Job		227		227	
Seat Cover							
	Demolish	Ea.		7.65		7.65	Includes material and labor to install a
	Install	Ea.	38.50	30.50		69	surface mounted toilet seat cover
	Demolish and Install	Ea.	38.50	38.15		76.65	dispenser.
	Reinstall	Ea.		24.21		24.21	
	Clean	Ea.	.41	10.30		10.71	
	Minimum Charge	Job		227		227	

Improvements / Appliances / Treatments

Bath Accessories

Bath Accessories		Unit	Material	Labor	Equip.	Total	Specification
Sanitary Napkin							
	Demolish	Ea.		19.30		19.30	Includes material and labor to install a
	Install	Ea.	38	48		86	surface mounted sanitary napkin
	Demolish and Install	Ea.	38	67.30		105.30	dispenser.
	Reinstall	Ea.		38.23		38.23	
	Clean	Ea.	.21	13.75		13.96	
	Minimum Charge	Job		227		227	
Partition							
Urinal							
	Demolish	Ea.		23		23	Includes material and labor to install a
	Install	Ea.	253	114		367	urinal screen.
	Demolish and Install	Ea.	253	137		390	
	Reinstall	Ea.		90.80		90.80	
	Clean	Ea.	.62	9.15		9.77	
	Minimum Charge	Job		227		227	
Toilet							
	Demolish	Ea.		68		68	Includes material and labor to install
	Install	Ea.	625	182		807	baked enamel standard size partition,
	Demolish and Install	Ea.	625	250		875	floor or ceiling mounted with one wall
	Reinstall	Ea.		145.28		145.28	and one door.
	Clean	Ea.	4.07	20.50		24.57	
	Minimum Charge	Job		227		227	
Handicap							
	Demolish	Ea.		68		68	Includes material and labor to install
	Install	Ea.	1025	182		1207	baked enamel handicap partition,
	Demolish and Install	Ea.	1025	250		1275	floor or ceiling mounted with one wall
	Reinstall	Ea.		145.28		145.28	and one door.
	Clean	Ea.	4.07	20.50		24.57	
	Minimum Charge	Job		227		227	

Fire Prevention / Protection

Fire Prevention / Protection		Unit	Material	Labor	Equip.	Total	Specification
Fire Extinguisher							
5# Carbon Dioxide							
	Install	Ea.	220			220	Includes labor and material to install one wall mounted 5 lb. carbon dioxide factory charged unit, complete with hose, horn and wall mounting bracket.
10# Carbon Dioxide							
	Install	Ea.	315			315	Includes labor and material to install one wall mounted 10 lb. carbon dioxide factory charged unit, complete with hose, horn and wall mounting bracket.
15# Carbon Dioxide							
	Install	Ea.	405			405	Includes labor and material to install one wall mounted 15 lb. carbon dioxide factory charged unit, complete with hose, horn and wall mounting bracket.

Improvements / Appliances / Treatments

Fire Prevention / Protection	Unit	Material	Labor	Equip.	Total	Specification
5# Dry Chemical						
Install	Ea.	58.50			58.50	Includes labor and material to install one wall mounted 5 lb. dry chemical factory charged unit, complete with hose, horn and wall mounting bracket.
20# Dry Chemical						
Install	Ea.	149			149	Includes labor and material to install one wall mounted 20 lb. dry chemical factory charged unit, complete with hose, horn and wall mounting bracket.

Gutters	Unit	Material	Labor	Equip.	Total	Specification
Galvanized						
Demolish	L.F.		1.53		1.53	Cost includes material and labor to install 5" galvanized steel gutter with enamel finish, fittings, hangers, and corners.
Install	L.F.	2.40	4.04		6.44	
Demolish and Install	L.F.	2.40	5.57		7.97	
Reinstall	L.F.		3.23		3.23	
Clean	L.F.		.44		.44	
Paint	L.F.	.24	1.16		1.40	
Minimum Charge	Job		252		252	
Aluminum						
Demolish	L.F.		1.53		1.53	Cost includes material and labor to install 5" aluminum gutter with enamel finish, fittings, hangers, and corners.
Install	L.F.	3.09	4.04		7.13	
Demolish and Install	L.F.	3.09	5.57		8.66	
Reinstall	L.F.		3.23		3.23	
Clean	L.F.		.44		.44	
Paint	L.F.	.24	1.16		1.40	
Minimum Charge	Job		252		252	
Copper						
Demolish	L.F.		1.53		1.53	Cost includes material and labor to install 5" half-round copper gutter with fittings, hangers, and corners.
Install	L.F.	8	4.04		12.04	
Demolish and Install	L.F.	8	5.57		13.57	
Reinstall	L.F.		3.23		3.23	
Clean	L.F.		.44		.44	
Minimum Charge	Job		252		252	
Plastic						
Demolish	L.F.		1.53		1.53	Cost includes material and labor to install 4" white vinyl gutter fittings, hangers, and corners.
Install	L.F.	1.43	3.95		5.38	
Demolish and Install	L.F.	1.43	5.48		6.91	
Reinstall	L.F.		3.16		3.16	
Clean	L.F.		.44		.44	
Paint	L.F.	.24	1.16		1.40	
Minimum Charge	Job		252		252	

Downspouts	Unit	Material	Labor	Equip.	Total	Specification
Galvanized						
Demolish	L.F.		1.05		1.05	Cost includes material and labor to install 4" diameter downspout with enamel finish.
Install	L.F.	2.77	3.48		6.25	
Demolish and Install	L.F.	2.77	4.53		7.30	
Reinstall	L.F.		2.78		2.78	
Clean	L.F.		.44		.44	
Paint	L.F.	.24	1.16		1.40	
Minimum Charge	Job		252		252	

Improvements / Appliances / Treatments

Downspouts

Downspouts	Unit	Material	Labor	Equip.	Total	Specification
Aluminum						
Demolish	L.F.		1.05		1.05	Includes material and labor to install
Install	L.F.	2.61	3.60		6.21	aluminum enameled downspout.
Demolish and Install	L.F.	2.61	4.65		7.26	
Reinstall	L.F.		2.88		2.88	
Clean	L.F.		.44		.44	
Paint	L.F.	.24	1.16		1.40	
Minimum Charge	Job		252		252	
Copper						
Demolish	L.F.		1.05		1.05	Includes material and labor to install
Install	L.F.	8.80	2.66		11.46	round copper downspout.
Demolish and Install	L.F.	8.80	3.71		12.51	
Reinstall	L.F.		2.12		2.12	
Clean	L.F.		.44		.44	
Paint	L.F.	.24	1.16		1.40	
Minimum Charge	Job		252		252	
Plastic						
Demolish	L.F.		1.05		1.05	Cost includes material and labor to
Install	L.F.	2.29	2.40		4.69	install 2" x 3" vinyl rectangular
Demolish and Install	L.F.	2.29	3.45		5.74	downspouts.
Reinstall	L.F.		1.92		1.92	
Clean	L.F.		.44		.44	
Paint	L.F.	.24	1.16		1.40	
Minimum Charge	Job		252		252	

Shutters

Shutters	Unit	Material	Labor	Equip.	Total	Specification
Exterior						
25" High						
Demolish	Ea.		11.45		11.45	Includes material and labor to install
Install	Pr.	193	45.50		238.50	pair of exterior shutters 25" x 16",
Demolish and Install	Pr.	193	56.95		249.95	unpainted, installed as fixed shutters
Reinstall	Pr.		45.40		45.40	on concrete or wood frame.
Clean	Ea.	.08	10.10		10.18	
Paint	Ea.	1.90	20.50		22.40	
Minimum Charge	Job		227		227	
39" High						
Demolish	Ea.		11.45		11.45	Includes material and labor to install
Install	Pr.	251	45.50		296.50	pair of shutters 39" x 16", unpainted,
Demolish and Install	Pr.	251	56.95		307.95	installed as fixed shutters on concrete
Reinstall	Pr.		45.40		45.40	or wood frame.
Clean	Ea.	.08	10.10		10.18	
Paint	Ea.	2.96	27.50		30.46	
Minimum Charge	Job		227		227	
51" High						
Demolish	Ea.		11.45		11.45	Includes material and labor to install
Install	Pr.	305	45.50		350.50	pair of shutters 51" x 16", unpainted,
Demolish and Install	Pr.	305	56.95		361.95	installed as fixed shutters on concrete
Reinstall	Pr.		45.40		45.40	or wood frame.
Clean	Ea.	.08	10.10		10.18	
Paint	Ea.	3.85	28.50		32.35	
Minimum Charge	Job		227		227	

Shutters	Unit	Material	Labor	Equip.	Total	Specification
59" High						
Demolish	Ea.		11.45		11.45	Includes material and labor to install
Install	Pr.	385	45.50		430.50	pair of shutters 59" x 16", unpainted,
Demolish and Install	Pr.	385	56.95		441.95	installed as fixed shutters on concrete
Reinstall	Pr.		45.40		45.40	or wood frame.
Clean	Ea.	.08	10.10		10.18	
Paint	Ea.	4.44	31		35.44	
Minimum Charge	Job		227		227	
67" High						
Demolish	Ea.		11.45		11.45	Includes material and labor to install
Install	Pr.	420	50.50		470.50	pair of shutters 67" x 16", unpainted,
Demolish and Install	Pr.	420	61.95		481.95	installed as fixed shutters on concrete
Reinstall	Pr.		50.44		50.44	or wood frame.
Clean	Ea.	.08	10.10		10.18	
Paint	Ea.	5.10	35		40.10	
Minimum Charge	Job		227		227	
Interior Wood						
23" x 24", 2 Panel						
Demolish	Ea.		11.45		11.45	Cost includes material and labor to
Install	Ea.	108	26.50		134.50	install 23" x 24", 4 panel interior
Demolish and Install	Ea.	108	37.95		145.95	movable pine shutter, 1-1/4" louver,
Reinstall	Ea.		21.36		21.36	knobs and hooks and hinges.
Clean	Ea.	.08	10.10		10.18	
Paint	Ea.	1.44	27		28.44	
Minimum Charge	Job		227		227	
27" x 24", 4 Panel						
Demolish	Ea.		11.45		11.45	Cost includes material and labor to
Install	Ea.	114	26.50		140.50	install 27" x 24", 4 panel interior
Demolish and Install	Ea.	114	37.95		151.95	movable pine shutter, 1-1/4" louver,
Reinstall	Ea.		21.36		21.36	knobs and hooks and hinges.
Clean	Ea.	.08	10.10		10.18	
Paint	Ea.	1.44	27		28.44	
Minimum Charge	Job		227		227	
31" x 24", 4 Panel						
Demolish	Ea.		11.45		11.45	Cost includes material and labor to
Install	Ea.	127	26.50		153.50	install 31" x 24", 4 panel interior
Demolish and Install	Ea.	127	37.95		164.95	movable pine shutter, 1-1/4" louver,
Reinstall	Ea.		21.36		21.36	knobs and hooks and hinges.
Clean	Ea.	.08	10.10		10.18	
Paint	Ea.	1.44	27		28.44	
Minimum Charge	Job		227		227	
35" x 24", 4 Panel						
Demolish	Ea.		11.45		11.45	Cost includes material and labor to
Install	Ea.	135	26.50		161.50	install 35" x 24", 4 panel interior
Demolish and Install	Ea.	135	37.95		172.95	movable pine shutter, 1-1/4" louver,
Reinstall	Ea.		21.36		21.36	knobs and hooks and hinges.
Clean	Ea.	.08	10.10		10.18	
Paint	Ea.	1.44	27		28.44	
Minimum Charge	Job		227		227	
39" x 24", 4 Panel						
Demolish	Ea.		11.45		11.45	Cost includes material and labor to
Install	Ea.	151	26.50		177.50	install 39" x 24", 4 panel interior
Demolish and Install	Ea.	151	37.95		188.95	movable pine shutter, 1-1/4" louver,
Reinstall	Ea.		21.36		21.36	knobs and hooks and hinges.
Clean	Ea.	.08	10.10		10.18	
Paint	Ea.	1.44	27		28.44	
Minimum Charge	Job		227		227	

Improvements / Appliances / Treatments

Shutters	Unit	Material	Labor	Equip.	Total	Specification
47" x 24", 4 Panel						
Demolish	Ea.		11.45		11.45	Cost includes material and labor to
Install	Ea.	171	26.50		197.50	install 47" x 24", 4 panel interior
Demolish and Install	Ea.	171	37.95		208.95	movable pine shutter, 1-1/4" louver,
Reinstall	Ea.		21.36		21.36	knobs and hooks and hinges.
Clean	Ea.	.08	10.10		10.18	
Paint	Ea.	1.50	29		30.50	
Minimum Charge	Job		227		227	

Awning	Unit	Material	Labor	Equip.	Total	Specification
Standard Aluminum						
36" Wide x 30" Deep						
Demolish	Ea.		11.45		11.45	Includes material and labor to install
Install	Ea.	185	38		223	one aluminum awning with
Demolish and Install	Ea.	185	49.45		234.45	weather-resistant finish.
Reinstall	Ea.		30.27		30.27	
Clean	Ea.		13.75		13.75	
Minimum Charge	Job		227		227	
48" Wide x 30" Deep						
Demolish	Ea.		11.45		11.45	Includes material and labor to install
Install	Ea.	216	45.50		261.50	one aluminum awning with
Demolish and Install	Ea.	216	56.95		272.95	weather-resistant finish.
Reinstall	Ea.		36.32		36.32	
Clean	Ea.		13.75		13.75	
Minimum Charge	Job		227		227	
60" Wide x 30" Deep						
Demolish	Ea.		23		23	Includes material and labor to install
Install	Ea.	252	50.50		302.50	one aluminum awning with
Demolish and Install	Ea.	252	73.50		325.50	weather-resistant finish.
Reinstall	Ea.		40.36		40.36	
Clean	Ea.		13.75		13.75	
Minimum Charge	Job		227		227	
72" Wide x 30" Deep						
Demolish	Ea.		23		23	Includes material and labor to install
Install	Ea.	288	57		345	one aluminum awning with
Demolish and Install	Ea.	288	80		368	weather-resistant finish.
Reinstall	Ea.		45.40		45.40	
Clean	Ea.		13.75		13.75	
Minimum Charge	Job		227		227	
Roll-up Type						
Demolish	S.F.		.83		.83	Includes material and labor to install
Install	S.F.	15.20	4.54		19.74	an aluminum roll-up awning.
Demolish and Install	S.F.	15.20	5.37		20.57	
Reinstall	S.F.		3.63		3.63	
Clean	S.F.	.04	.69		.73	
Minimum Charge	Job		227		227	
Canvas						
30" Wide						
Demolish	Ea.		6.90		6.90	Includes material and labor to install a
Install	Ea.	202	38		240	canvas window awning.
Demolish and Install	Ea.	202	44.90		246.90	
Reinstall	Ea.		30.27		30.27	
Clean	Ea.		4.13		4.13	
Minimum Charge	Job		227		227	

Improvements / Appliances / Treatments

Awning	Unit	Material	Labor	Equip.	Total	Specification
36" Wide						
Demolish	Ea.		6.90		6.90	Includes material and labor to install a
Install	Ea.	223	45.50		268.50	canvas window awning.
Demolish and Install	Ea.	223	52.40		275.40	
Reinstall	Ea.		36.32		36.32	
Clean	Ea.		4.52		4.52	
Minimum Charge	Job		227		227	
42" Wide						
Demolish	Ea.		6.90		6.90	Includes material and labor to install a
Install	Ea.	237	57		294	canvas window awning.
Demolish and Install	Ea.	237	63.90		300.90	
Reinstall	Ea.		45.40		45.40	
Clean	Ea.		5.35		5.35	
Minimum Charge	Job		227		227	
48" Wide						
Demolish	Ea.		6.90		6.90	Includes material and labor to install a
Install	Ea.	249	65		314	canvas window awning.
Demolish and Install	Ea.	249	71.90		320.90	
Reinstall	Ea.		51.89		51.89	
Clean	Ea.		6.25		6.25	
Minimum Charge	Job		227		227	
Cloth Canopy Cover						
8' x 10'						
Demolish	Ea.		385		385	Includes material and labor to install
Install	Ea.	385	227		612	one 8' x 10' cloth canopy patio cover
Demolish and Install	Ea.	385	612		997	with front bar and tension support
Reinstall	Ea.		181.60		181.60	rafters and 9" valance.
Clean	Ea.	3.26	55		58.26	
Minimum Charge	Job		455		455	
8' x 15'						
Demolish	Ea.		385		385	Includes material and labor to install
Install	Ea.	530	227		757	one 8' x 15' cloth canopy patio cover
Demolish and Install	Ea.	530	612		1142	with front bar and tension support
Reinstall	Ea.		181.60		181.60	rafters and 9" valance.
Clean	Ea.	3.26	82.50		85.76	
Minimum Charge	Job		455		455	

Vent	Unit	Material	Labor	Equip.	Total	Specification
Small Roof Turbine						
Demolish	Ea.		11.45		11.45	Includes material and labor to install
Install	Ea.	52.50	45.50		98	spinner ventilator, wind driven,
Demolish and Install	Ea.	52.50	56.95		109.45	galvanized, 4" neck diam, 180 CFM.
Reinstall	Ea.		36.32		36.32	
Clean	Ea.		7.35		7.35	
Minimum Charge	Job		258		258	
Large Roof Turbine						
Demolish	Ea.		9.15		9.15	Includes material and labor to install
Install	Ea.	61.50	65		126.50	spinner ventilator, wind driven,
Demolish and Install	Ea.	61.50	74.15		135.65	galvanized, 8" neck diam, 360 CFM.
Reinstall	Ea.		51.89		51.89	
Clean	Ea.		8.25		8.25	
Minimum Charge	Job		258		258	

Improvements / Appliances / Treatments

Vent		Unit	Material	Labor	Equip.	Total	Specification
Gable Louvered							
	Demolish	Ea.		23		23	Includes material and labor to install a
	Install	Ea.	49.50	15.15		64.65	vinyl gable end vent.
	Demolish and Install	Ea.	49.50	38.15		87.65	
	Reinstall	Ea.		12.11		12.11	
	Clean	Ea.		4.40		4.40	
	Paint	Ea.	1.20	5.80		7	
	Minimum Charge	Job		227		227	
Ridge Vent							
	Demolish	L.F.		1.18		1.18	Includes material and labor to install a
	Install	L.F.	4.27	3.25		7.52	mill finish aluminum ridge vent strip.
	Demolish and Install	L.F.	4.27	4.43		8.70	
	Minimum Charge	Job		227		227	
Foundation							
	Demolish	Ea.		8.20		8.20	Includes material and labor to install
	Install	Ea.	17.80	14.40		32.20	galvanized foundation block vent.
	Demolish and Install	Ea.	17.80	22.60		40.40	
	Reinstall	Ea.		11.52		11.52	
	Clean	Ea.	.21	8.25		8.46	
	Minimum Charge	Job		227		227	
Frieze							
	Demolish	L.F.		.81		.81	Includes material and labor to install
	Install	L.F.	1.32	1.82		3.14	pine frieze board.
	Demolish and Install	L.F.	1.32	2.63		3.95	
	Reinstall	L.F.		1.45		1.45	
	Clean	L.F.	.04	.28		.32	
	Minimum Charge	Job		227		227	
Water Heater Cap							
	Demolish	Ea.		17.90		17.90	Includes material and labor to install a
	Install	Ea.	39	63		102	galvanized steel water heater cap.
	Demolish and Install	Ea.	39	80.90		119.90	
	Reinstall	Ea.		50.44		50.44	
	Clean	Ea.	1.43	20.50		21.93	

Cupolas		Unit	Material	Labor	Equip.	Total	Specification
22" x 22"							
	Demolish	Ea.		219		219	Includes material and labor to install a
	Install	Ea.	395	114		509	cedar cupola with aluminum roof
	Demolish and Install	Ea.	395	333		728	covering.
	Reinstall	Ea.		90.80		90.80	
	Minimum Charge	Job		227		227	
29" x 29"							
	Demolish	Ea.		219		219	Includes material and labor to install a
	Install	Ea.	545	114		659	cedar cupola with aluminum roof
	Demolish and Install	Ea.	545	333		878	covering.
	Reinstall	Ea.		90.80		90.80	
	Minimum Charge	Job		227		227	
35" x 35"							
	Demolish	Ea.		219		219	Includes material and labor to install a
	Install	Ea.	920	151		1071	cedar cupola with aluminum roof
	Demolish and Install	Ea.	920	370		1290	covering.
	Reinstall	Ea.		121.07		121.07	
	Minimum Charge	Job		227		227	

Improvements / Appliances / Treatments

Cupolas

	Unit	Material	Labor	Equip.	Total	Specification
47" x 47"						Includes material and labor to install a cedar cupola with aluminum roof covering.
Demolish	Ea.		219		219	
Install	Ea.	1800	227		2027	
Demolish and Install	Ea.	1800	446		2246	
Reinstall	Ea.		181.60		181.60	
Minimum Charge	Job		227		227	
Weather Vane						Includes minimum labor and equipment to install residential type weathervane.
Install	Ea.	106	57		163	
Minimum Charge	Job		227		227	

Aluminum Carport

	Unit	Material	Labor	Equip.	Total	Specification
20' x 20' Complete						Includes material, labor and equipment to install an aluminum carport. Foundations are not included.
Demolish	Ea.		229		229	
Install	Ea.	8800	735	169	9704	
Demolish and Install	Ea.	8800	964	169	9933	
Reinstall	Ea.		587.52	135.26	722.78	
Clean	Ea.	.21	82.50		82.71	
Minimum Charge	Job		455		455	
24' x 24' Complete						Includes material, labor and equipment to install an aluminum carport. Foundations are not included.
Demolish	Ea.		229		229	
Install	Ea.	12700	980	225	13905	
Demolish and Install	Ea.	12700	1209	225	14134	
Reinstall	Ea.		783.36	180.35	963.71	
Clean	Ea.	.21	165		165.21	
Minimum Charge	Job		455		455	
Metal Support Posts						Includes material and labor to install metal support posts for a carport.
Demolish	Ea.		3.06		3.06	
Install	Ea.	29	57		86	
Demolish and Install	Ea.	29	60.06		89.06	
Reinstall	Ea.		45.40		45.40	
Clean	Ea.	.21	3.44		3.65	
Paint	Ea.	.19	7.55		7.74	
Minimum Charge	Job		455		455	

Storage Shed

	Unit	Material	Labor	Equip.	Total	Specification
Aluminum						
Pre-fab (8' x 10')						Includes material and labor to install an aluminum storage shed.
Demolish	Ea.		66.50		66.50	
Install	Ea.	1750	655		2405	
Demolish and Install	Ea.	1750	721.50		2471.50	
Clean	Ea.	1.35	41.50		42.85	
Minimum Charge	Ea.		525		525	
Pre-fab (10' x 12')						Includes material and labor to install an aluminum storage shed.
Demolish	Ea.		128		128	
Install	Ea.	2225	655		2880	
Demolish and Install	Ea.	2225	783		3008	
Clean	Ea.	1.35	46		47.35	
Minimum Charge	Ea.		525		525	

For customer support on your Contractor's Pricing Guide: Residential Repair & Remodeling, call 888.606.7279.

Improvements / Appliances / Treatments

Storage Shed

Storage Shed	Unit	Material	Labor	Equip.	Total	Specification
Wood						
Pre-fab (8′ x 10′)						
Demolish	Ea.		66.50		66.50	Includes material and labor to install a
Install	Ea.	1925	655		2580	pre-fabricated wood storage shed.
Demolish and Install	Ea.	1925	721.50		2646.50	
Clean	Ea.	1.35	41.50		42.85	
Minimum Charge	Ea.		525		525	
Pre-fab (10′ x 12′)						
Demolish	Ea.		128		128	Includes material and labor to install a
Install	Ea.	2425	655		3080	pre-fabricated wood storage shed.
Demolish and Install	Ea.	2425	783		3208	
Clean	Ea.	1.35	46		47.35	
Minimum Charge	Ea.		525		525	

Gazebo

Gazebo	Unit	Material	Labor	Equip.	Total	Specification
Good Grade						
Demolish	S.F.		3.67		3.67	Includes material and labor to install a
Install	S.F.	69.50	34		103.50	gazebo.
Demolish and Install	S.F.	69.50	37.67		107.17	
Reinstall	S.F.		27.24		27.24	
Clean	S.F.	.17	2.20		2.37	
Minimum Charge	Job		227		227	
Custom Grade						
Demolish	S.F.		3.67		3.67	Includes material and labor to install a
Install	S.F.	102	34		136	gazebo.
Demolish and Install	S.F.	102	37.67		139.67	
Reinstall	S.F.		27.24		27.24	
Clean	S.F.	.17	2.20		2.37	
Minimum Charge	Job		227		227	
Bench Seating						
Demolish	L.F.		6.10		6.10	Includes material and labor to install
Install	L.F.	2.28	22.50		24.78	pressure treated bench seating.
Demolish and Install	L.F.	2.28	28.60		30.88	
Reinstall	L.F.		18.16		18.16	
Clean	L.F.	.17	1.65		1.82	
Paint	L.F.	.66	3.02		3.68	
Minimum Charge	Job		227		227	
Re-screen						
Install	S.F.	1.77	1.82		3.59	Includes labor and material to
Minimum Charge	Job		227		227	re-screen wood frame.

Pool

Pool	Unit	Material	Labor	Equip.	Total	Specification
Screen Enclosure						
Re-screen						
Install	SF Wall	.03	5.25		5.28	Includes material and labor to install a
Minimum Charge	Job		227		227	screen enclosure for a pool.
Screen Door						
Demolish	Ea.		9		9	Includes material and labor to install
Install	Ea.	415	75.50		490.50	wood screen door with aluminum cloth
Demolish and Install	Ea.	415	84.50		499.50	screen. Frame and hardware not
Reinstall	Ea.		60.53		60.53	included.
Clean	Ea.	.15	12.15		12.30	
Paint	Ea.	4.54	15.10		19.64	
Minimum Charge	Job		227		227	

For customer support on your Contractor's Pricing Guide: Residential Repair & Remodeling, call 888.606.7279.

Improvements / Appliances / Treatments

Pool	Unit	Material	Labor	Equip.	Total	Specification
Heater / Motor						
Demolish	Ea.		46		46	Includes material and labor to install
Install	Ea.	2550	575		3125	swimming pool heater, not incl. base
Demolish and Install	Ea.	2550	621		3171	or pad, elec., 12 kW, 4,800 gal
Reinstall	Ea.		458.88		458.88	pool, incl. pump.
Clean	Ea.	4.07	41.50		45.57	
Minimum Charge	Job		264		264	
Filter System						
Demolish	Ea.		73.50		73.50	Includes material, labor and
Install	Total	1325	410		1735	equipment to install sand filter system
Demolish and Install	Total	1325	483.50		1808.50	tank including hook-up to existing
Reinstall	Total		329.73		329.73	equipment. Pump and motor not
Minimum Charge	Job		258		258	included.
Pump / Motor						
Demolish	Ea.		46		46	Includes material and labor to install a
Install	Ea.	985	103		1088	22 GPM pump.
Demolish and Install	Ea.	985	149		1134	
Reinstall	Ea.		82.43		82.43	
Clean	Ea.	4.07	41.50		45.57	
Minimum Charge	Job		264		264	
Pump Water						
Install	Day		465	95.50	560.50	Includes labor and equipment to pump
Minimum Charge	Job		207		207	water from a swimming pool.
Replaster						
Install	SF Surf	1.54	10.95		12.49	Includes labor and material to
Minimum Charge	Job		207		207	replaster gunite pool with a surface area of 500 to 600 S.F.
Solar Panels						
Demolish	S.F.		1.91		1.91	Includes material and labor to install
Install	S.F.	34.50	3.39		37.89	solar panel.
Demolish and Install	S.F.	34.50	5.30		39.80	
Reinstall	S.F.		2.71		2.71	
Minimum Charge	Ea.		525		525	

 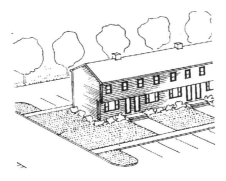 

Construction Clean-up	Unit	Material	Labor	Equip.	Total	Specification
Final						
Clean	S.F.		.29		.29	Includes labor for final site/job
Minimum Charge	Job		165		165	clean-up including detail cleaning.

Costs shown in *RSMeans Residential Cost Data* are based on national averages for materials and installation. To adjust these costs to a specific location, simply multiply the base cost by the factor for that city. The data is arranged alphabetically by state and postal zip code numbers. For a city not listed, use the factor for a nearby city with similar economic characteristics.

STATE	CITY	Residential
ALABAMA		
350-352	Birmingham	.88
354	Tuscaloosa	.89
355	Jasper	.85
356	Decatur	.88
357-358	Huntsville	.87
359	Gadsden	.79
360-361	Montgomery	.88
362	Anniston	.87
363	Dothan	.88
364	Evergreen	.84
365-366	Mobile	.89
367	Selma	.85
368	Phenix City	.87
369	Butler	.86
ALASKA		
995-996	Anchorage	1.23
997	Fairbanks	1.27
998	Juneau	1.25
999	Ketchikan	1.27
ARIZONA		
850,853	Phoenix	.85
851,852	Mesa/Tempe	.85
855	Globe	.82
856-857	Tucson	.84
859	Show Low	.86
860	Flagstaff	.87
863	Prescott	.82
864	Kingman	.83
865	Chambers	.84
ARKANSAS		
716	Pine Bluff	.79
717	Camden	.71
718	Texarkana	.75
719	Hot Springs	.71
720-722	Little Rock	.82
723	West Memphis	.78
724	Jonesboro	.76
725	Batesville	.73
726	Harrison	.75
727	Fayetteville	.70
728	Russellville	.76
729	Fort Smith	.81
CALIFORNIA		
900-902	Los Angeles	1.08
903-905	Inglewood	1.06
906-908	Long Beach	1.05
910-912	Pasadena	1.05
913-916	Van Nuys	1.08
917-918	Alhambra	1.09
919-921	San Diego	1.05
922	Palm Springs	1.05
923-924	San Bernardino	1.06
925	Riverside	1.07
926-927	Santa Ana	1.07
928	Anaheim	1.07
930	Oxnard	1.07
931	Santa Barbara	1.07
932-933	Bakersfield	1.05
934	San Luis Obispo	1.08
935	Mojave	1.07
936-938	Fresno	1.11
939	Salinas	1.14
940-941	San Francisco	1.24
942,956-958	Sacramento	1.12
943	Palo Alto	1.17
944	San Mateo	1.21
945	Vallejo	1.16
946	Oakland	1.21
947	Berkeley	1.23
948	Richmond	1.23
949	San Rafael	1.22
950	Santa Cruz	1.17
951	San Jose	1.22
952	Stockton	1.13
953	Modesto	1.11

STATE	CITY	Residential
CALIFORNIA (CONT'D)		
954	Santa Rosa	1.20
955	Eureka	1.16
959	Marysville	1.12
960	Redding	1.18
961	Susanville	1.16
COLORADO		
800-802	Denver	.90
803	Boulder	.92
804	Golden	.88
805	Fort Collins	.89
806	Greeley	.86
807	Fort Morgan	.90
808-809	Colorado Springs	.86
810	Pueblo	.86
811	Alamosa	.85
812	Salida	.88
813	Durango	.89
814	Montrose	.87
815	Grand Junction	.92
816	Glenwood Springs	.88
CONNECTICUT		
060	New Britain	1.11
061	Hartford	1.10
062	Willimantic	1.12
063	New London	1.11
064	Meriden	1.11
065	New Haven	1.11
066	Bridgeport	1.12
067	Waterbury	1.11
068	Norwalk	1.13
069	Stamford	1.13
D.C.		
200-205	Washington	.97
DELAWARE		
197	Newark	1.02
198	Wilmington	1.01
199	Dover	1.01
FLORIDA		
320,322	Jacksonville	.83
321	Daytona Beach	.86
323	Tallahassee	.82
324	Panama City	.78
325	Pensacola	.88
326,344	Gainesville	.82
327-328,347	Orlando	.85
329	Melbourne	.86
330-332,340	Miami	.85
333	Fort Lauderdale	.84
334,349	West Palm Beach	.84
335-336,346	Tampa	.85
337	St. Petersburg	.84
338	Lakeland	.83
339,341	Fort Myers	.82
342	Sarasota	.86
GEORGIA		
300-303,399	Atlanta	.90
304	Statesboro	.77
305	Gainesville	.79
306	Athens	.78
307	Dalton	.78
308-309	Augusta	.87
310-312	Macon	.77
313-314	Savannah	.82
315	Waycross	.79
316	Valdosta	.73
317,398	Albany	.76
318-319	Columbus	.78
HAWAII		
967	Hilo	1.22
968	Honolulu	1.25

295

Location Factors - Residential

STATE	CITY	Residential
STATES & POSS.		
969	Guam	.98
IDAHO		
832	Pocatello	.87
833	Twin Falls	.81
834	Idaho Falls	.85
835	Lewiston	.97
836-837	Boise	.87
838	Coeur d'Alene	.98
ILLINOIS		
600-603	North Suburban	1.20
604	Joliet	1.22
605	South Suburban	1.20
606-608	Chicago	1.21
609	Kankakee	1.13
610-611	Rockford	1.10
612	Rock Island	.98
613	La Salle	1.10
614	Galesburg	1.03
615-616	Peoria	1.06
617	Bloomington	1.03
618-619	Champaign	1.04
620-622	East St. Louis	1.02
623	Quincy	1.02
624	Effingham	1.04
625	Decatur	1.03
626-627	Springfield	1.04
628	Centralia	1.03
629	Carbondale	1.00
INDIANA		
460	Anderson	.91
461-462	Indianapolis	.93
463-464	Gary	1.05
465-466	South Bend	.91
467-468	Fort Wayne	.89
469	Kokomo	.91
470	Lawrenceburg	.86
471	New Albany	.85
472	Columbus	.91
473	Muncie	.91
474	Bloomington	.92
475	Washington	.90
476-477	Evansville	.90
478	Terre Haute	.91
479	Lafayette	.92
IOWA		
500-503,509	Des Moines	.89
504	Mason City	.75
505	Fort Dodge	.74
506-507	Waterloo	.83
508	Creston	.84
510-511	Sioux City	.86
512	Sibley	.73
513	Spencer	.75
514	Carroll	.79
515	Council Bluffs	.86
516	Shenandoah	.81
520	Dubuque	.87
521	Decorah	.79
522-524	Cedar Rapids	.92
525	Ottumwa	.87
526	Burlington	.88
527-528	Davenport	.97
KANSAS		
660-662	Kansas City	.98
664-666	Topeka	.81
667	Fort Scott	.89
668	Emporia	.83
669	Belleville	.81
670-672	Wichita	.80
673	Independence	.89
674	Salina	.80
675	Hutchinson	.80
676	Hays	.82
677	Colby	.84
678	Dodge City	.81
679	Liberal	.80
KENTUCKY		
400-402	Louisville	.89
403-405	Lexington	.86

STATE	CITY	Residential
KENTUCKY (CONT'D)		
406	Frankfort	.86
407-409	Corbin	.81
410	Covington	.90
411-412	Ashland	.93
413-414	Campton	.86
415-416	Pikeville	.90
417-418	Hazard	.86
420	Paducah	.88
421-422	Bowling Green	.89
423	Owensboro	.90
424	Henderson	.89
425-426	Somerset	.85
427	Elizabethtown	.86
LOUISIANA		
700-701	New Orleans	.86
703	Thibodaux	.82
704	Hammond	.77
705	Lafayette	.84
706	Lake Charles	.86
707-708	Baton Rouge	.84
710-711	Shreveport	.78
712	Monroe	.75
713-714	Alexandria	.78
MAINE		
039	Kittery	.92
040-041	Portland	.96
042	Lewiston	.96
043	Augusta	.92
044	Bangor	.95
045	Bath	.91
046	Machias	.95
047	Houlton	.96
048	Rockland	.95
049	Waterville	.90
MARYLAND		
206	Waldorf	.88
207-208	College Park	.86
209	Silver Spring	.88
210-212	Baltimore	.92
214	Annapolis	.89
215	Cumberland	.90
216	Easton	.85
217	Hagerstown	.89
218	Salisbury	.83
219	Elkton	.92
MASSACHUSETTS		
010-011	Springfield	1.06
012	Pittsfield	1.06
013	Greenfield	1.04
014	Fitchburg	1.13
015-016	Worcester	1.15
017	Framingham	1.18
018	Lowell	1.18
019	Lawrence	1.18
020-022, 024	Boston	1.23
023	Brockton	1.16
025	Buzzards Bay	1.14
026	Hyannis	1.13
027	New Bedford	1.16
MICHIGAN		
480,483	Royal Oak	1.01
481	Ann Arbor	1.02
482	Detroit	1.05
484-485	Flint	.95
486	Saginaw	.91
487	Bay City	.91
488-489	Lansing	.94
490	Battle Creek	.90
491	Kalamazoo	.90
492	Jackson	.92
493,495	Grand Rapids	.90
494	Muskegon	.88
496	Traverse City	.85
497	Gaylord	.88
498-499	Iron Mountain	.89
MINNESOTA		
550-551	Saint Paul	1.11
553-555	Minneapolis	1.12
556-558	Duluth	1.04

Location Factors - Residential

STATE	CITY	Residential
MINNESOTA (CONT'D)		
559	Rochester	1.03
560	Mankato	1.00
561	Windom	.92
562	Willmar	.95
563	St. Cloud	1.04
564	Brainerd	.95
565	Detroit Lakes	.94
566	Bemidji	.94
567	Thief River Falls	.93
MISSISSIPPI		
386	Clarksdale	.74
387	Greenville	.82
388	Tupelo	.75
389	Greenwood	.77
390-392	Jackson	.84
393	Meridian	.82
394	Laurel	.76
395	Biloxi	.80
396	McComb	.74
397	Columbus	.75
MISSOURI		
630-631	St. Louis	1.02
633	Bowling Green	.97
634	Hannibal	.92
635	Kirksville	.90
636	Flat River	.96
637	Cape Girardeau	.91
638	Sikeston	.88
639	Poplar Bluff	.89
640-641	Kansas City	1.03
644-645	St. Joseph	.98
646	Chillicothe	.96
647	Harrisonville	.97
648	Joplin	.90
650-651	Jefferson City	.93
652	Columbia	.93
653	Sedalia	.92
654-655	Rolla	.97
656-658	Springfield	.89
MONTANA		
590-591	Billings	.90
592	Wolf Point	.86
593	Miles City	.87
594	Great Falls	.90
595	Havre	.83
596	Helena	.86
597	Butte	.86
598	Missoula	.87
599	Kalispell	.86
NEBRASKA		
680-681	Omaha	.89
683-685	Lincoln	.89
686	Columbus	.86
687	Norfolk	.88
688	Grand Island	.87
689	Hastings	.89
690	McCook	.82
691	North Platte	.87
692	Valentine	.83
693	Alliance	.82
NEVADA		
889-891	Las Vegas	1.03
893	Ely	1.02
894-895	Reno	.91
897	Carson City	.91
898	Elko	.93
NEW HAMPSHIRE		
030	Nashua	.98
031	Manchester	.96
032-033	Concord	.97
034	Keene	.88
035	Littleton	.91
036	Charleston	.84
037	Claremont	.85
038	Portsmouth	.96

STATE	CITY	Residential
NEW JERSEY		
070-071	Newark	1.15
072	Elizabeth	1.17
073	Jersey City	1.14
074-075	Paterson	1.15
076	Hackensack	1.14
077	Long Branch	1.11
078	Dover	1.14
079	Summit	1.15
080,083	Vineland	1.12
081	Camden	1.13
082,084	Atlantic City	1.17
085-086	Trenton	1.15
087	Point Pleasant	1.12
088-089	New Brunswick	1.17
NEW MEXICO		
870-872	Albuquerque	.83
873	Gallup	.82
874	Farmington	.83
875	Santa Fe	.84
877	Las Vegas	.82
878	Socorro	.82
879	Truth or Consequences	.81
880	Las Cruces	.81
881	Clovis	.82
882	Roswell	.83
883	Carrizozo	.82
884	Tucumcari	.83
NEW YORK		
100-102	New York	1.35
103	Staten Island	1.30
104	Bronx	1.30
105	Mount Vernon	1.16
106	White Plains	1.19
107	Yonkers	1.22
108	New Rochelle	1.20
109	Suffern	1.17
110	Queens	1.30
111	Long Island City	1.33
112	Brooklyn	1.36
113	Flushing	1.32
114	Jamaica	1.31
115,117,118	Hicksville	1.23
116	Far Rockaway	1.31
119	Riverhead	1.23
120-122	Albany	1.00
123	Schenectady	1.02
124	Kingston	1.04
125-126	Poughkeepsie	1.21
127	Monticello	1.04
128	Glens Falls	.95
129	Plattsburgh	1.00
130-132	Syracuse	.98
133-135	Utica	.96
136	Watertown	.94
137-139	Binghamton	.98
140-142	Buffalo	1.07
143	Niagara Falls	1.03
144-146	Rochester	.99
147	Jamestown	.92
148-149	Elmira	.95
NORTH CAROLINA		
270,272-274	Greensboro	.90
271	Winston-Salem	.90
275-276	Raleigh	.89
277	Durham	.90
278	Rocky Mount	.89
279	Elizabeth City	.80
280	Gastonia	.94
281-282	Charlotte	.93
283	Fayetteville	.90
284	Wilmington	.91
285	Kinston	.91
286	Hickory	.89
287-288	Asheville	.91
289	Murphy	.83
NORTH DAKOTA		
580-581	Fargo	.83
582	Grand Forks	.78
583	Devils Lake	.81
584	Jamestown	.79
585	Bismarck	.81

297

STATE	CITY	Residential
NORTH DAKOTA (CONT'D)		
586	Dickinson	.79
587	Minot	.86
588	Williston	.79
OHIO		
430-432	Columbus	.93
433	Marion	.90
434-436	Toledo	.98
437-438	Zanesville	.90
439	Steubenville	.94
440	Lorain	.96
441	Cleveland	1.00
442-443	Akron	.98
444-445	Youngstown	.95
446-447	Canton	.94
448-449	Mansfield	.91
450	Hamilton	.92
451-452	Cincinnati	.92
453-454	Dayton	.93
455	Springfield	.94
456	Chillicothe	.97
457	Athens	.94
458	Lima	.92
OKLAHOMA		
730-731	Oklahoma City	.84
734	Ardmore	.79
735	Lawton	.82
736	Clinton	.80
737	Enid	.79
738	Woodward	.80
739	Guymon	.82
740-741	Tulsa	.80
743	Miami	.85
744	Muskogee	.77
745	McAlester	.76
746	Ponca City	.79
747	Durant	.77
748	Shawnee	.78
749	Poteau	.78
OREGON		
970-972	Portland	1.00
973	Salem	.99
974	Eugene	1.00
975	Medford	.98
976	Klamath Falls	.99
977	Bend	1.01
978	Pendleton	.99
979	Vale	.97
PENNSYLVANIA		
150-152	Pittsburgh	1.01
153	Washington	.97
154	Uniontown	.95
155	Bedford	.92
156	Greensburg	.98
157	Indiana	.96
158	Dubois	.92
159	Johnstown	.93
160	Butler	.94
161	New Castle	.94
162	Kittanning	.94
163	Oil City	.93
164-165	Erie	.95
166	Altoona	.89
167	Bradford	.92
168	State College	.91
169	Wellsboro	.93
170-171	Harrisburg	.96
172	Chambersburg	.90
173-174	York	.93
175-176	Lancaster	.94
177	Williamsport	.93
178	Sunbury	.94
179	Pottsville	.93
180	Lehigh Valley	1.00
181	Allentown	1.05
182	Hazleton	.94
183	Stroudsburg	.94
184-185	Scranton	.97
186-187	Wilkes-Barre	.95
188	Montrose	.92
189	Doylestown	1.09

STATE	CITY	Residential
PENNSYLVANIA (CONT'D)		
190-191	Philadelphia	1.17
193	Westchester	1.11
194	Norristown	1.11
195-196	Reading	.99
PUERTO RICO		
009	San Juan	.76
RHODE ISLAND		
028	Newport	1.09
029	Providence	1.09
SOUTH CAROLINA		
290-292	Columbia	.96
293	Spartanburg	.96
294	Charleston	.94
295	Florence	.93
296	Greenville	.95
297	Rock Hill	.83
298	Aiken	.93
299	Beaufort	.81
SOUTH DAKOTA		
570-571	Sioux Falls	.77
572	Watertown	.73
573	Mitchell	.74
574	Aberdeen	.76
575	Pierre	.78
576	Mobridge	.73
577	Rapid City	.78
TENNESSEE		
370-372	Nashville	.85
373-374	Chattanooga	.84
375,380-381	Memphis	.84
376	Johnson City	.71
377-379	Knoxville	.81
382	McKenzie	.73
383	Jackson	.77
384	Columbia	.78
385	Cookeville	.72
TEXAS		
750	McKinney	.81
751	Waxahackie	.81
752-753	Dallas	.84
754	Greenville	.82
755	Texarkana	.80
756	Longview	.80
757	Tyler	.82
758	Palestine	.79
759	Lufkin	.81
760-761	Fort Worth	.82
762	Denton	.85
763	Wichita Falls	.85
764	Eastland	.83
765	Temple	.81
766-767	Waco	.84
768	Brownwood	.79
769	San Angelo	.79
770-772	Houston	.84
773	Huntsville	.81
774	Wharton	.82
775	Galveston	.83
776-777	Beaumont	.86
778	Bryan	.80
779	Victoria	.83
780	Laredo	.81
781-782	San Antonio	.82
783-784	Corpus Christi	.85
785	McAllen	.85
786-787	Austin	.80
788	Del Rio	.82
789	Giddings	.81
790-791	Amarillo	.80
792	Childress	.82
793-794	Lubbock	.81
795-796	Abilene	.83
797	Midland	.85
798-799,885	El Paso	.80
UTAH		
840-841	Salt Lake City	.82
842,844	Ogden	.81
843	Logan	.81

Location Factors - Residential

STATE	CITY	Residential
UTAH (CONT'D)		
845	Price	.78
846-847	Provo	.82
VERMONT		
050	White River Jct.	.89
051	Bellows Falls	.96
052	Bennington	.99
053	Brattleboro	.95
054	Burlington	.94
056	Montpelier	.92
057	Rutland	.93
058	St. Johnsbury	.90
059	Guildhall	.90
VIRGINIA		
220-221	Fairfax	1.07
222	Arlington	1.10
223	Alexandria	1.11
224-225	Fredericksburg	1.08
226	Winchester	1.02
227	Culpeper	1.07
228	Harrisonburg	.86
229	Charlottesville	.89
230-232	Richmond	.92
233-235	Norfolk	.94
236	Newport News	.94
237	Portsmouth	.89
238	Petersburg	.92
239	Farmville	.83
240-241	Roanoke	.95
242	Bristol	.83
243	Pulaski	.82
244	Staunton	.86
245	Lynchburg	.92
246	Grundy	.80
WASHINGTON		
980-981,987	Seattle	1.02
982	Everett	1.03
983-984	Tacoma	1.00
985	Olympia	.98
986	Vancouver	.96
988	Wenatchee	.93
989	Yakima	.98
990-992	Spokane	.97
993	Richland	.97
994	Clarkston	.94
WEST VIRGINIA		
247-248	Bluefield	.94
249	Lewisburg	.93
250-253	Charleston	.97
254	Martinsburg	.90
255-257	Huntington	.98
258-259	Beckley	.94
260	Wheeling	.94
261	Parkersburg	.93
262	Buckhannon	.94
263-264	Clarksburg	.94
265	Morgantown	.94
266	Gassaway	.93
267	Romney	.91
268	Petersburg	.91
WISCONSIN		
530,532	Milwaukee	1.07
531	Kenosha	1.03
534	Racine	1.01
535	Beloit	1.00
537	Madison	1.00
538	Lancaster	.97
539	Portage	.96
540	New Richmond	.97
541-543	Green Bay	1.03
544	Wausau	.97
545	Rhinelander	.95
546	La Crosse	.96
547	Eau Claire	.99
548	Superior	.96
549	Oshkosh	.95
WYOMING		
820	Cheyenne	.82
821	Yellowstone Nat. Pk.	.82
822	Wheatland	.77

STATE	CITY	Residential
WYOMING (CONT'D)		
823	Rawlins	.85
824	Worland	.80
825	Riverton	.80
826	Casper	.79
827	Newcastle	.84
828	Sheridan	.83
829-831	Rock Springs	.86
CANADIAN FACTORS (reflect Canadian currency)		
ALBERTA		
	Calgary	1.07
	Edmonton	1.06
	Fort McMurray	1.10
	Lethbridge	1.08
	Lloydminster	1.03
	Medicine Hat	1.03
	Red Deer	1.03
BRITISH COLUMBIA		
	Kamloops	1.01
	Prince George	1.01
	Vancouver	1.02
	Victoria	1.02
MANITOBA		
	Brandon	1.08
	Portage la Prairie	.98
	Winnipeg	.95
NEW BRUNSWICK		
	Bathurst	.90
	Dalhousie	.92
	Fredericton	.95
	Moncton	.92
	Newcastle	.91
	St. John	.99
NEWFOUNDLAND		
	Corner Brook	1.02
	St. John's	1.03
NORTHWEST TERRITORIES		
	Yellowknife	1.12
NOVA SCOTIA		
	Bridgewater	.94
	Dartmouth	1.03
	Halifax	.97
	New Glasgow	1.02
	Sydney	1.01
	Truro	.94
	Yarmouth	1.02
ONTARIO		
	Barrie	1.11
	Brantford	1.10
	Cornwall	1.09
	Hamilton	1.06
	Kingston	1.09
	Kitchener	1.02
	London	1.05
	North Bay	1.17
	Oshawa	1.07
	Ottawa	1.06
	Owen Sound	1.10
	Peterborough	1.08
	Sarnia	1.10
	Sault Ste. Marie	1.04
	St. Catharines	1.04
	Sudbury	1.01
	Thunder Bay	1.06
	Timmins	1.07
	Toronto	1.08
	Windsor	1.05
PRINCE EDWARD ISLAND		
	Charlottetown	.90
	Summerside	.97
QUEBEC		
	Cap-de-la-Madeleine	1.08
	Charlesbourg	1.08
	Chicoutimi	1.11
	Gatineau	1.08
	Granby	1.08

299

Location Factors - Residential

STATE	CITY	Residential
QUEBEC (CONT'D)		
	Hull	1.08
	Joliette	1.09
	Laval	1.08
	Montreal	1.05
	Quebec City	1.06
	Rimouski	1.11
	Rouyn-Noranda	1.08
	Saint-Hyacinthe	1.08
	Sherbrooke	1.08
	Sorel	1.09
	Saint-Jerome	1.08
	Trois-Rivieres	1.19
SASKATCHEWAN		
	Moose Jaw	.91
	Prince Albert	.90
	Regina	1.05
	Saskatoon	1.03
YUKON		
	Whitehorse	1.03

For customer support on your Contractor's Pricing Guide: Residential Repair & Remodeling, call 888.606.7279.

Abbreviations

A	Area
ASTM	American Society for Testing and Materials
B.F.	Board feet
Carp.	Carpenter
C.F.	Cubic feet
CWJ	Composite wood joist
C.Y.	Cubic yard
Ea.	Each
Equip.	Equipment
Exp.	Exposure
Ext.	Exterior
F	Fahrenheit
Ft.	Foot, feet
Gal.	Gallon
Hr.	Hour
in.	Inch, inches
Inst.	Installation
Int.	Interior
Lb.	Pound
L.F.	Linear feet
LVL	Laminated veneer lumber
Mat.	Material
Max.	Maximum
MBF	Thousand board feet
MBM	Thousand feet board measure
MSF	Thousand square feet
Min.	Minimum
O.C.	On center
O&P	Overhead and profit
OWJ	Open web wood joist
Oz.	Ounce
Pr.	Pair
Quan.	Quantity
S.F.	Square foot
Sq.	Square, 100 square feet
S.Y.	Square yard
V.L.F.	Vertical linear feet
'	Foot, feet
"	Inch, inches
o	Degrees

302

Index

Online

Book

CD

eBook

Online add-on available
for additional fee

Annual Cost Guides

Unit prices according to the latest MasterFormat

RSMeans Building Construction Cost Data 2016

Offers you unchallenged unit price reliability in an easy-to-use format. Whether used for verifying complete, finished estimates or for periodic checks, it supplies more cost facts better and faster than any comparable source. More than 24,700 unit prices have been updated for 2016. The City Cost Indexes and Location Factors cover more than 930 areas for indexing to any project location in North America. Order and get *RSMeans Quarterly Update Service* FREE.

RSMeans Green Building Cost Data 2016

Estimate, plan, and budget the costs of green building for both new commercial construction and renovation work with this sixth edition of *RSMeans Green Building Cost Data*. More than 10,000 unit costs for a wide array of green building products plus assemblies costs. Easily identified cross references to LEED and Green Globes building rating systems criteria.

RSMeans Mechanical Cost Data 2016

Total unit and systems price guidance for mechanical construction—materials, parts, fittings, and complete labor cost information. Includes prices for piping, heating, air conditioning, ventilation, and all related construction.

Plus new 2016 unit costs for:

- Thousands of installed HVAC/controls, sub-assemblies and assemblies
- "On-site" Location Factors for more than 930 cities and towns in the U.S. and Canada
- Crews, labor, and equipment

RSMeans Facilities Construction Cost Data 2016

For the maintenance and construction of commercial, industrial, municipal, and institutional properties. Costs are shown for new and remodeling construction and are broken down into materials, labor, equipment, and overhead and profit. Special emphasis is given to sections on mechanical, electrical, furnishings, site work, building maintenance, finish work, and demolition.

More than 49,800 unit costs, plus assemblies costs and a comprehensive Reference Section are included.

RSMeans Square Foot Costs 2016

Accurate and Easy-to-Use

- Updated price information based on nationwide figures from suppliers, estimators, labor experts, and contractors.
- Green building models
- Realistic graphics, offering true-to-life illustrations of building projects
- Extensive information on using square foot cost data, including sample estimates and alternate pricing methods

RSMeans Commercial Renovation Cost Data 2016

Commercial/Multi-family Residential

Use this valuable tool to estimate commercial and multi-family residential renovation and remodeling.

Includes: Updated costs for hundreds of unique methods, materials, and conditions that only come up in repair and remodeling, PLUS:

- Unit costs for more than 19,600 construction components
- Installed costs for more than 2,700 assemblies
- More than 930 "on-site" localization factors for the U.S. and Canada

RSMeans Electrical Cost Data 2016

Pricing information for every part of electrical cost planning. More than 13,800 unit and systems costs with design tables; clear specifications and drawings; engineering guides; illustrated estimating procedures; complete labor-hour and materials costs for better scheduling and procurement; and the latest electrical products and construction methods.

- A variety of special electrical systems, including cathodic protection
- Costs for maintenance, demolition, HVAC/mechanical, specialties, equipment, and more

RSMeans Electrical Change Order Cost Data 2016

RSMeans Electrical Change Order Cost Data provides you with electrical unit prices exclusively for pricing change orders. Analyze and check your own change order estimates against those prepared when using RSMeans cost data. It also covers productivity analysis and change order cost justifications. With useful information for calculating the effects of change orders and dealing with their administration.

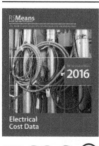

RSMeans Assemblies Cost Data 2016

RSMeans Assemblies Cost Data takes the guesswork out of preliminary or conceptual estimates. Now you don't have to try to calculate the assembled cost by working up individual component costs. We've done all the work for you.

Presents detailed illustrations, descriptions, specifications, and costs for every conceivable building assembly—over 350 types in all—arranged in the easy-to-use UNIFORMAT II system. Each illustrated "assembled" cost includes a complete grouping of materials and associated installation costs, including the installing contractor's overhead and profit.

RSMeans Open Shop Building Construction Cost Data 2016

The latest costs for accurate budgeting and estimating of new commercial and residential construction, renovation work, change orders, and cost engineering.

RSMeans Open Shop "BCCD" will assist you to:

- Develop benchmark prices for change orders.
- Plug gaps in preliminary estimates and budgets.
- Estimate complex projects.
- Substantiate invoices on contracts.
- Price ADA-related renovations.

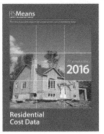

RSMeans Residential Cost Data 2016

Contains square foot costs for 28 basic home models with the look of today, plus hundreds of custom additions and modifications you can quote right off the page. Includes more than 3,900 costs for 89 residential systems. Complete with blank estimating forms, sample estimates, and step-by-step instructions.

Contains line items for cultured stone and brick, PVC trim, lumber, and TPO roofing.

RSMeans Site Work & Landscape Cost Data 2016

Includes unit and assemblies costs for earthwork, sewerage, piped utilities, site improvements, drainage, paving, trees and shrubs, street openings/repairs, underground tanks, and more. Contains more than 60 types of assemblies costs for accurate conceptual estimates.

Includes:

- Estimating for infrastructure improvements
- Environmentally-oriented construction
- ADA-mandated handicapped access
- Hazardous waste line items

RSMeans Facilities Maintenance & Repair Cost Data 2016

RSMeans Facilities Maintenance & Repair Cost Data gives you a complete system to manage and plan your facility repair and maintenance costs and budget efficiently. Guidelines for auditing a facility and developing an annual maintenance plan. Budgeting is included, along with reference tables on cost and management, and information on frequency and productivity of maintenance operations.

The only nationally recognized source of maintenance and repair costs. Developed in cooperation with the Civil Engineering Research Laboratory (CERL) of the Army Corps of Engineers.

RSMeans Light Commercial Cost Data 2016

Specifically addresses the light commercial market, which is a specialized niche in the construction industry. Aids you, the owner/designer/contractor, in preparing all types of estimates—from budgets to detailed bids. Includes new advances in methods and materials.

The Assemblies section allows you to evaluate alternatives in the early stages of design/planning.

More than 15,500 unit costs ensure that you have the prices you need, when you need them.

RSMeans Concrete & Masonry Cost Data 2016

Provides you with cost facts for virtually all concrete/masonry estimating needs, from complicated formwork to various sizes and face finishes of brick and block—all in great detail. The comprehensive Unit Price Section contains more than 8,000 selected entries. Also contains an Assemblies [Cost] Section, and a detailed Reference Section that supplements the cost data.

RSMeans Labor Rates for the Construction Industry 2016

Complete information for estimating labor costs, making comparisons, and negotiating wage rates by trade for more than 300 U.S. and Canadian cities. With 46 construction trades in each city, and historical wage rates included for comparison. Each city chart lists the county and is alphabetically arranged with handy visual flip tabs for quick reference.

RSMeans Construction Cost Indexes 2016

What materials and labor costs will change unexpectedly this year? By how much?

- Breakdowns for 318 major cities
- National averages for 30 key cities
- Expanded five major city indexes
- Historical construction cost indexes

RSMeans Interior Cost Data 2016

Provides you with prices and guidance needed to make accurate interior work estimates. Contains costs on materials, equipment, hardware, custom installations, furnishings, and labor for new and remodel commercial and industrial interior construction, including updated information on office furnishings, as well as reference information.

RSMeans Heavy Construction Cost Data 2016

A comprehensive guide to heavy construction costs. Includes costs for highly specialized projects such as tunnels, dams, highways, airports, and waterways. Information on labor rates, equipment, and materials costs is included. Features unit price costs, systems costs, and numerous reference tables for costs and design.

RSMeans Plumbing Cost Data 2016

Comprehensive unit prices and assemblies for plumbing, irrigation systems, commercial and residential fire protection, point-of-use water heaters, and the latest approved materials. This publication and its companion, *RSMeans Mechanical Cost Data*, provide full-range cost estimating coverage for all the mechanical trades.

Contains updated costs for potable water, radiant heat systems, and high efficiency fixtures.

Unit prices according to the latest MasterFormat

Annual Cost Guides

RSMeans Online

Competitive Cost Estimates Made Easy

RSMeans Online is a web-based service that provides accurate and up-to-date cost information to help you build competitive estimates or budgets in less time.

Quick, intuitive, easy to use, and automatically updated, RSMeans Online gives you instant access to hundreds of thousands of material, labor, and equipment costs from RSMeans' comprehensive database, delivering the information you need to build budgets and competitive estimates every time.

With RSMeans Online, you can perform quick searches to locate specific costs and adjust costs to reflect prices in your geographic area. Tag and store your favorites for fast access to your frequently used line items and assemblies and clone estimates to save time. System notifications will alert you as updated data becomes available. RSMeans Online is automatically updated throughout the year.

RSMeans Online's visual, interactive estimating features help you create, manage, save and share estimates with ease! It provides increased flexibility with customizable advanced reports. Easily edit custom report templates and import your company logo onto your estimates.

	Standard	Advanced	Professional
Unit Prices	✓	✓	✓
Assemblies	X	✓	✓
Sq Ft Models	X	X	✓
Editable Sq Foot Models	X	X	✓
Editable Assembly Components	X	✓	✓
Custom Cost Data	X	✓	✓
User Defined Components	X	✓	✓
Advanced Reporting & Customization	X	✓	✓
Union Labor Type	✓	✓	✓
Open Shop*	Add on for additional fee	Add on for additional fee	✓
History (3-year lookback)**	Add on for additional fee	Add on for additional fee	Add on for additional fee
Full History (2007)	Add on for additional fee	Add on for additional fee	Add on for additional fee

*Open shop included in Complete Library & Bundles
** 3-year history lookback included with complete library

Estimate with Precision

Find everything you need to develop complete, accurate estimates.

- Verified costs for construction materials
- Equipment rental costs
- Crew sizing, labor hours and labor rates
- Localized costs for U.S. and Canada

Save Time & Increase Efficiency

Make cost estimating and calculating faster and easier than ever with secure, online estimating tools.

- Quickly locate costs in the searchable database
- Create estimates in minutes with RSMeans cost lines
- Tag and store favorites for fast access to frequently used items

Improve Planning & Decision-Making

Back your estimates with complete, accurate and up-to-date cost data for informed business decisions.

- Verify construction costs from third parties
- Check validity of subcontractor proposals
- Evaluate material and assembly alternatives

Increase Profits

Use RSMeans Online to estimate projects quickly and accurately, so you can gain an edge over your competition.

- Create accurate and competitive bids
- Minimize the risk of cost overruns
- Reduce variability
- Gain control over costs

30 Day Free Trial

Register for a free trial at www.RSMeansOnline.com

- Access to unit prices, building assemblies and facilities repair and remodeling items covering every category of construction
- Powerful tools to customize, save and share cost lists, estimates and reports with ease
- Access to square foot models for quick conceptual estimates
- Allows up to 10 users to evaluate the full range of collaborative functionality used items.

2016 RSMeans Seminar Schedule

Note: call for exact dates and details.

Location	Dates	Location	Dates
Seattle, WA	January and August	Bethesda, MD	June
Dallas/Ft. Worth, TX	January	El Segundo, CA	August
Austin, TX	February	Jacksonville, FL	September
Anchorage, AK	March and September	Dallas, TX	September
Las Vegas, NV	March	Philadelphia, PA	October
New Orleans, LA	March	Houston, TX	October
Washington, DC	April and September	Salt Lake City, UT	November
Phoenix, AZ	April	Baltimore, MD	November
Kansas City, MO	April	Orlando, FL	November
Toronto	May	San Diego, CA	December
Denver, CO	May	San Antonio, TX	December
San Francisco, CA	June	Raleigh, NC	December

📞 877-620-6245

Beginning early 2016, RSMeans will provide a suite of online, self-paced training offerings. Check our website www.RSMeans.com for more information.

Professional Development

eLearning Training Sessions

Learn how to use *RSMeans Online*® or the *RSMeans CostWorks*® CD from the convenience of your home or office. Our eLearning training sessions let you join a training conference call and share the instructors' desktops, so you can view the presentation and step-by-step instructions on your own computer screen. The live webinars are held from 9 a.m. to 4 p.m. or 11 a.m. to 6 p.m. eastern standard time, with a one-hour break for lunch.

For these sessions, you must have a computer with high speed Internet access and a compatible Web browser. Learn more at www.RSMeansOnline.com or call for a schedule: 781-422-5115. Webinars are generally held on selected Wednesdays each month.

RSMeans Online	RSMeans CostWorks CD
$299 per person	$299 per person

Professional Development

RSMeans Online Training

Construction estimating is vital to the decision-making process at each state of every project. RSMeansOnline works the way you do. It's systematic, flexible and intuitive. In this one day class you will see how you can estimate any phase of any project faster and better.

Some of what you'll learn:
- Customizing RSMeansOnline
- Making the most of RSMeans "Circle Reference" numbers
- How to integrate your cost data
- Generate reports, exporting estimates to MS Excel, sharing, collaborating and more

Also available: RSMeans Online training webinar

Facilities Construction Estimating

In this *two-day* course, professionals working in facilities management can get help with their daily challenges to establish budgets for all phases of a project.

Some of what you'll learn:
- Determining the full scope of a project
- Identifying the scope of risks and opportunities
- Creative solutions to estimating issues
- Organizing estimates for presentation and discussion
- Special techniques for repair/remodel and maintenance projects
- Negotiating project change orders

Who should attend: facility managers, engineers, contractors, facility tradespeople, planners, and project managers.

Construction Cost Estimating: Concepts and Practice

This introductory course to improve estimating skills and effectiveness starts with the details of interpreting bid documents and ends with the summary of the estimate and bid submission.

Some of what you'll learn:
- Using the plans and specifications for creating estimates
- The takeoff process—deriving all tasks with correct quantities
- Developing pricing using various sources; how subcontractor pricing fits in
- Summarizing the estimate to arrive at the final number
- Formulas for area and cubic measure, adding waste and adjusting productivity to specific projects
- Evaluating subcontractors' proposals and prices
- Adding insurance and bonds
- Understanding how labor costs are calculated
- Submitting bids and proposals

Who should attend: project managers, architects, engineers, owners' representatives, contractors, and anyone who's responsible for budgeting or estimating construction projects.

Maintenance & Repair Estimating for Facilities

This *two-day* course teaches attendees how to plan, budget, and estimate the cost of ongoing and preventive maintenance and repair for existing buildings and grounds.

Some of what you'll learn:
- The most financially favorable maintenance, repair, and replacement scheduling and estimating
- Auditing and value engineering facilities
- Preventive planning and facilities upgrading
- Determining both in-house and contract-out service costs
- Annual, asset-protecting M&R plan

Who should attend: facility managers, maintenance supervisors, buildings and grounds superintendents, plant managers, planners, estimators, and others involved in facilities planning and budgeting.

Practical Project Management for Construction Professionals

In this *two-day* course, acquire the essential knowledge and develop the skills to effectively and efficiently execute the day-to-day responsibilities of the construction project manager.

Some of what you'll learn:
- General conditions of the construction contract
- Contract modifications: change orders and construction change directives
- Negotiations with subcontractors and vendors
- Effective writing: notification and communications
- Dispute resolution: claims and liens

Who should attend: architects, engineers, owners' representatives, and project managers.

Mechanical & Electrical Estimating

This *two-day* course teaches attendees how to prepare more accurate and complete mechanical/electrical estimates, avoid the pitfalls of omission and double-counting, and understand the composition and rationale within the RSMeans mechanical/electrical database.

Some of what you'll learn:
- The unique way mechanical and electrical systems are interrelated
- M&E estimates—conceptual, planning, budgeting, and bidding stages
- Order of magnitude, square foot, assemblies, and unit price estimating
- Comparative cost analysis of equipment and design alternatives

Who should attend: architects, engineers, facilities managers, mechanical and electrical contractors, and others who need a highly reliable method for developing, understanding, and evaluating mechanical and electrical contracts.

Unit Price Estimating

This interactive *two-day* seminar teaches attendees how to interpret project information and process it into final, detailed estimates with the greatest accuracy level.

The most important credential an estimator can take to the job is the ability to visualize construction and estimate accurately.

Some of what you'll learn:
- Interpreting the design in terms of cost
- The most detailed, time-tested methodology for accurate pricing
- Key cost drivers—material, labor, equipment, staging, and subcontracts
- Understanding direct and indirect costs for accurate job cost accounting and change order management

Who should attend: corporate and government estimators and purchasers, architects, engineers, and others who need to produce accurate project estimates.

Conceptual Estimating Using the RSMeans CostWorks CD

This *two-day* class uses the leading industry data and a powerful software package to develop highly accurate conceptual estimates for your construction projects. All attendees must bring a laptop computer loaded with the current year *Square Foot Costs* and the *Assemblies Cost Data* CostWorks titles.

Some of what you'll learn:
- Introduction to conceptual estimating
- Types of conceptual estimates
- Helpful hints
- Order of magnitude estimating
- Square foot estimating
- Assemblies estimating

Who should attend: architects, engineers, contractors, construction estimators, owners' representatives, and anyone looking for an electronic method for performing square foot estimating.

RSMeans CostWorks CD Training

This *one-day* course helps users become more familiar with the functionality of the *RSMeans CostWorks* program. Each menu, icon, screen, and function found in the program is explained in depth. Time is devoted to hands-on estimating exercises.

Some of what you'll learn:
- Searching the database using all navigation methods
- Exporting RSMeans data to your preferred spreadsheet format
- Viewing crews, assembly components, and much more
- Automatically regionalizing the database

This training session requires you to bring a laptop computer to class.

When you register for this course you will receive an outline for your laptop requirements.

Also offering web training for the RSMeans CostWorks CD!

Assessing Scope of Work for Facility Construction Estimating

This *two-day* practical training program addresses the vital importance of understanding the scope of projects in order to produce accurate cost estimates for facility repair and remodeling.

Some of what you'll learn:
- Discussions of site visits, plans/specs, record drawings of facilities, and site-specific lists
- Review of CSI divisions, including means, methods, materials, and the challenges of scoping each topic
- Exercises in scope identification and scope writing for accurate estimating of projects
- Hands-on exercises that require scope, take-off, and pricing

Who should attend: corporate and government estimators, planners, facility managers, and others who need to produce accurate project estimates.

Facilities Estimating Using the RSMeans CostWorks CD

Combines hands-on skill building with best estimating practices and real-life problems. Brings you up-to-date with key concepts and provides tips, pointers, and guidelines to save time and avoid cost oversights and errors.

Some of what you'll learn:
- Estimating process concepts
- Customizing and adapting RSMeans cost data
- Establishing scope of work to account for all known variables
- Budget estimating: when, why, and how
- Site visits: what to look for—what you can't afford to overlook
- How to estimate repair and remodeling variables

This training session requires you to bring a laptop computer to class.

Who should attend: facility managers, architects, engineers, contractors, facility tradespeople, planners, project managers and anyone involved with JOC, SABRE, or IDIQ.

Unit Price Estimating Using the RSMeans CostWorks CD

Step-by-step instructions and practice problems to identify and track key cost drivers—material, labor, equipment, staging, and subcontractors—for each specific task. Learn the most detailed, time-tested methodology for accurately "pricing" these variables, their impact on each other and on total cost.

Some of what you'll learn:
- Unit price cost estimating
- Order of magnitude, square foot, and assemblies estimating
- Quantity takeoff
- Direct and indirect construction costs
- Development of contractors' bill rates
- How to use *RSMeans Building Construction Cost Data*

This training session requires you to bring a laptop computer to class.

Who should attend: architects, engineers, corporate and government estimators, facility managers, and government procurement staff.

Professional Development

319

Visit RSMeans.com/Online for more details on data titles in online format.

For more information visit our website at www.RSMeans.com

Professional Development *(vertical text, left margin)*

Registration Information

Register early and save up to $100!
Register 30 days before the start date of a seminar and save $100 off your total fee. Note: This discount can be applied only once per order. It cannot be applied to team discount registrations or any other special offer.

How to register
Register by phone today! The RSMeans toll-free number for making reservations is 781-422-5115.

Two-day seminar registration fee - $935. One-day RSMeans CostWorks® or RSMeans Online training registration fee - $375.
To register by mail, complete the registration form and return, with your full fee, to: RSMeans Seminars, 1099 Hingham Street, Suite 201, Rockland, MA 02370.

One-day Construction Cost Estimating - $575.

Government pricing
All federal government employees save off the regular seminar price. Other promotional discounts cannot be combined with the government discount.

Team discount program
For over five attendee registrations. Call for pricing: 781-422-5115

Multiple course discounts
When signing up for two or more courses, call for pricing.

Refund policy
Cancellations will be accepted up to ten business days prior to the seminar start. There are no refunds for cancellations received later than ten working days prior to the first day of the seminar. A $150 processing fee will be applied for all cancellations. Written notice of the cancellation is required. Substitutions can be made at any time before the session starts. No-shows are subject to the full seminar fee.

AACE approved courses
Many seminars described and offered here have been approved for 14 hours (1.4 recertification credits) of credit by the AACE International Certification Board toward meeting the continuing education requirements for recertification as a Certified Cost Engineer/Certified Cost Consultant.

AIA Continuing Education
We are registered with the AIA Continuing Education System (AIA/CES) and are committed to developing quality learning activities in accordance with the CES criteria. Many seminars meet the AIA/CES criteria for Quality Level 2. AIA members may receive 14 learning units (LUs) for each two-day RSMeans course.

Daily course schedule
The first day of each seminar session begins at 8:30 a.m. and ends at 4:30 p.m. The second day begins at 8:00 a.m. and ends at 4:00 p.m. Participants are urged to bring a hand-held calculator since many actual problems will be worked out in each session.

Continental breakfast
Your registration includes the cost of a continental breakfast and a morning and afternoon refreshment break. These informal segments allow you to discuss topics of mutual interest with other seminar attendees. (You are free to make your own lunch and dinner arrangements.)

Hotel/transportation arrangements
RSMeans arranges to hold a block of rooms at most host hotels. To take advantage of special group rates when making your reservation, be sure to mention that you are attending the RSMeans seminar. You are, of course, free to stay at the lodging place of your choice. (Hotel reservations and transportation arrangements should be made directly by seminar attendees.)

Important
Class sizes are limited, so please register as soon as possible.

Note: Pricing subject to change.

Registration Form   ADDS-1000

Call 781-422-5115 to register or fax this form to 800-632-6732. Visit our website: www.RSMeans.com

Please register the following people for the RSMeans construction seminars as shown here. We understand that we must make our own hotel reservations if overnight stays are necessary.

☐ Full payment of $_____ enclosed.

☐ Bill me.

Please print name of registrant(s).

(To appear on certificate of completion)

_____

_____

_____

P.O. #: _____
GOVERNMENT AGENCIES MUST SUPPLY PURCHASE ORDER NUMBER OR TRAINING FORM.

Firm name_____

Address_____

City/State/Zip_____

Telephone no._____ Fax no._____

E-mail address_____

Charge registration(s) to: ☐ MasterCard ☐ VISA ☐ American Express

Account no._____ Exp. date_____

Cardholder's signature_____

Seminar name_____

Seminar City_____

Please mail check to: 1099 Hingham Street, Suite 201, Rockland, MA 02370 USA